History, Philosophy and Theory of the Life Sciences

Volume 33

Stuart A. Newman, New York Medical College, Valhalla, NY, USA
Frederik Nijhout, Duke University, Durham, NC, USA
Samir Okasha, University of Bristol, Bristol, UK
Susan Oyama, The City University of New York, New York, USA
Kevin Padian, University of California, Berkeley, CA, USA
David Queller, Washington University in St. Louis, St. Louis, MO, USA
Stephane Schmitt, Archives Poincaré, Nancy, France
Phillip Sloan, University of Notre Dame, Notre Dame, IN, USA
Jacqueline Sullivan, Western University, London, ON, Canada
Giuseppe Testa, University of Milan, Milan, Italy
J. Scott Turner, SUNY College of Environmental Science and Forestry, Syracuse, NY, USA
Denis Walsh, University of Toronto, Toronto, ON, Canada
Marcel Weber, University of Geneva, Geneva, Switzerland

History, Philosophy and Theory of the Life Sciences is a space for dialogue between life scientists, philosophers and historians – welcoming both essays about the principles and domains of cutting-edge research in the life sciences, novel ways of tackling philosophical issues raised by the life sciences, as well as original research about the history of methods, ideas and tools, which constitute the genealogy of our current ways of understanding living phenomena.

The series is interested in receiving book proposals that • are aimed at academic audience of graduate level and up • combine historical and/or philosophical and/or theoretical studies with work from disciplines within the life sciences broadly conceived, including (but not limited to) the following areas: • Anatomy & Physiology • Behavioral Biology • Biochemistry • Bioscience and Society • Cell Biology • Conservation Biology • Developmental Biology • Ecology • Evolution & Diversity of Life • Genetics, Genomics & Disease • Genetics & Molecular Biology • Immunology & Medicine • Microbiology • Neuroscience • Plant Science • Psychiatry & Psychology • Structural Biology • Systems Biology • Systematic Biology, Phylogeny Reconstruction & Classification • Virology The series editors aim to make a first decision within 1 month of submission. In case of a positive first decision the work will be provisionally contracted: the final decision about publication will depend upon the result of the anonymous peer review of the complete manuscript. The series editors aim to have the work peer-reviewed within 3 months after submission of the complete manuscript. The series editors discourage the submission of manuscripts that contain reprints of previously published material and of manuscripts that are below 150 printed pages (75,000 words). For inquiries and submission of proposals prospective authors can contact one of the editors: Charles T. Wolfe: ctwolfe1 @gmail.com Philippe Huneman: huneman@wanadoo.fr Thomas A.C. Reydon: reydon@ww.uni-hannover.de

Matteo Mossio
Editor

Organization in Biology

 Springer

Editor
Matteo Mossio
Institut d'Histoire et de Philosophie
des Sciences et des Techniques
(IHPST, CNRS/Paris1)
Paris, France

ISSN 2211-1948   ISSN 2211-1956   (electronic)
History, Philosophy and Theory of the Life Sciences
ISBN 978-3-031-38967-2   ISBN 978-3-031-38968-9   (eBook)
https://doi.org/10.1007/978-3-031-38968-9

This Springer imprint is published by the registered company Springer Nature Switzerland AG
The registered company address is: Gewerbestrasse 11, 6330 Cham, Switzerland

Paper in this product is recyclable.

Contents

Chapter 1
Introduction: Organization as a Scientific Blind Spot

Matteo Mossio

Abstract For most of the twentieth century, biology forgot or largely neglected organization. By this term, I mean a certain mode of interaction among the parts of a system, which is by hypothesis distinctively realized by biological systems. While a systemic trend is progressively pervading various biological fields – notably Evolutionary Biology, Systems Biology and Origins of Life – I suggest that organization still remains a blind spot of biological thinking. Therefore, I submit, biology should be enriched by an explicit and specific notion of organization, drawing in particular on the theory of autonomy, of which I recall some central tenets. I conclude with a brief overview of the scientific and philosophical tradition which has explicitly elaborated on biological organization, and of the more recent literature to which this book aims to contribute.

1.1   The Neglect of Organization

For most of the twentieth century, biology forgot or largely neglected organization. Since the establishment of the Modern Synthesis in evolutionary biology in the 1930s and 1940s, and the flourishing of molecular biology in the 1950s and 1960s, biological research has focused almost exclusively on entities described at the populational or molecular level, by adopting what is usually called "genocentrism," the perspective that places strong emphasis on genes as the fundamental determinants of biological phenomena (Rosenberg, 2007; Fox Keller, 2000).

As Gilbert and Sarkar have pointed out, "for most of this century, the major project of biology has been to reinterpret living properties as being epiphenomena of genes" (Gilbert & Sarkar, 2000: 5).

M. Mossio (✉)
Institut d'Histoire et de Philosophie des Sciences et des Techniques (IHPST, CNRS/Paris1),
Paris, France
e-mail: matteo.mossio@univ-paris1.fr

© The Author(s) 2024
M. Mossio (ed.), *Organization in Biology*, History, Philosophy and Theory of
the Life Sciences 33, https://doi.org/10.1007/978-3-031-38968-9_1

The Modern Synthesis reconciled Darwin's theory of natural selection with Mendelian genetics through population genetics and put forward a reconceptualization of evolution as the change in allele frequencies in a population (Dobzhansky, 1937). Later, with the discovery of the structure of DNA in 1953, molecular biology underwent a comprehensive research program aimed at studying how the genetic program – a notion introduced simultaneously by Ernst Mayr on the one hand, and François Jacob and Jacques Monod on the other (Mayr, 1961; Monod & Jacob, 1961) – governs the synthesis of macromolecules and their interactions and, therefore, cell activity and functionality. The so-called "central dogma of molecular biology," i.e., the idea that biological information flows unidirectionally from DNA to proteins, was thereby interpreted as the idea that genes are the primary (if not sole) determinants of form, function, and behavior of organisms (Crick, 1966). In both evolutionary and molecular biology, genocentrism has therefore consisted in a reductionist stance, resulting in the neglect of biological organization as such.

By the term "organization" I mean a certain mode of interaction among the parts of a system, distinctively realized by biological systems, when compared to other kinds of natural systems, or to artifacts. Broadly speaking (I will return to this below), organization refers to a regime in which a set of entities happen to be related to each other so as to constitute a system that displays both functional differentiation and integration. Moreover, the activity of the whole system plays a role in maintaining its constituents over time: organized systems self-maintain. Let me point out right away that organization is typically, but not exclusively realized by organisms. For instance, it might be argued that colonies, symbioses, or, at a higher level of description, ecosystems can be described as organized systems, although they would not necessarily count as organisms. Accordingly, the notion of "organization" and that of "organism" should not be straightforwardly conflated, although they are closely related: organisms are organized systems, but organized systems are not necessarily organisms.

1.2   Organization as an Explanandum and an Explanans of Biology

Before reductionist and genocentric approaches became mainstream, many eminent biologists and philosophers put emphasis on the centrality of organization in biology. For instance, in his *Modern Theories of Development* (1933), Ludwig von Bertalanffy argued that "all vital processes are so organized that they are directed to the maintenance, production, or restoration of the wholeness of the organism." Therefore, he writes, "there is no 'living substance' because the characteristic of life is the organization of substances" (von Bertalanffy, 1933: 8).

At first sight, one might think that any biologist would easily agree with Bertalanffy. In a sense, it seems obvious that biological systems are organized and, hence, that biology should deal with their organization, its general principles and its various concrete manifestations. After all, biology emerged at the turn of the nineteenth century precisely as the science of "organized beings" (Gambarotto, 2018). Yet, several authors have noted that twentieth-century genocentric biology has lost sight of organisms (Laublicher, 2000; Huneman & Wolfe, 2010; Nicholson, 2014; Walsh, 2015) and thereby, I hold, of organization. What does this alleged neglect of organization consist in? Philosophically, I submit – following Huneman (2010) – that it has taken two different forms, related to the *explanandum* (what must be explained) and the *explanans* (what explains) of biology, respectively.

Firstly, organization is not the *explanandum* of genocentric biology; at best, the latter takes organization as a vague object of explanation. Indeed, the Modern Synthesis has been classically criticized because of its *atomization* of organisms, i.e., the fact of explaining biological traits separately as adaptations, without accounting for their integration as organized systems (Gould & Lewontin, 1979). As Huxley famously put it "every organism cannot be other than a bundle of adaptations" (Huxley, 1942: 20). Even though some advocates of the Modern Synthesis such as Mayr and Dobzhansky did disagree with such a radical view, it seems only fair to claim that the organized complexity of biological systems is not what this framework was designed to explain. Modern Synthesis' main explanandum is the evolution of adaptive traits as the result of differential selection acting on genes. In the case of molecular biology, it could certainly be argued that organization is the ultimate explanandum, insofar as the study of the parts taken separately is supposed to lead, in the long run, to an explanation of the whole system. Nevertheless, such a remote objective remained largely out of reach throughout the actual development of the discipline, which has seemed de facto unable to provide a molecular understanding of biological organization in all its characteristic complexity.

Secondly, organization does not play any explanatory role in genocentric biology. If it were the case, explanations would presuppose the fact that biological phenomena occur *because* biological systems are organized, and would not aim at explaining organization as such, by relying on more fundamental principles. Biology would explain phenomena in the light of their organizational nature. In evolutionary biology, the theoretical framework of the Modern Synthesis identifies natural selection as the main (and often sole) *explanans*, while the impact of organization in the evolutionary dynamics of biological systems is seldom taken into account. As for molecular biology, the rejection of organization as the explanans relies on more explicit theoretical reasons. Indeed, molecular genocentrism has advocated the idea that the specificity of biological phenomena is the fact that they are the result of the expression of a genetic program. Genes code for protein synthesis which, in turn, contributes to the realization of biological functions and, ultimately, of the whole organization. Accordingly, organization itself is explained by the expression of the genetic program (Mayr, 1961; Jacob, 1973).

It is worth noting that genocentrism has also influenced another research domain, that of the origins of life. Influenced by the flourishing of molecular biology, many researchers in this field have defended the idea that life on earth was initiated by the appearance of the first replicators, i.e., the first self-replicative molecules (the "first genes") (Pereto, 2005). On the one hand, organization should constitute – as it does for molecular biology – the ultimate explanandum, to the extent that self-replicative molecules are supposed to trigger an evolutionary process leading to the emergence of living systems endowed with organized complexity. Here again, however, the "replication first" approach has had a hard time in providing satisfactory explanations of the emergence of organized complexity. On the other hand, the "replication first" approach has explicitly excluded organization from the explanans, which seems at first sight an obvious move for a discipline that aims at explaining the *emergence* of life. Organized complexity cannot be adequately explained by appealing to an explanans that already implies organization. Rather, what could explain the emergence of organization would be the inherent capacity of (a population of) self-replicative molecules to generate more and more complex systems by evolving through natural selection – assuming that natural selection can operate before organized systems exist.

In recent times, genocentrism has been increasingly criticized. As it has been previously highlighted (Gilbert & Sarkar, 2000; Bateson, 2005), biology is paying more and more attention to the fact that biological phenomena should be studied by taking into account their distinctive complexity. Accordingly, a general trend toward a more organicist – or at least systemic – perspective can be easily perceived, not only in biology but also in related fields, like prebiotic chemistry. Importantly, recent organicist approaches challenge genocentrism not only on the side of the explanandum but also, and somehow more radically, on the side of the explanans: biology should be more inclusive and ambitious with the kind of phenomena to be explained and, at the same time, equip itself with more adequate explanatory tools.

1.3   The Anti-reductionist Trend in Evolutionary Biology

One of the fields in which an anti-reductionist reaction to genocentrism has been explicitly invoked is evolutionary biology, in the context of a lively debate about whether evolutionary theory needs a rethink (Laland et al., 2014). The debate puts into question the mainstream conceptual framework of the modern synthesis, leading to the call for an "extended evolutionary synthesis" (Pigliucci & Müller, 2010; Laland et al., 2015; Huneman & Walsh, 2017). The extended evolutionary synthesis puts emphasis on four key research themes whose implications, in the view of its advocates, converge on a different, "extended" understanding of the main factors that determine evolutionary change: constructive development, phenotypic plasticity, niche construction, and inclusive inheritance.

Constructive development refers to the idea that the development of organisms does not result from the execution of a genetic program but, rather, from multiple interactions between many factors within the developing system, as well as between the system and the environment (Oyama, 1985). Such a view has evolutionary implications, insofar as development is conceived as a process that facilitates the emergence of phenotypic variation (within and between individuals, populations, species, etc.) on which selection may operate (Minelli & Fusco, 2008). The classical conception of "developmental bias" that reduces the range of possible variations is replaced by a vision according to which development produces phenotypic novelty and affects evolutionary trajectories (Gerhart & Kirschner, 2007).

Phenotypic (or developmental) plasticity is the capacity of a given organism to generate different phenotypes to adapt to different environmental conditions. Plasticity can be seen as a generalization of constructive development: as such, it is meant to lead to phenotypic innovations and, thereby, to play a role in modulating adaptive evolution. In particular, a lively debate exists on whether – and if so, how – phenotypic plasticity can be consolidated by means of genetic accommodation, beyond phenotypic accommodation. As a consequence, adaptive phenotypic novelties would induce genetic changes in populations, and not vice versa (West-Eberhard, 2003).

Niche construction refers to all those processes through which the organism systematically modifies the environment and, thereby, the selective pressures acting on itself. Typical examples are the construction of dams, nests or webs, or the active intervention in the composition or distribution of nutrients in the environment (Odling-Smee et al., 2003).

Inclusive (or extended) inheritance challenges the idea according to which genes are the sole factors involved in biological inheritance. Rather, increasing experimental evidence shows that trans-generational similarities also rely on a variety of transmission processes that do not seem to require a genetic ground. Examples that are often mentioned include epigenetic, symbiotic, ecological, and cultural inheritance. The main implication for evolutionary biology consists in the fact that nongenetic inheritance would allow the transmission of acquired characteristics, thus being responsible for adaptive variations (Jablonka & Lamb, 1995, 2005).

Taken together, these recent research themes are promoting a shift from a genetic- to an organismic-centered view of evolution, insofar as each of them focus on phenomena with respect to which explanations relying on genetic determinism do not appear relevant or adequate. The general idea is that a better understanding of evolutionary trajectories requires taking into account a number of capacities and processes that cannot be described and explained by appealing only to genetic factors. Rather, biological systems as wholes become the relevant object with respect to which phenomena as development, plasticity, niche construction, and extended inheritance can be described and explained. When compared to the conceptual framework of the Modern Synthesis, it seems therefore clear that the organization of biological systems is being brought back to the foreground by the extended evolutionary synthesis, both as an explanandum and as an explanans (Nicholson, 2014). On the one hand, the EES pays more attention to phenomena that reflect the integrated activity of organized systems and their complexity; on the other hand, the explanatory strategy does not consist in deriving organization from more fundamental principles but, rather, in relying on organized systems (including, of course, organisms) and their characteristics in order to elaborate more adequate explanations of these phenomena.

There is therefore no doubt that the EES is more attentive to the organized complexity of the entities which are involved in evolutionary processes, like other authors remarked (Callebaut et al., 2007;[1] Wagner & Laublicher, 2000). Nevertheless, I submit, a conceptual characterization of biological organization as such has not yet been elaborated. In other words, the EES does not rely on a characterization of what is an organized system, which rather consists in a particular *level of description* at which phenomena are described and explanations are provided. The fact that the systemic shift is not accompanied by a characterization of organization has important implications for the explanatory scope and power of evolutionary biology. Some of the contributions to this book explore these implications. Here, I want to emphasize that, in spite of the increasing focus on it, organization remains a blind spot from the current EES views.

1.4   The Anti-reductionist Trend in Systems Biology

The systemic trend is also explicit in molecular biology with the emergence, in the early 2000s, of systems biology. Broadly speaking, systems biology was promoted in reaction to the genocentrism of classic molecular biology, as an approach that aims at shifting the view from genes to the larger systems in which genes are embedded. Rather than studying how gene expression generates biological complexity, systems biology focuses on how the biological system (typically, a cell) works and, in particular, how it regulates gene expression itself. It is now quite common to make a distinction, initially proposed by O'Malley and Dupré (2005), between two attitudes within systems biology: a "pragmatic" one and a "theoretic" one.

The first attitude includes the great majority of research in the field. Pragmatic systems biology can be described as an extension of molecular biology, which studies the interactions and dynamics of large molecular networks. Pragmatic (or "molecular") systems biology challenges genocentrism by considering large systems within which genes are a (crucial) component among many others. Different sets of molecular components (typically biopolymers) belonging to specific cellular organisms, under specific environmental conditions, are studied by different sub-fields: e.g., full cell genomes, as studied by genomics, the whole pool of RNA transcripts (awaiting translation, at a given time) by transcriptomics, the diverse proteins operating in the system by proteomics, metabolites by metabolomics, membrane lipids by lipidomics, and so on. Collectively, these sub-fields generate huge datasets that systems biologists are nowadays trying to interpret and integrate by using mathematical models and computer simulations (Nicholson, 2014: 352).

In spite of these innovations, systems biology can be said to place itself in continuity with the reductionist bottom-up strategies of classical molecular biology, insofar as the general objective consists in obtaining knowledge about the structure, function, organization, and dynamics of whole biological systems by elaborating

---

[1] Callebaut, Müller, and Newman, for instance, propose an "Organismic Systems Approach" to Evo-Devo, which explicitly elaborates on biological organization as a core notion.

models that integrate the parts and their relations (Saetzler et al., 2011: 3). Accordingly, the explanatory strategy operative in the field assumes organization as something that must be explained, but that does not play any explanatory role. More generally, as it has been pointed out by O'Malley and Dupré, molecular systems biology relies on a generic and theoretically ungrounded notion of "system," which designates the network of interacting molecules of different kinds. The central question is whether molecular systems biology can succeed in providing an adequate understanding of organized complexity through such a bottom-up explanatory strategy. Recently, several researchers have expressed skepticism in this respect (Mesarović & Sreenath, 2006; Bertolaso, 2011; Saetzler et al., 2011; Noble, 2017), arguing that the quantity and complexity of available data make more and more difficult their interpretation and integration. Data – the criticism points out – do not speak by themselves, and biologists are in trouble in asking them the relevant questions so as to get an adequate understanding of the whole organization.

This is where theoretical systems biology steps in. Instead of producing models that include more and more experimental data, theoretical systems biology looks for what Green and Wolkenhauer call "organizing principles" (Green & Wolkenhauer, 2013), which are used to select relevant data. One may say that while pragmatic systems biology aims at getting knowledge by adding more details, theoretical systems biology pursues the same objective by abstracting from details. The principles on which theoretical systems biology focuses are mathematical descriptions of recurrent constraints, relations, and patterns that are similar ("isomorphic") in different systems, not necessarily or exclusively biological. A classic example is homeostasis, the capacity to maintain an internal steady state in spite of external perturbation (Cannon, 1929). Other organizing principles have been applied to explain phenomena as flows and oscillations (see for instance the classical work of Goodwin, 1963, and recent developments). A milestone in this respect is the first mathematical model of the heart rhythm (Noble, 1962).

When compared to the pragmatic approach, theoretical systems biology makes a further step in challenging the reductionist perspective. Systemic principles are understood as general hypotheses, which means that they are supposed to explain the data (top-down) and not be explained by them (bottom-up). For instance, if one elaborates a model for accounting for a homeostatic behavior of a biological system, the question would not be "why is the system homeostatic?", but rather "how does the system manage to maintain that specific variable steady, given its (hypothetical) homeostatic capacities"? Accordingly, the notion of "system" is theoretically enriched, and its explanatory role enhanced by the expression of the principles.

Yet those principles are not meant to be distinctive to the biological realm. They are usually elaborated and formalized in other domains, as engineering and graph theory. For instance, negative feedbacks were formalized by cybernetics to account for homeostatic behavior in both animal and machines (Wiener, 1948). The transdisciplinary application of the principles is taken to be a fruitful explanatory strategy, which allows getting insights into the properties of biological systems by looking for analogies with physical systems or machines.

The strategy has proven successful and could certainly foster the acquisition of further knowledge. Nevertheless, it raises the question whether a full-fledged theoretical systems biology can rely on principles that are not designed to capture specific biological features, but more general systemic ones. While it is certainly true that biological systems are a class of physical systems, and therefore share with them (including artifacts and machines) common principles, a satisfactory theoretical framework of biological complexity should also look for *distinctive* biological principles.

One of the objectives of this book consists precisely in promoting the idea that theoretical biology should be endowed with a distinctive principle of organization, which would characterize biological complexity as such, above and beyond the different systemic principles that account for some of its features.

1.5   The Anti-reductionist Trend in the Origins of Life

In the origins of life field, the anti-reductionist alternative to the "replication first" view, which assumes that the individuals of an evolving population can be bare molecules, has taken the form of what is usually called the "metabolism first" view (Pereto, 2005). According to this perspective, the relevant starting point of the emergence of life is the spontaneous appearance of primitive self-catalytic metabolic networks, which would be a condition for the subsequent synthesis of replicators and genes.

The central objection of the "metabolism first" perspective to the "replication first" one is that a process of evolution by natural selection faces "bottlenecks" when it starts from a population of "naked" molecular species. In sharp contrast to the underlying assumptions of the competing view, the anti-reductionist approach argues that molecular replicators alone cannot generate the relevant kind of complexity leading to the appearance of life in the form of organized chemical systems.

Accordingly, the anti-reductionist view challenges the reductionist one with respect to the *explanans* that is adopted. Although it might seem obvious, the idea that we should not presuppose organization to explain the emergence of life (given that organization is taken as an inherent characteristic of life) leads to an impasse: if the explanans is too simple (in terms of its complexity), it cannot generate entities that are complex enough for our explanatory purposes. That is why the "metabolism first" approach does presuppose organization as an *explanans*, under the general hypothesis that some degree of organized complexity is actually required to bootstrap an evolutionary process leading to the appearance of living systems as we know them (Hordijk et al., 2011).

The general strategy consists in characterizing chemical systems (usually referred to as "protocells" or "proto-organisms," Rasmussen et al., 2008) that can appear spontaneously in plausible prebiotic environments and are endowed with the capacity of dynamically self-maintain and increase their functional complexity. Of course, organization cannot be the explanans of this very initial phase, in which organized systems emerge from the integration of different kinds of preexisting processes and components. Once these systems have appeared, however, the process toward primitive living cells implies a set of intermediate forms of organization, each playing a role in the emergence of the next, more complex one. During that long process, thus, each form of organization plays both the role of *explanans* of the next one and of *explanandum* of the previous one.

Although the systemic perspective on the origins of life is taking momentum, after the flourishing of "systems chemistry" as a research field important criticisms have also been addressed to it (Vasas et al., 2010), which explains that the debate is still lively in the field. One important issue is that – here again – the very notion of organization is not expressed in explicit conceptual and theoretical terms and, therefore, does not provide a sufficiently precise guidance in the elaboration of relevant protocells models (see Ruiz-Mirazo et al., 2017, for a discussion of this issue and a proposal).

1.6   What Is Organization?

The central objective of this book is to make a contribution to the current anti-reductionist trend in biology, by putting organization to the foreground. I submit that the systemic, or even organicist-thinking that is progressively pervading various biological fields should be enriched by an explicit and specific notion of organization, understood as both a fundamental explanandum and explanans in biology.

Adopting organization as an explanandum means that the object of biological explanation should be the phenomenon of organization itself in its various *realizations*, as well as its evolution. A satisfactory biological explanation should aim at making explicit how any specific phenomenon under scrutiny is to be understood as a manifestation or an aspect of organized complexity and, for this precise reason, biological. Beyond the fragmentation of reductionist approaches, anti-reductionist ones should hence aim at explaining biological phenomena by locating them into integrated organized wholes.

Adopting organization as an explanans, in turn, means conceiving it as a *theoretical principle* (Mossio et al., 2016). A theoretical principle is an overarching hypothesis that frames the intelligibility of the objects within a scientific domain. All biological systems, in all their diversity and richness of forms and kinds, comply with the principle, and are therefore organized. The crucial implication here is that theoretical principles enable explanations, but are not themselves the object of an explanation. Accordingly, the adoption of organization as an explanans means that biology presupposes the principle without trying to deduce it from something else, which would be precisely the reductionist stance.

A recent example of the adoption of organization as both an explanandum and an explanans is given by the model of a self-maintaining metabolic system, inspired by Robert Rosen's idea of (M,R)-systems (Piedrafita et al., 2010). The model is a computational simulation of a chemical network, made of three interlocking catalytic cycles, in which catalysts are produced by the very network that they contribute to maintain. Here, organization is the explicit explanandum in the sense that the model explores the conditions under which the system *as a whole* exhibits properties of organized – and therefore biologically relevant – systems, such as steady autocatalysis, robustness, and bi-stability. At the same time, organization is an explanans to the extent that the mutual dependence between the catalysts is presupposed as a background hypothesis and not obtained as a result. The model does not try to explain how and why catalysts get organized in the first place (i.e., mutually dependent); rather, it takes the very fact of being organized as an unexplained premise (an explanatory principle) and aims at explaining the properties of a system that realizes a specific instantiation of biological organization.

It is worth reflecting on this apparent contradiction, stemming from the double role of organization as both the explanandum and the explanans of biology. In the two cases, the term designates in fact different conceptual entities: as an explanandum, organization refers to its various concrete realizations in nature; as an explanans, it designates the general "regime" or "set of relations" that are common to all realizations. To make a (somehow very perilous) analogy, the dual role of organization as explanans and explanandum can be grasped by thinking to the role played by Newton's laws of motion in Classical Mechanics. On the one hand, the laws of motion are principles that are presupposed (i.e., not explained within the field), so as to provide explanation of physical phenomena; on the other hand, Classical Mechanics provide explanations of phenomena which realize (are in conformity with) the laws. Classical Mechanics presuppose the laws of motion so as to explain specific instantiations of these laws. In this very general sense, I submit that organization could play an analogous role in the biological domain.

Needless to say, for organization to play such a role, it should be characterized in a way that is more precise than the general and intuitive notions of "system" or even "organism" as they are currently employed in evolutionary and systems biology. So, the main question is: what does organization mean? To answer this question, let me spell out some guidelines that will be explored in the book. These guidelines are mostly inspired by the theory of biological autonomy, one of the contemporary heirs of the organicist tradition, to which I have myself contributed (Moreno and Mossio, 2015).

As already mentioned, organization designates a specific kind of complexity, a specific set of relations among elements. To a first approximation, organization refers to the differentiation of functional roles (i.e., division of labor) among the parts of a system and, at the same time, to their integration and coordination as a whole. Furthermore, and crucially, organization involves a generative dimension in the form of a mutual dependence, such that the very activity and existence of each organized part depends on its mutual relationship with the others. Organized parts have functions (Mossio et al., 2009; Saborido et al., 2011), which means that organization most basically designates *functional* complexity. Overall, biological organization is capable of self-determination, insofar as functional constraints collectively contribute to determine their conditions of existence. As I have argued elsewhere (Mossio & Bich, 2017), the capacity of self-determination provides a naturalized ground for purposiveness: biological organization can be legitimately said to be cause and effect of itself and thereby an intrinsically purposive regime.

Biological organization goes along with thermodynamic openness, which is the fact that organized systems continuously exchange energy and matter with the surroundings. The connection is theoretically deep, insofar as only thermodynamically open systems can possibly comply with the organization principle, although not any thermodynamically open system does. As all open systems, indeed, be they physical or chemical, biological systems are traversed by a flow of energy and matter, which takes the form of processes and reactions occurring in open thermodynamic conditions. In this respect, biological systems do not differ from other natural open systems. Yet, unlike "self-organizing" dissipative structures, they constrain and canalize the thermodynamic flow through the collective activity of their functional parts, which realize a specific form of mutual relationship, i.e., organization.

Because of their distinctive functional complexity, furthermore, organized systems (usually) do not appear spontaneously when some specific boundary conditions are met, as self-organizing structures do. Rather, organized systems are the result of a long historical evolutionary process of increase and preservation of complexity, which means in particular that any individual biological system is generated through the reproduction of other biological systems. In spite of their common thermodynamic grounding, hence, (biological) organization cannot be conflated with (physical) self-organization.

As the result of an historical process, biological complexity raises the central question of understanding how biological systems manage to maintain their stability while continuously undergoing variation (Montévil et al., 2016a; Longo et al., 2012, argue that biological variation is theoretically unprestatable). In this respect, I submit that the notion of organization plays a twofold explanatory role: on the one hand, organization allows explaining the stability of biological systems (both at the individual and evolutionary scale), and the maintenance of their constitutive dynamics over time; on the other hand, it provides a ground to understand how quantitative and qualitative innovations can be produced, and then preserved though its functional integration.

Lastly, let me emphasize again that biological organization should be conceived as a broader notion than that of "organism." Although organisms are – by hypothesis – organized systems, not all organized systems are necessarily organisms. For instance, some contributions to this book explore the idea that ecosystems might also be described as organized systems, without implying that they constitute a kind of organisms. Reciprocally, additional specifications should be added to characterize organisms among the broader set of organized systems. In this respect, a possible route is traced by the theory of autonomy, according to which organisms are autonomous systems, i.e., organized systems endowed with agential and adaptive capacities (see for instance Moreno and Mossio, 2015, section 4.4). As autonomous systems, in a word, organisms are organized adaptive agents.

1.7 Historical Overview

The historical roots of the notion of biological organization I refer to can be traced back to Immanuel Kant. In his *Critique of the Power of Judgment* (1790/1987, Kant argues that, unlike any other kind of system, the parts of biological systems do not and cannot exist by themselves, but only insofar as they constitute an organized whole which, in turn, is itself a condition of their own existence and functioning. In this sense, biological systems display self-organizing features that are absent in machines. In a watch, for example, every part is organically arranged in relation to the others, but the watch does not *produce* them. It "is certainly present for the sake of the other but not because of it." Hence the producing cause of the watch is the watchmaker, not the watch itself: "one wheel in the watch does not produce the other, and even less does one watch produce another, using for that purpose other matter (organizing it); hence it also cannot by itself replace parts that have been taken from it, or make good defects in its original construction by the addition of other parts, or somehow repair itself when it has fallen into disorder: all of which, by contrast, we can expect from organized nature." Based on these considerations, Kant claims that "an organized being is thus not a mere machine, for that has only a *motive* force, while the organized being possesses in itself a *formative* force (*Bildungskraft*), and indeed one that it communicates to the matter, which does not have it (it organizes the latter): thus it has a self-propagating formative power, which cannot be explained through the capacity for movement alone (that is, mechanism)" (Kant, 1790/1987: §65).

The Kantian focus on biological organization had continuity in the (mostly Continental) Biology of the nineteenth century, notably in the work of Goethe (1995) and Cuvier (1817). Cuvier's principle of the "condition of existence," for instance, claims that "the different parts of each being must be coordinated in such a way as to render possible the existence of the being as a whole" (1817 i., 6, quoted and translated by Reiss, 2009). By implying that the different parts are linked and coordinated, Cuvier's principle grounds and guides his empirical investigations in comparative anatomy and paleontology (Cuvier, 1805; see also Huneman, 2006, for an analysis).

Kant's and Cuvier's perspectives further influenced German organicist tradition leading to Johannes Müller's physiology (1837–1840) and Karl von Baer's embryology (1828). They both consider that, as Huneman writes "the proper object of life sciences should be a set of parts organizing itself as a whole, the development and the functioning of this specific kind of entity being the proper field of, respectively, embryology and physiology" (Huneman, 2010: 342).

Claude Bernard explicitly invokes Cuvier's view and claims that biological systems are to be conceived as organized entities, whose parts are interdependent and mutually generative. In his words, "The physiologist and the physician must never forget that the living being comprises an organism and an individuality... If we decompose the living organism into its various parts, it is only for the sake of experimental analysis, not for them to be understood separately. Indeed, when we wish to ascribe to a physiological quality its value and true significance, we must always refer to this whole and draw our final conclusions only in relation to its effects in the whole" (Bernard, 1865/1984, II, ii, x 1, 137, quoted and translated by Wolfe, 2010). Bernard's main focus is on the contribution of the organized parts – that must be investigated through the experimental method to the conservation of the internal milieu, in spite of the continuous variations taking place in the external milieu.

An important moment in the history of the scientific treatment of biological organization is represented by the so-called Theoretical Biology Club, that refers to a group of researchers including Woodger, Needham, and Waddington (Etxeberria & Umerez, 2006; Peterson, 2010). The Theoretical Biology Club promoted a scientific organicist perspective for biology and underwent a rigorous conceptual and theoretical treatment of various dimensions of the very idea of organization, including the analysis of internal relations (Woodger, 1929) and hierarchies (Needham, 1937). Another particularly relevant contribution is due to Ludwig von Bertalanffy (1952), who was one of the first authors that made explicit the fact that biological systems as thermodynamically open systems. Initially used by Bertalanffy as an argument against both vitalism and mechanism, the thermodynamic openness of biological systems remains central aspect role in the subsequent elaborations on organization.

Later on, the notion of organization played a central role in the organicist perspective that permeated embryology in the first half of the twentieth century. In particular, Paul Weiss refers to organization as the "coordinating principle" (Weiss, 1963: 190) that characterizes biological systems beyond local components and processes and that grounds their stability in the face of internal or external perturbations (Rosslenbroich, 2011; Bich & Arnellos, 2013; Nicholson & Gawne, 2015).

In the second half of the twentieth century, the conceptualization and scientific treatment of biological organization entered into a new phase, characterized by an increasing coherence and theoretical refinement. A milestone in this tradition is the account put forward by Jean Piaget (Piaget, 1967), whose core idea was to integrate into a single coherent picture thermodynamic openness and organizational closure. On the one hand, as emphasized by Bertalanffy, organisms are thermodynamically open systems, traversed by a continuous flow of matter and energy. On the other hand, they realize "closure," i.e., the mutual dependence between a set of constituents which maintain each other through their interactions and which could not exist in isolation.

In Piaget's view, closure captures a fundamental aspect of the very idea of "organization," through the association between division of labor and mutual dependence that it implies. In other words, biological organisms are organized precisely because they realize closure. The centrality of closure and its connection to organization, as well as its distinction from (and, yet, complementarity to) thermodynamic openness have become givens in most subsequent accounts of biological organization (Letelier et al., 2011).

One of the best-known accounts of biological organization is the one centered on the concept of autopoiesis (Varela et al., 1974; Varela, 1979) which, among other aspects, places heavy emphasis on the generative dimension of closure: biological systems determine themselves in the sense that they "make themselves" (auto-poiein). Precisely because of their dissipative nature, the components of biological organisms undergo degradation over time; the whole system preserves its coherence and identity only insofar as it maintains and stabilizes not just some internal states or processes but the autopoietic system itself as an organized unity. In spite of its qualities, however, a central weakness of the concept of autopoiesis is that it does not provide a sufficiently explicit characterization of closure (Montévil & Mossio, 2015). Biological systems are at the same time thermodynamically open and orga-nizationally closed, but no details are given regarding how the two dimensions are interrelated, what constituents are involved in closure, and at what level of descrip-tion. In the absence of such specifications, it remains unclear in what precise sense closure would constitute a causal regime that distinctively characterizes biological organization.

A concerted attempt to answer this question was made by Robert Rosen. In *Life Itself* (Rosen, 1991), Rosen reinterprets the Aristotelian categories of causality and claims that the distinction between closure and openness should be grounded on a distinction between efficient cause and material cause. By relying on this distinc-tion, Rosen's central thesis is that: "a material system is an organism [a living sys-tem] if, and only if, it is closed to efficient causation" (Rosen, 1991: 244). In turn, a natural system is closed to efficient causation if, and only if, all components having the status of efficient causes within the system are materially produced by the sys-tem itself. What matters here is that closure is located at the level of efficient causes: what constitutes the organization is the set of efficient causes subject to closure, and its maintenance (and stability) is the maintenance of the closed network of effi-cient causes.

Although Rosen's account represents a crucial step forward in the theoretical understanding of organization, I think that it still remains too abstract, and therefore hardly applicable as a guiding principle for biological theorizing, modeling, and experimentation. Rosen defines closure as involving efficient causes, but, without additional specifications, it might be difficult to identify efficient causes in the sys-tem: what entities actually play the role of efficient causes in a biological system? To deal with this issue, decisive insights have emerged from more recent literature that elaborates more explicitly on the "thermodynamic grounding" of biological systems (Bickhard, 2000; Christensen & Hooker, 2000; Moreno & Ruiz Mirazo, 1999) and the relations between closure and openness. In particular, Stuart Kauffman (2000) argues that biological organization implies a circular relationship between work and constraints, in the form of what he labels a "work-constraint (W-C) cycle." When a (W-C) cycle is realized, constraints that apply to the system are produced and maintained by the system itself. Hence, the system needs to use the work gener-ated by the constraints in order to generate those very constraints, by establishing a mutual relationship – a cycle – between constraints and work.

More recently, a characterization of biological organization as "closure of con-straints," which puts together many of the central ideas evoked above, has been proposed (Montévil & Mossio, 2015). Most of the contributions to this book actu-ally rely on this characterization to further develop its implications.

1.8 The Current Context, and The Place of the Book

Nowadays, what is generically called "organicism" is undergoing resurgence, as an increasing number of philosophical, theoretical, and even formal accounts have advocated it as an integrative and fecund framework for biology (Gilbert & Sarkar, 2000). Among these accounts, the aforementioned theory of biological autonomy – originally elaborated by Varela (Varela, 1979) – is gaining momentum (Moreno & Mossio, 2015).

Recently, several studies have relied on the pioneering work mentioned above, and tried to further elaborate on the central notion of organization. Some have investigated its philosophical (Mossio & Moreno, 2010) and theoretical (Letelier et al., 2011; Wolkenhauer & Hofmeyr, 2007) implications, while others have developed applications to various domains, such as the already mentioned metabolic networks (see also Cornish-Bowden et al., 2013), physiological regulation (Bich et al., 2020), the transition from unicellular to multicellular organisms (Arnellos et al., 2014), organogenesis (Montévil et al., 2016b), ecology (Nunes et al., 2014), agency (Barandiaran et al., 2009), cognition (Barandiaran & Moreno, 2006), and the origins of life (Ruiz-Mirazo & Moreno, 2004). These are just a few examples showing the existence of a scientific perspective that aims at establishing an organizational framework for biology – whether or not they stem from the theory of autonomy – and elaborates on some recurrent theoretical themes such as openness, closure, constraints, agency, and circularities, as well as their connections with philosophical issues as teleology, functionality, normativity, historicity, and individuation.

The main objective of the book is to assess the prospects and the fecundity of the concept of organization in biological research, both as a philosophical foundation and as a theoretical principle able to generate models and experimental protocols. The various chapters deal with a variety of issues with respect to which an organizational perspective can be adopted and discussed. Collectively, they show that the notion of organization can nourish the current anti-reductionist trend, by guiding the elaboration of models and the connection with experimental biology.

In the second chapter (Chap. 2), Georg Toepfer recounts the history of the concept of organization, as used in relation to organic bodies. Toepfer underscores that organization becomes a defining feature of life from the seventeenth century and plays a central role in the establishment of biology as an autonomous scientific discipline. During the nineteenth and twentieth centuries, then, it has been supplemented by the concepts of evolution and regulation, which refer to the transformation and stabilization of organized systems, respectively. In its more recent formulations (notably in terms of constraints closure) – Toepfer argues – the specificity of organization is more explicitly tied to the specificity of *forms* that enable its realization. As he writes: "The only life-forces that exist are life-forms."

Charles Wolfe (Chap. 3) discusses the challenges that a naturalistic and non-foundationalist – and thereby scientifically workable – organicist project should take up. In his view, some versions of organicism suffer from at least three main hesitations or "instabilities," which relate to the interpretation of organismal properties (epistemological vs. ontological, irreducible vs. empirical), as well as the opposition with mechanism. Wolfe argues that "one more effort" should be made to overcome these instabilities, without giving in to the symmetrical temptations of objectification and subjectification of organisms. The very concept of organization is likely to play a crucial role in making this effort successful.

In Chap. 4, Gertrudis Van de Vijver and Levi Haeck focus on one of the instabilities noted by Wolfe, that between the epistemological and ontological interpretations of organisms as organized systems. They put forward an original transcendental stance, inspired by Kant's treatment of biological organization, according to which both the subject and the object involved in our understanding of organisms should be treated as organized living systems. Thereby, the enquiry about the properties of living organization is simultaneously an enquiry about the subject and the object of biological knowledge. Insofar as our rational capacities are a manifestation of life, studying the latter "folds back onto" the former and reveals that our cognition shares fundamental organizational properties with biological phenomena, starting with their purposiveness.

Cliff Hooker (Chap. 5) shares with Van de Vijver and Haeck the idea that cognition and life display common properties. As life, cognition is organized in a narrow, theoretically precise sense, which implies in particular the realization of agency, purposiveness, and anticipation. In a word, both life and cognition realize autonomy. Yet, instead of focusing on the epistemic loop between cognition and life, Hooker emphasizes that autonomy comes in degrees: in particular, cognition (and notably human cognition) relies on much more sophisticated anticipatory and adaptive capacities, when compared to noncognitive biological autonomy. The take home message is "unification without reduction": the concept of organization can be usefully put to work to provide a general understanding of cognition and life, while preserving their specific features.

Chapter 6 provides a counterpoint to the general message of the book. Olivier Sartenaer argues that organicism does not need organization to remain "chauvinist about organisms and autonomist about biology." Organicism can vindicate the irreducibility of organisms – and justify its epistemological autonomy – by showing that they comply with the requirements of transformational emergence. Sartenaer's argument is not that organization is an illegitimate concept, but that organicism could stand while discarding it. To the objection that, without organization, transformational emergence does not capture what distinguishes organisms from other emergents, Sartenaer replies that their specificity could be grounded in their being the outcome of specific transformational transitions during biological evolution. Yet, this solution begs the question whether biological evolution can be the evolution of anything else than organized systems.

Philippe Huneman (Chap. 7) offers a comparison between organizational and evolutionary approaches of organisms, that he labels "Kantian" and "liberal," respectively. While the former aim at characterizing organisms by appealing to a distinctive set of organizational properties, the latter situate organisms in a larger, continuous spectrum of biological individuals understood as units of selection. In liberal approaches, organisms are not the only biological individuals and, in addition, "being an organism comes by degrees." The comparison raises the question of the connections between the two conceptions of individuality, as well as between the two underlying theoretical frameworks. Huneman addresses this question and explores the prospects of a fruitful reconciliation between Kantian and liberal approaches.

In Chap. 8, Johannes Jaeger provides an answer to the question raised by Huneman. He advocates the radical idea according to which biological evolution cannot be but the evolution of organized systems. Evolutionary theory should move toward a fourth perspective, which would complement and succeed existing structural, functional, and processual perspectives. In particular, the fourth perspective is an agential perspective, centered on the hypothesis that evolving organisms are organized purposive agents. As such, not only organisms are what evolutionary processes operate on, but they also modulate such processes. As Jaeger, following Walsh, puts it: "Some things in evolution happen because organisms make them happen." The agential perspective relies on a naturalized understanding of purposiveness, provided by recent characterizations of organization as constraints closure. Importantly, Jaeger underscores that such an organizational purposiveness applies to individual organisms exclusively, and not to evolutionary processes as such. The existence of macroevolutionary trends is a different issue that should be explored separately.

Sharing Jaeger's perspective on the role of organization in shaping evolution – one might argue – still leaves room to the assumption that, while looking at the origins of life, organization initially emerged from evolutionary processes. In Chap. 9, however, Kepa Ruiz-Mirazo and Alvaro Moreno argue that things are more complicated. They put forward an account of biogenesis that also ascribes an explanatory role to the concept of organization. Again, the emphasis is placed on the interplay between individual organization and evolutionary processes, which take place at a different spatial and temporal scale. Their main thesis is that, to result in the emergence of complex biological organisms (as we observe them), biogenetic trends require complex enough, organized self-maintaining systems as a starting point. Complexity begets complexity, in the sense of generating functional variety and more sophisticated forms of control. In particular, Ruiz-Mirazo and Moreno emphasize the evolutionary significance of forms of regulation and heredity relying on dynamical decoupling, whose emergence has drastically enhanced individual adaptivity and cross-generation stability.

Gaëlle Pontarotti (Chap. 10) specifically deals with the connection between organization and heredity, a key ingredient of evolution. Pontarotti argues that the general trend beyond genocentrism implies a shift from a heuristic of replication, which sees evolution in terms of a competition among self-replicating objects, to a heuristic of collaboration, which emphasizes the mutual dependence of objects belonging to integrated wholes. The heuristic of collaboration can be applied to elaborate an organizational account of heredity, which characterizes the latter as the "trans-generational conservation of functional networks." Pontarotti submits that the organizational account allows expanding heredity beyond genes, while keeping the concept conceptually bounded. The extension avoids then the dilution of heredity into a too general concept of biological cross-generation stability. The chapter also discusses how the organizational account of heredity impacts some of the central tenets of evolutionary theory.

Chapter 11 shifts to the individual scale and explores the explanatory role that the concept of organization can have with regard to development. Leonardo Bich and Derek Skillings put forward an organizational view on development that, here again, makes determination reciprocal: not only development determines the establishment of biological organization but, reciprocally, organization enables developmental processes. As Bich and Skillings emphasize, the organizational view "favors a switch in perspective," whereby each stage of development is understood as an organized system aiming at its own maintenance, rather than being an intermediate step of a process tending to a final state (typically identified with the achievement of reproductive capabilities). Bich and Skillings argue that the organizational view accounts not only for maintenance but also for change, which is genuinely developmental only if it is controlled by regulatory functions exerted by the organized system. Regulation also draws the boundaries of development, which starts when regulatory functions appear and ends when the organized system ceases to undergo regulated change.

Maël Montévil and Ana Soto (Chap. 12) further explore developmental processes by discussing their recent efforts to model morphogenesis, and more specifically mammary ductal morphogenesis. In their model, Montévil and Soto have applied two principles: cells default state and organization. According to the default state, cells move and proliferate when unconstrained, in the presence of sufficient nutrients and space, while the organization of the multicellular system that they constitute exerts the constraints that canalize or inhibit the default state. In particular, Montévil and Soto show that the formation of mammal ducts is determined by the interplay between the constraints exerted by proliferating cells on the extracellular matrix (notably on collagen fibers), which in turn constrains cells proliferation and motility. The chapter also addresses important issues raised by the modeling practice relying on organization. These include the choice of those organized parts and constraints expected to play a role in determining the target phenomenon, and their insertion into a description of the whole organismal organization.

The last chapter (Chap. 13) shifts again to a different scale and discusses some theoretical and ethical implications stemming from the application of the organizational framework to the ecological domain. Charbel El-Hani, Felipe Lima, and Nei Nunes-Neto argue that the concept of organization provides a relevant tool to individuate ecosystems and to ascribe functions to their parts (both to items of biodiversity and abiotic items). In particular, they provide a detailed reply to some objections recently raised against the organizational account of ecosystemic functions. El-Hani, Lima and Nunes-Neto's contribution is particularly important because it shows that the concept of organization may be pertinently used to characterize biological systems in general, and not only organisms. Also, it opens the way to future research directions, which would explore the relations and interplay between nested levels of organization. Lastly, El-Hani, Lima, and Nunes-Neto argue that the concept of organization, by naturalizing intrinsic purposiveness, provides a ground to ascribe intrinsic value to ecosystems. In turn, this supports an original conception of sustainability, which is alternative to the usual anthropocentric interpretation.

Acknowledgments I am indebted to many colleagues and friends who contributed to this project in different ways. First of all, I thank all authors for their contributions, as well as for taking the time to read and to comment on other chapters. The cross-reading effort certainly helped to improve the global coherence of the book. I also warmly thank the external reviewers for their detailed and constructive remarks: Argyris Arnellos, Mike Beaton, James DiFrisco, Antoine Dussault, James Griesemer, Francesca Merlin, Guglielmo Militello, Auguste Nahas, Dan Nicholson, Juli Peretó, Cyril Rauch, and Denis Walsh.

I am thankful to my research unit, the Institut d'Histoire et de Philosophie des Sciences et des Techniques (IHPST), and to the Université Paris 1 Panthéon – Sorbonne for funding and hosting, a few years ago, the workshop from which this book has been subsequently elaborated. A special mention goes to Andrea Gambarotto, who co-organized the workshop, and encouraged me to undertake the project of this book.

I would also like to thank the Labex "Who Am I? Exploring identity: from molecules to individuals" for its support, and to express my gratitude to the John Templeton Foundation for generously funding the open access publication of the book through the Agency, Directionality and Function program (#62220), led by Alan Love. The opinions expressed in this book are those of the authors and not those of the John Templeton Foundation.

References

Arnellos, A., Moreno, A., & Ruiz-Mirazo, K. (2014). Organizational requirements for multicellular autonomy: Insights from a comparative case study. *Biology and Philosophy, 29*, 851–884.

Barandiaran, X., & Moreno, A. (2006). On what makes certain dynamical systems cognitive. *Adaptive Behavior, 14*, 171–185.

Barandiaran, X., Di Paolo, E., & Rohde, M. (2009). Defining agency. Individuality, normativity, asymmetry and spatio-temporality in action. *Adaptive Behavior, 17*(5), 367–386.

Bateson, P. (2005). The return of the whole organism. *Journal of Biosciences, 30*(1), 31–39.

Bernard, C. (1865/1984). *Introduction á l'étude de la médecine expérimentale*. Baillière.

Bertolaso, M. (2011). Hierarchies and causal relationships in interpretative models of the neoplastic process. *History and Philosophy of the Life Sciences, 33*, 515–536.

Bich, L., & Arnellos, A. (2013). Autopoiesis, autonomy and organizational biology: Critical remarks on "Life after Ashby". *Cybernetics and Human Knowing, 19*(4), 75–103.

Bich, L., Mossio, M., & Soto, A. (2020). Glycemia regulation: From feedback loops to organizational closure. *Frontiers in Physiology, 11*, 69.

Bickhard, M. H. (2000). Autonomy, function, and representation. *Communication and Cognition Artificial Intelligence, 17*(3–4), 111–131.

Callebaut, W., Müller, G. B., & Newman, S. A. (2007). The organismic systems approach: Evo-devo and the streamlining of the naturalistic agenda. In R. Sansom & R. N. Brandon (Eds.), *Integrating evolution and development: From theory to practice* (pp. 25–92). MIT Press.

Cannon, W. B. (1929). Organization for physiological homeostasis. *Physiological Reviews, 9*(3), 399–431.

Christensen, W. D., & Hooker, C. A. (2000). An interactivist-constructivist approach to intelligence: Self-directed anticipative learning. *Philosophical Psychology, 13*, 5–45.

Cornish-Bowden, A., Piedrafita, G., Morán, F., Cárdenas, M.-L., & Montero, F. (2013). Simulating a model of metabolic closure. *Biological Theory, 8*(4), 383–390.

Crick, F. (1966). *Of molecules and man*. University of Washington Press.

Cuvier, G. (1805). *Leçons d'anatomie comparée*. Baudoin.

Cuvier, G. (1817). *Le règne animal distribué d'après son organization, pour servir de base à l'histoire naturelle des animaux et d'introduction à l'anatomie compare*. Déterville.

Dobzhansky, T. (1937). *Genetics and the origin of species*. Columbia University Press.

Etxeberria, A., & Umerez, J. (2006). Organización y organismo en la Biología Teórica ¿Vuelta al organicismo? *Ludus Vitalis, 26*, 3–38.

Fox Keller, E. (2000). *The century of the gene*. Harvard University Press.

Gambarotto, A. (2018). *Vital forces, teleology and organization: Philosophy of nature and the rise of biology in Germany*. Springer.

Gerhart, J. C., & Kirschner, M. W. (2007). The theory of facilitated variation. *Proceedings of the National Academy of Sciences of the United States of America, 104*, 8582–8589.

Gilbert, S. F., & Sarkar, S. (2000). Embracing complexity: Organicism for the 21st century. *Developmental Dynamics, 219*(1), 1–9.

Goethe, J. W. (1995). *Collected works, XII. Scientific studies*. Princeton University Press.

Goodwin, B. C. (1963). *Temporal Organization in Cells*. Academic.

Gould, S. J., & Lewontin, R. C. (1979). The spandrels of San Marco and the Panglossian paradigm: A critique of the adaptationist programme. *Proceedings of the Royal Society of London - Series B: Biological Sciences, 205*(1161), 581–598.

Green, S., & Wolkenhauer, O. (2013). Tracing organizing principles – Learning from the history of systems biology. *History and Philosophy of the Life Sciences, 35*, 553–576.

Hordijk, W., Kauffman, S. A., & Steel, M. (2011). Required levels of catalysis for emergence of autocatalytic sets in models of chemical reaction systems. *International Journal of Molecular Sciences, 12*(5), 3085–3101.

Huneman, P. (2006). Naturalizing purpose: From comparative anatomy to the "adventures of reason". *Studies in History and Philosophy of Life Sciences, 37*(4), 621–656.

Huneman, P. (2010). Assessing the prospects for a return of organisms in evolutionary biology. *History and Philosophy of the Life Sciences, 32*, 341–372.

Huneman, P., & Walsh, D. M. (2017). *Challenging the modern synthesis: Adaptation development, and inheritance*. Oxford University Press.

Huneman, P., & Wolfe, C. T. (Eds.). (2010). The concept of organism: Historical, philosophical, scientific perspectives. *History and Philosophy of the Life Sciences, 32*(2–3), 145–424.

Huxley, J. (1942). *Evolution: The modern synthesis*. Allen & Unwin.

Jablonka, E., & Lamb, M. J. (1995). *Epigenetic inheritance and evolution: The Lamarckian dimension*. Oxford University Press.

Jablonka, E., & Lamb, M. J. (2005). *Evolution in four dimensions*. MIT Press.

Jacob, F. (1973). *The logic of life: A history of heredity*. Princeton University Press.

Kant, I. (1790/1987). *Critique of judgment*. Hackett Publishing.

Kauffman, S. (2000). *Investigations*. Oxford University Press.

Laland, K., Uller, T., Feldman, M., Sterelny, K., Muller, G. B., Moczek, A., & Odling-Smee, J. (2014). Does evolutionary theory need a rethink? Yes, urgently. *Nature, 514*, 161–164.

Laland, K. N., Uller, T., Feldman, M. W., Sterelny, K., Muller, G. B., Moczek, A., & Odling-Smee, J. (2015). The extended evolutionary synthesis: Its structure, assumptions, and predictions. *Proceedings of the Royal Society B, 282*(1813).

Laublichler, M. D. (Ed.). (2000). The organism in philosophical focus. *Philosophy of Science, 67*(3), S256–S321.

Letelier, J. C., Cárdenas, M., & Cornish-Bowden, A. (2011). From L'Homme machine to metabolic closure: Steps towards understanding life. *Journal of Theoretical Biology, 286*(1), 100–113.

Longo, G., Montévil, M., & Kauffman, S. (2012). No entailing laws, but enablement in the evolution of the biosphere. In *Proceedings of the 14th international conference on genetic and evolutionary computation conference*, pp. 1379–1392.

Mayr, E. (1961). Cause and effect in biology. *Science, 134*(3489), 1501–1506.

Mesarović, M., & Sreenath, S. N. (2006). Beyond the flat earth perspective in systems biology. *Biological Theory, 1*, 33–34.

Minelli, A., & Fusco, G. (Eds.). (2008). *Evolving pathways. Key themes in evolutionary developmental biology.* Cambridge University Press.

Monod, J., & Jacob, F. (1961). Genetic regulatory mechanisms in the synthesis of proteins. *Journal of Molecular Biology, 3*(3), 318–356.

Montévil, M., & Mossio, M. (2015). Biological organization as closure of constraints. *Journal of Theoretical Biology, 372,* 179–191.

Montévil, M., Mossio, M., Pocheville, A., & Longo, G. (2016a). Theoretical principles for biology: Variation. *Progress in Biophysics and Molecular Biology, 122,* 36–50.

Montévil, M., Speroni, L., Sonnenschein, C., & Soto, A. M. (2016b). Modelling mammary organogenesis from biological first principles: Cells and their physical constraints. *Progress in Biophysics and Molecular Biology, 122*(1), 58–69.

Moreno, M., & Mossio, M. (2015). *Biological autonomy. A philosophical and theoretical enquiry.* Springer.

Moreno, A., & Ruiz Mirazo, K. (1999). Metabolism and the problem of its universalization. *Biosystems, 49*(1), 45–61.

Mossio, M., & Bich, L. (2017). What makes biological organization teleological? *Synthese, 194,* 1089–1114.

Mossio, M., & Moreno, A. (2010). Organizational closure in biological organisms. *History and Philosophy of Life Sciences, 32,* 269–288.

Mossio, M., Saborido, C., & Moreno, A. (2009). An organizational account of biological functions. *British Journal for the Philosophy of Science, 60,* 813–841.

Mossio, M., Montévil, M., & Longo, G. (2016). Theoretical principles for biology: Organization. *Progress in Biophysics and Molecular Biology, 122,* 24–35.

Müller, J. (1837–1840). *Handbuch der Physiologie des Menschenfür Vorlesungen* (Vol. 2 Vols). Verlag von J. Hölscher.

Needham, J. (1937). *Integrative levels: A revaluation of the idea of progress.* Clarendon Press.

Nicholson, D. J. (2014). The return of the organism as a fundamental explanatory concept in biology. *Philosophy Compass, 9,* 347–359.

Nicholson, D. J., & Gawne, R. (2015). Neither logical empiricism nor vitalism, but organicism: What the philosophy of biology was. *History and Philosophy of the Life Sciences, 37,* 345–381.

Noble, D. (1962). A modification of the Hodgkin-Huxley equations applicable to Purkinje fibre action and pace-maker potentials. *Journal of Physiology, 160,* 317–352.

Noble, D. (2017). Systems biology beyond the genome. In S. Green (Ed.), *Philosophy of systems biology.* Springer.

Nunes, N., Moreno, A., & El Hani, C. (2014). Function in ecology: An organizational approach. *Biology and Philosophy, 29*(1), 123–141.

O'Malley, M. A., & Dupré, J. (2005). Fundamental issues in systems biology. *BioEssays, 27,* 1270–1276.

Odling-Smee, J., Laland, K., & Feldman, M. (2003). *Niche construction: The neglected process in evolution.* Princeton University Press.

Oyama, S. (1985). *The ontogeny of information.* Cambridge University Press.

Pereto, J. (2005). Controversies on the origin of life. *International Microbiology, 8,* 23–31.

Peterson, E. (2010). *Finding mind, form, organism, and person in a reductionist age.* PhD Dissertation, 2 vols., Program in History and Philosophy of Science, University of Notre Dame.

Piaget, J. (1967). *Biologie et connaissance.* Éditions de la Pléiade.

Piedrafita, G., Montero, F., Morán, F., Cárdenas, M.-L., & Cornish-Bowden, A. (2010). A simple self-maintaining metabolic system: Robustness, autocatalysis, Bistability. *PLOS Computational Biology, 6*(8): e1000872.

Pigliucci, M., & Müller, G. B. (Eds.). (2010). *Evolution: The extended synthesis.* MIT Press.

Rasmussen, S., Bedau, M. A., Liaohai, C., Deamer, D., Krakauer, D. C., Packhard, N. H., & Stadler, P. F. (Eds.). (2008). *Protocells: Bridging nonliving and living matter.* MIT Press.

Reiss, J. O. (2009). *Not by design: Retiring Darwin's watchmaker.* University of California Press.

Rosen, R. (1991). *Life itself. A comprehensive enquiry into the nature, origin and fabrication of lif.* Columbia University Press.

Rosenberg, A. (2007). Reductionism (and anti-reductionism) in biology. In D. L. Hull & M. Ruse (Eds.), *The Cambridge companion to the philosophy of biology* (pp. 120–138). Cambridge University Press.

Rosslenbroich, B. (2011). Outline of a concept for organismic systems biology. *Seminars in Cancer Biology, 21*, 156–164.

Ruiz-Mirazo, K., & Moreno, A. (2004). Basic autonomy as a fundamental step in the synthesis of life. *Artificial Life, 10*(3), 235–259.

Ruiz-Mirazo, K., Briones, C., & Escoura, A. (2017). Chemical roots of biologicalevolution: The origins of life as a process of development of autonomous functional systems. *Open Biology, 7*, 170050.

Saborido, C., Mossio, M., & Moreno, A. (2011). Biological organization and cross-generation functions. *The British Journal for the Philosophy of Science, 62*, 583–606.

Saetzler, K., Sonnenschein, C., & Soto, A. M. (2011). Systems biology beyond networks: Generating order from disorder through self-organization. *Seminars in Cancer Biology, 21*, 165–174.

Varela, F. J. (1979). *Principles of biological autonomy*. North Holland.

Varela, F. J., Maturana, H., & Uribe, R. (1974). Autopoiesis: The organization of living systems, its characterization and a model. *Biosystems, 5*, 187–196.

Vasas, V., Szathmary, E., & Santos, M. (2010). Lack of evolvability in self-sustaining autocatalytic networks constraints metabolism-first scenarios for the origin of life. *Proceedings of the National Academy of Sciences of the United States of America, 107*, 1470–1475.

von Baer, K. E. (1828). *Über Entwickelungsgeschichte der Thiere: Beobachtung und Reflexion*. Bornträger.

von Bertalanffy, L. (1933). *Modern theories of development*. Oxford University Press.

von Bertalanffy, L. (1952). *Problems of life: An evaluation of modern biological thought*. Watts & Co.

Wagner, G. P., & Laublicher, M. D. (2000). Character identification: The role of the organism. *Theory in Biosciences, 119*, 20–40.

Walsh, D. M. (2015). *Organisms, agency, and evolution*. Cambridge University Press.

Weiss, P. (1963). The cell as unit. *ICSU Review, 5*, 185–193.

West-Eberhard, M. J. (2003). *Developmental plasticity and evolution*. Oxford University Press.

Wiener, N. (1948). *Cybernetics: Or control and communication in the animal and the machine*. The MIT Press.

Wolfe, C. (2010). Do organisms have an ontological status? *History and Philosophy of the Life Sciences, 32*(2–3), 195–232.

Wolkenhauer, O., & Hofmeyr, J.-H. (2007). An abstract cell model that describes the self-organization of cell function in living systems. *Journal of Theoretical Biology, 246*, 461–476.

Woodger, J. H. (1929). *Biological principles. A critical study*. Routledge and Kegan.

Chapter 2
"Organization": Its Conceptual History and Its Relationship to Other Fundamental Biological Concepts

Georg Toepfer

Abstract The conceptual history of the term "organization" begins in Medieval times with the reception and transformation of Aristotle's philosophy of life. It designates the corporeal structure and conditions of identity of natural "organic bodies," a term that had been used to refer to living beings since antiquity. The term played an important role in specifying the ontological status of living beings. At the same time, it offered a basis for their mechanistic understanding. Starting with mechanistic models of life in the second half of the seventeenth century, "organization" and "life" were increasingly used interchangeably. This conjunction of meaning transformed "living beings" into "organisms." Within physiological accounts of the eighteenth century, the living organization was compared to a causal cycle of interdependency. Philosophically, this conjunction was adapted at the end of the century in Kant's philosophy of "organized beings of nature" in which he located the idea of causal cyclicity within a teleological framework and specified an "organized being" in causal terms as a system of interacting and interdependent parts characterized by functional closure. Thus, "organization" refers to the constitution of living beings as a particular kind of causal system. In the nineteenth century, the term achieves the status of a signal word for the life sciences and starts being applied in a wide variety of contexts, from comparative anatomy to physiology and ecology. It was supplemented by two other fundamental notions, namely, "regulation" and "evolution," the first referring to the stabilization and the second to the long-term transformation of natural organizations. The twentieth century saw a further intensification of the complementarity of the perspectives associated with these three terms. Finally, in recent years, a substantial improvement in understanding the causal structure of "organization" was achieved by analyzing it in terms of the "closure of constraints."

G. Toepfer (✉)
Berlin, Germany
e-mail: toepfer@zfl-berlin.org

© The Author(s) 2024
M. Mossio (ed.), *Organization in Biology*, History, Philosophy and Theory of the Life Sciences 33, https://doi.org/10.1007/978-3-031-38968-9_2

23

2.1 Introduction

For a very long time, "organization" has been a central concept in biology. Since antiquity, the material basis of a living being has been called "organic body" ("corpus organicum"). For Aristotle, this meant that the body is an instrument ("organon") of the soul (Bos, 2003). However, since late antiquity, especially since the writings of Galen, an "organic body" was understood as an integrated system in which the parts mutually depend on one another. Thus, they were seen as instruments not only for the soul but also for their own interdependent activity. Galen compared the working order of the organic body to a "symphony" and explained it as "sympathy" or "synergy" in the sense of a "functional organization" (Siegel, 1973, p. 129; Toepfer, 2011a, vol. 2, p. 779). However, the term "organization" was not applied to this functional organization prior to the Middle Ages. Surprisingly, there was apparently no original semantic connection between the Greek expressions for "organization" and "organ": whereas the former was used in the sense of "forming," the latter had a functional meaning from the beginning. Both expressions have therefore been described as "semantically autonomous" (Wolf, 1971, p. 31). Only in the course of their later development were the two notions semantically unified. Starting with mechanistic models of vital processes in the second half of the seventeenth century, "organization" and "life" were increasingly used interchangeably. Since the end of the eighteenth century, "organization" has thus predominantly been understood as a characteristic of living beings, becoming a signal word for the animate world and its scientific analysis. The assumption that it was the organization of their body that constituted the defining characteristic of living beings led to the transformation of "living beings" into "organisms." However, on the level of individual organisms, this did not happen before the end of the eighteenth century; before that, "organism" was used (in parallel to "mechanism") in the sense of "organization," referring to the abstract structure of organisms, not specific individuals (Cheung, 2006). In the nineteenth century, it became common practice to equate life with organization: in the words of the early Neo-Kantian philosopher Kuno Fischer, "Life is organization, self-organization."(Fischer, 1865, p. 534) A few decades later, at the beginning of the twentieth century, "biology," the label that had been given to the science of life, was defined as the "study of the organization of the living" with "organization" being the name for "the association of different elements according to a uniform plan for a common effect" (Uexküll, 1903, p. 269). However, for the longer part of its history, "organization" has functioned as a dummy concept or placeholder for a theory of the living still under negotiation. It was embedded in an explanatory approach in which living beings were seen as functional systems composed of interacting parts. It was only during the last decades that a full-fledged theory of organization was proposed, enabling the term to fulfill the theoretical role it was meant to fulfill since the 1800s.

2.2   The Conceptual History of "Organization"

The term "organization" first appeared in medieval Latin. The word is related to the Greek expression "ὀργάνωσις" meaning "formation, arrangement." The Greek term was used, for example, by Sextus Empiricus and Porphyrios in late antiquity, in the second and third century AD, respectively. Whereas Sextus Empiricus used the Greek expression in his treatise *Against the Logicians* in the more general sense of "arrangement" or "conjunction" (Adversus dogmaticos, 7, 126; Engl. transl. Bett, 2005, p. 114), Porphyrios applied it to the more specific context of bodily structures that living beings of distinct categories have in common (in order to argue against the practice of eating meat): "Almost everyone agrees that animals are like us in perception and in organisation generally with regard both to sense-organs and to the flesh" (De abstinentia 3, 7; Engl. transl. Clark, 2000, p. 84).

Later, in the High Middle Ages, the Latin term "organization" appears in several texts by Thomas Aquinas, especially in reference to Aristotle's second book of *On the Soul*. Here, Thomas closely associated the term with the formation of organic bodies ("formatio et organizatio corporis"; In III Sententiarum distinctio (1254–56): dist. 3, qu. 2, art. 1; dist. 4, qu.2, art. 1) and claimed that, for Aristotle, "organization" was the basic principle of living bodies ("de ratione corporis vivi est organizatio"; In IV Sententiarum distinctio (1254–56) dist. 10, qu. 1, art. 2, quaestiuncula 3, sed contra 2). Furthermore, he stated that the term referred to a multitude of parts and was relevant to the form of the body. In these and other passages, Thomas attributed the concept of an "organic body" to Aristotle, in the sense of a body consisting of a diversity of organs; only those bodies which feature this inner diversity are "organic bodies"; this diversity was said to be "necessary" for living bodies: "dicitur corpus organicum, quod habet diversitatem organorum" (Commentarius in libros de anima II et III: 2, 1, 20 (No 230)). Following these lines of thought, future authors even referred to Aristotle as the "father of animal organization" (Schiller, 1978, p. 84).

However, there is some debate as to whether Aristotle actually saw living beings as organisms (Bolton, 1978; Bos, 2003). In *On the Soul*, the book referred to by Thomas Aquinas, Aristotle claimed that the soul is the form of a natural body that potentially bears life. However, according to Aristotle, the soul as the principle of life was not directly related to the disposition of the organs in the body. For Aristotle, the body was an organ or an instrument of the soul but not necessarily an organized being consisting of a diversity of interacting organs. He did not elucidate the relationship of the soul to the diversity of its organs. "Organic," a term Aristotle apparently introduced into the Greek language, was used by him in reference to something "instrumental." Here, the term does not refer to a diversity of organs and does not even imply an "endowment with organs." Consequently, in the modern and contemporary sense, the concept of organization was not an inherent part of Aristotle's philosophy of living beings.

Aristotle had no specific term for the arrangement or disposition of the organs that provided essence and unity to living beings. In Aristotle's terminology, this function is fulfilled by the soul. However, there are good reasons to assume that the Aristotelean "soul" corresponds well with the medieval and modern idea of organization since both refer to the essence and unity of a living system. There are four particular parallels between the concepts of "soul" and "organization" (Quarantotto, 2010): (1) like the soul, the organization is not itself a body, but a property belonging to the body; (2) soul (or organization) and body do not exist independently of what possesses them, they are not self-sufficient autonomous entities; (3) both are considered the principle of unity and identity of the body that they organize/endow with soul; and (4) soul and organization are both explanatory principles for fundamental organic activities such as movement—the reason why animals have the capacity to move is found in their organization or endowment with a soul.

Thus, prior to the scientific revolution in the seventeenth century, "soul" was a concept perfectly equipped to fulfill the explanatory role that was later taken on by "organization." In fact, the latter term only came into frequent use during the seventeenth century. Until then, it was employed nearly exclusively in the context of scholastic discussions on the changing arrangement of parts in embryonic development.

That situation changed in the early seventeenth century, when, in a commentary on Aristotle's *On the Soul* (1600), "organized" was equated with the state of "potentially having life" (Collegium Conimbricense, 1600, p. 55; for the context, see Des Chene, 2000). Henceforth, the term "organization" entered the academic language, especially as a result of debates on the value of mechanistic models for living beings.

At the beginning of this debate, "organization" was not yet in the position to become the fundamental principle of life. For the Cambridge Platonist Henry More, for example, "organization" was not equivalent to the living state of a being. He postulated a "Plastical Power" that "organized" "duly-prepared Matter," as he called it, "into life" (More, 1659, p. 46). Thus, the "mere organization of the Body" (ibid., p. 107) was not enough to constitute life; this was a merely *mechanical* organization—or, in More's terms, "matter mechanically organized" (ibid., p. 109). Thus, More still differentiated between life as "being ensouled" and life as "organization."

Three years later, in 1662, Joachim Jungius, a mathematician and philosopher of science from Hamburg, announced that "true organization alone" was at least sufficient for plants to perform their life functions of nutrition, growth, and reproduction. Jungius followed Descartes as he denied plants a soul, arguing that their life functions could be explained by the mere disposition and arrangement of their parts: "vero organisatio sola sufficiat" (Jungius, 1662, part. 2, sect. 3). To my knowledge, this is the first instance in which "organization" and "life" were used interchangeably. The exact wording was later resumed by La Mettrie in his description of the relationship between the mental and the material aspect of the brain: "The organization, is it really sufficient for everything? Once again, yes" (La Mettrie, 1747, p. 180).

In the period of dominant mechanistic thinking from the mid-1660s onward, several authors accepted the equation of life and organization. Robert Hooke, for example, used the term "organization" in English. Presumably inspired by his microscopic observations of plant tissues, he claimed that there was an organization common to all vegetables ("the same Schematism or Organization that is common to all Vegetables"; Hooke, 1665, p. 116). In a similar vein, Francis Glisson argued in 1672 that the difference between plants and animals and other bodies could be deduced from their "organization." Thus, life had no cause other than the organization of bodies (Glisson, 1672, p. 226; for the context, see Hartbecke, 2006, p. 165). Additionally, in an important and well-known ontological argument, John Locke reflected on the conditions of identity of living beings whose parts are constantly being exchanged whereas the entirety of the system remains the same. Locke used the term "organization" to refer to this bodily property that always persists even as its parts are exchanged (Locke, 1689, p. 331).

In the 150 years between 1650 and 1800, which could be viewed as the formative period of biology, the ancient principle of life, the "soul," was gradually replaced by "organization." Organization became the central explanatory concept for biology. This was the conceptual revolution at the beginning of biology, which at the same time maintained the ontological specificity of life phenomena, and their mechanistic explainability starting from uniform principles transformed the study of life into an explanatory endeavor that maintained rests on a unifying principle that provided: life was equated with being organized (see also Jacob, 1973, Chap. 2). An important element of this revolution was a reversal of the relationship between the concepts of "life" and "organization": in the seventeenth century, "life" was the more fundamental notion and the living state was thought to somehow determine the organization of the body. However, during the eighteenth century, it was established that it actually works the other way around, a notion that persists until today: "organization" now forms the basic concept from which the determination and consequently the analysis starts; "life" becomes a phenomenon derived from "organization" (see Schiller, 1978, p. 24).

In the first half of the eighteenth century, however, this equation was not accepted by all authors. In order to integrate the living world into the nonliving (and thus advocate for the possibility of a spontaneous generation of living beings), some considered all parts of nature to be organized. This was the position held by Leibniz, for example. He reasoned that since everything is arranged by God, every piece of matter is organized ("la matiere arrangée par une sagesse divine doit estre essentiellement organisée partout"; Leibniz, 1705, p. 342). In the 1720s, this view is supported by microscopic investigations and descriptions of the regular crystalline structure in minerals (see, e.g., Bourguet, 1729, p. 58: "tout est organisé dans la matière"). Thus, until the middle of the eighteenth century, there were influential authors for whom the concept of organization served to unify rather than to separate realms of nature.

For the life sciences, "organization" increasingly served as an important explanatory concept. This is especially true for mechanical approaches toward the generation and transformation of living beings. In fact, in the mid-eighteenth century, it was preformationism, i.e., the idea that the forms of living beings are already existent in their germs, that mainly contributed to the diffusion of the concept (Schiller, 1978, p. 40). In reference to the preexistent structures in the germ, its "organization," it was possible to explain the emergence of complex adult forms as "development."

In the eighteenth century, the idea of organization was mainly associated with mechanical understandings of life and preformationist accounts of individual development. In this context, the term mainly referred to the specific body plan of organisms and could therefore be used not only in developmental studies but also in natural history for the classification of organisms into taxonomic groups. Linnaeus, the master of this approach, defended the view that "organization" was a specific concept within the life sciences, presumably because matter in living beings, in contrast to nonliving ones, was specifically arranged into recurrent, taxonomically significant structures. In the tenth edition of his *Systema Naturae*, the association of "organized" and "living" was made explicit and formalized typographically by characterizing plants and animals as "organized and living" (*organisata & viva*), whereas stones were seen as merely "composite" (*congesta*) (Linnaeus 1758, p. 6). In the mid-eighteenth century, several authors stressed the explanatory value of "organization" in different fields within the life sciences: in 1750, John Turberville Needham argued that vitality, sensation, and thinking appear to be an immediate consequence of "organization" (Needham, 1750, p. 375). In 1772, Voltaire famously defined life as organization: "La vie est organisation avec capacité de sentir" (Voltaire, 1772, p. 55). Some years later, Diderot claimed that the soul is nothing but organization and life: "L'organisation et la vie, voilà l'âme" (Diderot, 1778, p. 358). At the end of the century, Christoph Girtanner also directly connected the living state with being organized: "Les mots organisé & vivant sont, selon moi, des synonimes" (Girtanner, 1790, p. 150). Even in Kant, in his *Opus postumum*, one can find this equation of "being alive" and "having an organization" (Kant, OP., AA XXI, p. 66). In his major work on the epistemology of biology, the *Critique of Judgement*, Kant (AA V) called living beings "organized beings in nature," although he avoided the term "living" because it had a terminological use in his ethical writings.[1]

Since the beginning of the nineteenth century, "organization" has been regarded as one of life's most essential aspects. It proved to be valuable for the self-understanding of biology as an independent natural science. This was particularly evident in situations where life was either reduced to the merely mechanical or explained with additional supernatural "life forces." In the mid-nineteenth century, Claude Bernard essentially relied on the concept of organization when he rejected vitalistic approaches. According to Bernard, all manifestations of life are not to be derived from a mysterious life force, but from "the phenomena of organization" (Bernard, 1867, p. 138). At the beginning of the twentieth century, in light of the growing struggle between mechanism and vitalism, the study of life was simultaneously confronted with the postulation of mysterious vital principles and reductionist views that denied any peculiarity of vital phenomena and their scientific explanations. In this situation, "organization" was propagated as a concept that offered a

---

[1] However, things are complicated in Kant as he also has the concept of nonliving natural purposes which are the plants. Hence, for Kant, not all organized beings are living. I thank Gertrudis Van de Vijver for pointing to this; see also Piché, 2001.

third way between these two metaphysical positions and, thus, was seen as a way out of the fundamental dispute in theoretical biology. On the one hand, the assumption of a central guiding vital force was considered unnecessary, because, if seen as functional organizations, living beings and the orderly processes within them could be described as the outcome of a decentralized structure of interacting parts. On the other hand, "organization" was understood (in the Kantian sense) as an additional principle that is not part of a purely mechanical approach, because it added the aspect of integrating isolated causal relations into a coherent functional whole. In 1900, Oscar Hertwig argued that the explanation of life should neither introduce mysterious forces nor follow the "mechanistic dogma," according to which "life with all its complicated phenomena is nothing than a physico-chemical problem" (Hertwig, 1900, p. 24). Instead, Hertwig argued for recognizing "that life is based on a peculiar organization of the substance" (ibid, p. 4). Via the concept of organization, biology could thus take a third path and navigate between the approaches of vitalism and mechanism, thus securing biology's status as a natural science and its methodological autonomy from physics (see also Wolfe's contribution to this volume). In the twentieth century, biological research programs that aimed to find life on other planets and create life in the lab found the organizational approach to be more stable than any purely material or molecular characterization of life: "The peculiarity of life is not due to some chemical mystery but to organization" (Bertalanffy, 1928, p. 68–9).

Hence, at least for 250 years, "organization" has been one of biology's basic if not one of its most fundamental concepts that which explains what life is. This poses a question: What is "organization"? What does the term actually mean?

2.3 The Meaning of "Organization"

A fairly good, but still very open, definition can be found in the *Encylopédie*: "Organization" is defined as "the arrangement of parts that constitute a living body" (Anonymous, 1765, p. 629). Thirty years later, Kant, in a letter to Sömmering, contributed an important amendment: he expanded this definition by including purposiveness. For him, "organization" was "the purposeful and in its form persistent arrangement of parts" (Kant, 1795, p. 33). In his works on natural philosophy, Kant had a very specific understanding of purposefulness. As is well known, Kant stated that in a thing as a natural purpose, the parts are "reciprocally the cause and effect of their form" (Kant, 1790, p. 373). This means in a thing as a natural purpose the parts' *very existence* depends on the system's other parts. Kant stresses this ontological dependency in another passage: "For a body [...] which is to be judged as a natural end in itself and in accordance with its internal possibility, it is required that its parts reciprocally produce each other, as far as both their form and their combination is concerned, and thus produce a whole out of their own causality" (ibid.).

Kant did not indicate the origin of his idea of reciprocity as a condition for a thing to be a natural purpose. However, similarities in the wording and his personal contacts suggest that the idea was inspired by the Leiden physiologist Herman Boerhaave (see Toepfer, 2011b). In 1727, Boerhaave provided a definition for the concept of an "organic body" in which the decisive moment is the interdependence of the parts (*harum partium actiones ab invicem dependent*; Boerhaave, 1727, p. 3). To be sure, the emphasis on reciprocity as a hallmark of organic systems has its roots in Antiquity (see Toepfer, 2011a, vol. 3, pp. 738–763). However, only with the physiological theories since the end of the seventeenth century did it acquire a fundamental role in the identification and definition of living beings. This process took place in parallel with the introduction of the concept of "organism." Georg Ernst Stahl, who proposed this notion in 1684, already described the relationship of the parts in an organism as an "adaptation of forms" (*aptatio configurationis*) and a dynamic interaction of a single part with the others (*cum aliis partibus cohaerens, conspirans, atque communicans*) (Stahl, 1707, p. 17). The parts in an organism would act "reciprocally and together" (*mutua & socia*) and thus be interrelated (ibid.). According to Stahl, the whole complex of the diverse organs in an organism forms a functional unity since the ultimate purpose of all movements is to preserve the body. The concept of "organism" thus establishes a causal model for a functionally closed and self-referential system of heterogeneous components. During the first half of the eighteenth century, Boerhaave and other mechanistically minded physiologists invoked the image of a "circle" (*circulo quasi*) for the causal pattern of organic systems, consequently firmly anchoring any discussions of causal reciprocity in physiological language (*mutuas causæ vices & effectuum gerant*) (Boerhaave, 1708, p. 11).

This physiological understanding of the interactions within an organic body formed the background for Kant's understanding of "organization" in terms of causal reciprocity. Kant's philosophical contribution was the integration of teleology into this causal understanding of natural systems of interdependent parts as well as the clarification of the peculiar metaphysical and ontological status of organisms or, in his terms, "things as natural purposes," with respect to the explanatory level of causal mechanisms. In doing so, in combining teleology and cyclicity, Kant gave a justification of teleology within biology as the science of cyclical organized systems: the teleological way of thinking by focusing on outcomes of processes is justified in biology because biology is the study of systems consisting of interdependent parts in which the final state of one process is important for the existence of the other parts of the system and ultimately for its own maintenance (as a type of process or part).

Kant's philosophy of the organic was widely received in the years around 1800. Disciples of Kant gave explicit definitions of "organization" with Kant's philosophy in mind, for example: "Organization is the disposition of a body in which every part is at the same time means and ends to all the others" (Schmid, 1799, p. 274). Not only philosophers but also practicing biologists accepted this foundational role of reciprocity and teleology for the specification of their objects of study. One example by a famous author: "A living body is a natural organized body composed of different kinds of parts which act and react upon each other" (Lamarck, 1797, pp. 249–50). Lamarck repeatedly formulated concise sentences which express the close connection between the state of being alive and organization or order. For him, life constitutes a "physical phenomenon resulting from the order of things and from the state of the parts," their "organization" itself being a "physical phenomenon" (Lamarck, 1815, p. 60; 122; see Schiller, 1978, p. 70). "Life" was explained as an "ensemble of functions" with the functions being nothing but "acts of the organization and its pars" (ibid., p. 59). As these short quotations make clear, Lamarck had a dynamical understanding of "organization"; in his view, it is linked to movements caused by the arrangement of the parts within a body.

In the first half of the nineteenth century, however, the static interpretation of "organization" proved to be at least as important as the dynamic view.[2] In comparative anatomy, one of the dominant research areas at that time, "organization" was understood as the "disposition" of the parts in an organic body; it referred to the configuration of the parts, the "body plan." Anatomy with its focus on the spatial arrangement of parts within a body has even been called "the science of organization" (Schiller, 1978, p. 88). In this context, the analysis of the "organization" of body plans formed the foundation for the classification of animals into larger groups. For Georges Cuvier, one of its main representatives, comparative anatomy is the study of "the laws of organization of animals and of the modifications this organization shows in different species" (Cuvier, 1817, vol. 1, p. v). In Cuvier's taxonomic system, the arrangement of the nervous system was of particular importance; it provided the basis for the classification of all animals into four main "branches" (see Figlio, 1976; Guillo, 2003). Here, "organization" was an important notion because it stressed the interdependence of the parts. In comparative anatomy, this interdependence was not primarily a causal notion but referred to the observation that traits of the body plan covary and do not exist independently from one another. Besides that, "organization" was also used as a measure for the complexity or "degree of perfection" of body plans. Even Charles Darwin, who was generally skeptical of the idea of progress in the history of life on earth, held the view that natural selection would result in an "improvement" that inevitably led to "the gradual advancement of the organization" (Darwin, 1860, p. 117).

For the philosophy of biology and its reflection on the ontological peculiarity of living beings, the important aspect was not this morphological concept of organization but the physiological meaning of the term. It was in the years around 1800 that the essential aspects of the concept were established—those that have persisted ever since. "Organization" now referred to the disposition of the parts in a certain type of causal system; the pattern of causal interactions has the form of a cycle because the parts mutually depend on one another's influence, resulting in a functional "closure."[3]

---

[2] Therefore, most of the historical accounts of the concept of "organization" in the history of biology focus on this aspect (see Figlio, 1976, Schiller 1978, and Guillo, 2003).

[3] This currently prominent term that is most often derived from Piaget (1967, p. 182) also has its roots in early nineteenth-century reflections on the ontological status of organisms. Georges Cuvier, for example, put it this way: "Tout être organisé forme un ensemble, un système unique et clos, dont toutes les parties se correspondent mutuellement, et concourent à la meme action définitive par une réaction réciproque" (Cuvier 1812, vol. 1, p. 58).

Later definitions elaborated on these points, especially by highlighting the self-referential character of organizations. In 1928, Helmuth Plessner explained: "Organization is the mode of existence of the living body, which must differentiate itself and through which it generates the inner teleology according to which it is formed and functions" (Plessner, 1928, p. 170). Thus, an organization consists of differentiated functional parts, which, through their activity, generate and permanently regenerate the entire system. The same point was expressed in the theory of autopoiesis since the early 1970s, in which the "living organization" was characterized as a perpetual self-regeneration (Maturana et al., 1974). In this tradition, organization was defined as "the complex of interaction and properties of structure that make the perpetuation of structure possible" (Kolasa & Pickett, 1989, p. 8837).

Many authors have stressed the close connection between the concept of organization and teleology (except Maturana and his co-workers). Following Kant, one could say that "organization" and "function" or "purpose" go hand in hand: wherever there is organization in nature there is function and vice versa. As John von Neumann once said in conversation with Colin Pittendrigh, "Organization has purpose; order does not" (Pittendrigh, 1993, p. 20). Since functional reasoning and exploring purposes are essential to the domain of biology, it makes sense that "organization" has become a fundamental concept for that science—in contrast to physics, as "the physical sciences don't deal in function" (Wicken, 1987, p. 40). Consequently, "organization" and "function" are frequently regarded as crucial to any attempt to justify the autonomy of biology as a science. Since Kant, this position has been defended by many authors, and teleology was even defined as the "philosophy of biology" because "the organism requires teleological consideration" (Kühnemann, 1924, p. 494).

It is a striking feature of biology that functions have been ascribed to living systems long before the causal pattern of their working order had been understood. Surprisingly, the basic inventory and supposition of functions have changed very little throughout the long history of biology. In fact, Aristotle had already named them: *nutrition*, *growth*, *movement*, *sensation*, and *reproduction*. However, it took more than 2000 years before biology began to understand the way in which they are realized in living beings. Functional knowledge is therefore a one-way kind of knowledge; it reduces the complexity of a system without necessarily having a complete understanding of it. Or, in other words: "organization [and hence function] emerges as a problem when there is too much knowledge in one direction and too little in another" (Beckner, 1959, p. 10).

The integrative power of the concept of "organization" in biology can also be demonstrated by explicating the fundamental properties of living beings as based on this concept. Metabolism, reproduction, development, metamorphosis, and evolution are fundamental aspects of life that can be described and analyzed, respectively, as maintenance, transmission, expansion, individual transformation, modification, or as the supraindividual transformation of organization. Based on this universal applicability, from cell theory to evolution, "organization" has been called "the key concept at all levels" for the life sciences (Figlio, 1976, p. 34).

2.4 "Organization" as One of the Three Basic Principles of Biology

"Organization" refers to the constitution of a system of interdependent parts. Two other important aspects of such a system that are related but not solely reducible to its constitution refer to the *permanence* and the *transformation* of the system. In well-known biological terms, they are called *regulation* and *evolution*. The three concepts refer to related but different aspects of organized systems: the causal pattern of their constitution, the capacity to control their relationship to the environment, and the potential for long-term transformation. The general meaning of all three principles operates on the same level of abstraction.[4]

Organization, in the systems-theoretical, Kantian tradition, essentially refers to the mutual dependence of parts in a system. *Regulation* refers to the stabilization of an organization by controlling environmental influences. Basically, regulation covers three processes: (1) supplying the system with necessary materials and other factors from its surroundings, (2) protecting it from detrimental influences, and (3) coordinating and integrating all the processes within the organized system. Taken together, they ensure the maintenance of the system, its preservation, and its perpetuation through time by managing the system's relationship with the environment. However, controlling the relation to the environment is not a conceptually necessary feature of organized systems. We can think of organized systems that are not regulated. Ecosystems might be an example for systems that are most certainly organized as their parts depend on each other. However, at least conceptually, we may think of them as not being controlled but more vulnerable to disturbances than organisms. In simple terms, *evolution* can be defined as the transgenerational transformation of organizations due to differential reproduction that is due to selection or genetic drift. The distinctness of "evolution" as a fundamental concept might be less controversial. Of course, we can imagine organisms that do not evolve, and most biologists did so until 1859.

The introduction of these three fundamental concepts in biology, organization, evolution, and regulation can be related to three conceptual turns or even "revolutions" in biology. They took place in the eighteenth, nineteenth, and twentieth century, respectively. The first is the revolution that established biology as a distinct scientific discipline at the end of the eighteenth century. It resulted in the conception

---

[4] There have been several attempts to separate the aspects of organization, regulation, and evolution of systems. Especially the relation between organization and regulation has been investigated from different angles, i.e., from the angle of economics with the distinction between internal order (organization) and external interventions (regulation) (Sombart, 1925) or the attempt to distinguish general systems theory from cybernetics (Bertalanffy 1951) or the efforts of autopoiesis theory to differentiate between (internal) self-organization and (external) control (Varela 1979). In all these cases, organization concerns the system-constituting internal structure of a dynamic entity, the regulation of its relation to the environment, especially the mechanisms of maintenance in the face of disturbances. For recent attempts to connect organization, closure, and regulation, see Di Paolo (2005) and Bich et al. (2016).

of living beings as organisms. The second revolution was connected to evolution, to the insight that all life on earth is united in one all-encompassing process of transformation. The third revolution, the regulation revolution, took place mainly in the middle of the twentieth century and resulted in the description of organisms as cybernetic systems of control and information flow comparable to man-made machines.

One may think of organization, regulation, and evolution as a rather symmetrical trio: "organization" as the central category concerns the constitution of a system, "regulation" its stabilization, and "evolution" its transformation. However, it is also possible to derive the concepts from one linear argument. This argument begins with "organization" as the fundamental descriptive term for the constitution of living systems. It basically identifies a cycle of interdependent processes. Apart from this internal cycle, which grounds Kant's internal purposiveness, there is an external cycle, a cybernetic feedback cycle that relates the system to its environment and stabilizes the system—the fundamental point of regulation. Regulation is directed toward the perpetuation of the system in time. This can be realized through two mechanisms: by stabilizing the individual system or by its multiplication, by the production of similar systems. Thus, we have two forms of self-preservation in organized systems: One is the regulation that consists in the preservation of individual systems, the maintenance of the dynamic state of an individual organism by devices for nutrition and protection. The other is preservation by multiplication of organizational types, which biologists call *reproduction*, the maintenance of the organizational structure of an individual by its multiplication in new individuals with a similar organization. In this view, reproduction is a preservation strategy by means of perpetuating an organizational type. Ironically, this most effective way of preservation has resulted in the vast process of transformation we call evolution.[5] Therefore, reproduction leads to two contradictory consequences: On the one hand, it emerged as the most efficient means of self-preservation ("preservation by multiplication"). On the other hand, however, since it allows for variation (as mutations are inevitable and often even functional), this eventually results in the transformation of these systems. Ironically, preserving organization therefore means to transform it. As Paul Valéry surmised in an elegant and paradoxical formula: "Bios. Se transformer et transformer pour conserver" (Valéry, 1933, p. 755).

Thus, according to this argument, evolution is a derived feature of organized systems and regulated with respect to their maintenance. By seeking to maintain their systems through the most effective means at their disposal—reproduction—they initiate the process of transformation, which, in the long run, will erode their organization, at least in its original form.

---

[5] For the analysis of the relationship between organization and evolution, see also Ruiz-Mirazo et al. (2004) and Walsh (2006).

2.5 Organization, Constraints, and Morphology

However, for evolution to begin in the first place, there must be an organized system of interdependent parts directed toward its own (or its type's) maintenance (for a critique of this view, see Ruiz-Mirazo et al., 2017). To understand the embodiment of such a system, the central concept of "constraints" has proven useful. Its conceptual history goes back to Gauss' principle of least constraint in classical mechanics. In the context of organization, it refers to the material structure or configuration of parts in a system that has a harnessing or channeling influence on the flow of energy within the system, for example, the structure of an organism's body, which serves as a boundary condition for physical laws. This general idea was described by Franz Reuleaux in his *Theoretical Kinematics* (1875).

Reuleaux is considered the founder of what has later been called "machine morphology." According to Reuleaux, a machine is a "compound of resistant bodies, which is disposed in such a way that mechanical laws of nature are constrained to be effective under certain conditions" (Reuleaux, 1875, p. 38). This means that the effectiveness of a machine depends on the disposition of its parts, the structure of the whole, or its morphology. Morphology works by constraining the laws of nature. The machine does not introduce additional laws of nature; it simply channels or harnesses general laws through its morphology.

A hundred years later, Michael Polanyi applied this line of reasoning to biological systems. According to Polanyi, an organism has the same general makeup as a machine: its bodily structure serves as a boundary condition harnessing physical-chemical processes. In the case of the organism, this harnessing serves its organic functions (1968, p. 1308) Thus, quite surprisingly, life's irreducible structure rests on the very thing living beings have in common with machines: their specific structure (or morphology) that functions as a boundary condition in constraining laws of nature.

Starting in the late 1960s, Howard Pattee elaborated on this by stressing that it is not just the possession of functional constraints that is unique to living beings but the production and coordination of these constraints *by* the system itself. As he put it in 1971: "Life is distinguished from inanimate matter by the co-ordination of its constraints" (Pattee, 1971, p. 273). Organisms are embodied structures that produce their own structure, which feeds back on itself by maintaining and regenerating it. Thus, the boundary conditions or "constraints" of the system are, in the case of organisms, self-imposed. The structure of organismic bodies channels the energy in such a way that the body is preserved or rebuilt. The basic pattern is that of a cycle.

In recent years, a more detailed and precise rendering of this pattern has been provided by Alvaro Moreno and his collaborators. Since the early 1990s, his group has been describing the circular organization of systems on the basis of their components functioning as constraints. In their view, local constraints within the system are generated by the activity of components of the system (Moreno et al., 1994, p. 17; see also Ruiz-Mirazo & Moreno in this volume). They call the resulting system an "autonomous organization" since the system itself generates and regenerates its constraints.

Stuart Kauffman called this circular organization "a virtuous cycle," a cycle of works and constraints: The work of the system generates constraints, namely, precisely those constraints required for the work to be done. According to Kaufman, this cycle is "the heart of a new concept of 'organization'" (Kauffman, 2000, p. 4).

In the last few years, Matteo Mossio has further elaborated on these matters. According to Mossio and his colleagues, biological organization is characterized by the fact that it realizes a specific kind of causal regime. This regime is based on nothing but the material structure of an organism acting as constraints for the physical laws. Because the constraints within the organization are mutually determined, the "organizational closure" of organisms consists in a "closure of constraints" and therefore in a biological self-determination (Montévil & Mossio, 2015, p. 180).

An important consequence of this account is that biological autonomy, in the double sense of biology as a distinct discipline and of organisms as self-determining systems, is entirely based on structures. Nothing but the structure of an organism embodies the constraints that effectively control the boundary conditions for the laws of nature. Biological autonomy is grounded in the material form of living bodies. Biology's distinct causal regimes, once referred to as "life-forces," are embodied in the forms of organisms.

With this emphasis on form, the biological subdiscipline of *morphology*, which has been marginalized for over a century now, is again taking center stage. Ultimately, it is morphology that provides the basis for the organism as an integrated autonomous system because it provides the only factor beyond the laws of nature that is specific to organisms. Insofar as organisms are considered to be autonomous, they are determined by their form. Form is the only additional factor that distinguishes organisms from inorganic bodies. This is true, at least at the explanatory level, because forms provide the only specific biological causal factor. To put it bluntly: The only life forces that exist are life forms.

Thus, morphology, the study of forms, is the fundamental explanatory principle of biology. It has always been fairly easy for biologists to identify functions in living beings. Aristotle was famously prolific at it, and his well-known functional categories, such as nutrition, growth, and reproduction, are still being used today. However, although they may still define what it means to be alive, functions do not necessarily provide causal understanding of processes. In biology, this is done by identifying mechanisms; and mechanisms are based on morphology, because it is morphology that identifies the structures that function as (self-)constraints within natural organizations, establishing them as a distinct type of material bodies.

Organic forms, then, are the mediators for the realization of biological organizations. They instantiate these organizations in specific living bodies; in their function as particular "constraints," they enable the causal interdependence of the components and the self-referentiality of the whole system. Thus, "organization" and "form" are two complementary aspects of living bodies: the first refers to the causal pattern that constitutes the unity of the system and the second to the individuality of the system and the specific material "constraints" by which this causal pattern is realized and instantiated in concrete living beings. Or, in other words, "organization" provides the law-like universal feature of all living beings (existent and potential), "forming" the physical realization of its causal pattern in distinct instances. Biological explanations demonstrate how forms are effective as functions, i.e., how they are integrated in functional closure. The fixed form of the heart chambers (in a particular individual or in the type of individuals of a certain class) explains how this form constrains the general laws in order to achieve closure.

"Organization" and "form" both fulfill *descriptive* and *explanatory* functions in biology, albeit at different levels: The first visible feature of living beings is, of course, their form; forms are *described* and are the basis for biological classifications (as they indicate genealogical relationships). However, forms also *explain* (and only they can do that) how the forces and energy flows in an organism are channeled to realize the functional closure that is characteristic of every living being. "Organization," on the other hand, is *descriptive* with respect to the causal pattern that all living beings have in common; this pattern, causal "cyclicity" or functional "closure," is identified when an entity is described as being organized.[6] However, on a more abstract level, "organization" gives the *explanation* to the fact that all living beings share a certain functional order.[7] This explanation consists in the specification of the causal pattern or the working order of every organism, namely, the self-referentiality of all their activities, their orientation toward self-maintenance (what the developmental biologist Wilhelm Roux once called the "autophely" or "self-utility" of organisms; Roux, 1895, p. 58). Since "organization" specifies this causal and functional pattern common to all living beings, it grounds the biological approach toward nature. At the same time, as there are living beings on earth that do not always behave autophelically, "organization" also marks the end of the biological perspective focused on autophely: "Man has reached a level of existence that stands above purpose. It is his distinctive value that he can act without purpose" (Simmel, 1918, p. 28).

References

Anonymous. (1765). Organisation. In *Encyclopédie ou Dictionnaire raisonné des sciences, des arts et des métiers* (Vol. 11, p. 629). Briasson.
Beckner, M. (1959). *The biological way of thought*. Columbia University Press.

---

[6] For Kant, this identification procedure takes place in reflective judgments, and he is very clear about his view that they are not explanatory, but descriptive: "positing ends of nature in its products, insofar as it constitutes a system in accordance with teleological concepts, belongs only to the description of nature" (Kant, 1790, p. 417).

[7] "Form," however, is also a concept that involves abstractions. In most biological contexts, it does not refer to individual bodies but to structural aspects that numerous individuals (of one species or of another taxonomic unit) have in common. Nevertheless, for every specific causal interaction, it is the realization of one form in an individual organism that serves as constraint.

Bernard, C. (1867). *Rapport sur les progrès et la marche de la physiologie générale en France.* L'Imprimerie Impériale.

Bich, L., Mossio, M., Ruiz-Mirazo, K., & Moreno, A. (2016). Biological regulation: Controlling the system from within. *Biology and Philosophy, 31,* 237–265.

Boerhaave, H. (1708). *Institutiones medicae.* Linden.

Boerhaave, H. (1727). *Historia plantarum.* Gonzaga.

Bolton, R. (1978). Aristotle's definitions of the soul: *De anima* II, 1–3. *Phronesis, 23,* 258–278.

Bos, A. P. (2003). *The soul and its instrumental body. A reinterpretation of Aristotle's philosophy of living nature.* Brill.

Bourguet, L. (1729). *Lettres philosophiques sur la formation des sels et des crystaux et sur la génération & le méchanisme organique des plantes et des animaux.* Honore.

Cheung, T. (2006). From the organism of a body to the body of an organism: Occurrence and meaning of the word 'organism' from the seventeenth to the nineteenth centuries. *The British Journal for the History of Science, 39,* 319–339.

Collegium Conimbricense. (1600). *Commentarii Collegii Conimbricensis Societatis Iesv, in tres libros De anima Aristotelis Stagiritae Coimbra.* Zetzner.

Cuvier, G. (1812). *Recherches sur les ossemens fossils de quadrupèdes* (4 vols.). Deterville.

Cuvier, G. (1817). *Le règne animal, distribué après son organisation* (4 vols.). Deterville.

Darwin, C. (1860). *On the origin of species* (rev ed.). Appleton.

de La Mettrie, J. O. (1747). L'Homme machine (A. Vartanian, Ed.). Princeton University Press 1960.

de Lamarck, J. B. (1815). *Histoire naturelle des animaux sans vertèbres* (Vol. 1). Verdière.

Des Chene, D. (2000). *Life's form. Late Aristotelian conceptions of the soul.* Cornell University Press.

Di Paolo, E. A. (2005). Autopoiesis, adaptivity, teleology, agency. *Phenomenology and the Cognitive Sciences, 4*(4), 429–452.

Diderot, D. (1778). *Éléments de physiologie* (P. Quintili, Ed.). Champion 2004.

Figlio, K. M. (1976). The metaphor of organization: an historiographical perspective on the biomedical sciences of the early nineteenth century. *History of Science, 14,* 17–53.

Fischer, K. (1865). *System der Logik und Metaphysik oder Wissenschaftslehre* (2nd ed.). Bassermann.

Girtanner, C. (1790). Mémoires sur l'irritabilité, considérée comme principe de vie dans la nature organisée. *Observ Phys History Natural Arts, 37,* 139–154.

Glisson, F. (1672). *Tractatus de natura substantiae energetica, seu de vita naturae.* Flesher.

Guillo, D. (2003). *Les figures de l'organisation. Sciences de la vie et sciences sociales au XIXe siècle.* Presses universitaires de France.

Hartbecke, K. (2006). Metaphysik und Naturphilosophie im 17. Jahrhundert. Francis Glissons Substanztheorie in ihrem ideengeschichtlichen Kontext. Niemeyer.

Hertwig, O. (1900). *Die Entwicklung der Biologie im 19. Jahrhundert.* Fischer.

Hooke, R. (1665). *Micrographia.* Martyn.

Jacob, F. (1973). *The logic of life* (1970), transl. by Betty E. Spillmann. Pantheon.

Jungius, J. (1662). *Doxoscopiae physicae minores.* Naumann.

Kant, I. (1790). *Critique of the power of judgment* (P. Guyer, Ed.). Cambridge University Press 2004.

Kant, I. (1795). [Letter to Samuel Thomas Soemmerring, 10 Aug. 1795]. In Akademie-Ausgabe, vol. XII. de Gruyter 1922, pp. 30–35.

Kant, I. (OP). Opus postumum. In: A. Buchenau (Ed.), Kant's Opus postumum, vol. 1 (= Akademie Ausgabe, vol. XXI). de Gruyter 1936.

Kauffman, S. (2000). *Investigations.* Oxford University Press.

Kolasa, J., & Pickett, S. T. A. (1989). Ecological systems and the concept of biological organization. *Proceedings of the National Academy of Sciences of the United States of America, 86,* 8837–8841.

Kühnemann, E. (1924). Kant (Vol. 2). Das Werk Kants und der europäische Gedanke. Beck.

Lamarck, J. B. (1797). *Mémoires de physique et d'histoire naturelle*. Auteur.

Leibniz, G. W. (1705). Considérations sur les principes de vie, et sur les natures plastiques. Philosophische Schriften, Vol. 4. Suhrkamp 1996, pp. 327–347.

Locke, J. (1689). *An essay concerning human understanding*. Clarendon Press 1979.

Maturana, H. R., Varela, F. J., & Uribe, R. (1974). Autopoiesis: The organisation of living systems, its characterization and a model. *Biosystems, 5*, 187–196.

Montévil, M., & Mossio, M. (2015). Biological organization as closure of constraints. *Journal of Theoretical Biology, 372*, 179–191.

More, H. (1659). *The immortality of the soul*. Nijhoff 1987.

Moreno, A., Umerez, J., & Fernandez, J. (1994). Definition of life and the research program in artificial life. *Ludus Vitalis, 2*, 15–33.

Needham, J. T. (1750). *Nouvelles observations microscopiques, avec des découvertes intéressantes sur la composition et la décomposition des corps organisés*. Ganeau.

Pattee, H. H. (1971). Physcial theories of biological co-ordination. *Quartery Review of Biophysics, 4*, 255–276.

Piaget, J. (1967). *Biologie et connaissance*. Gallimard.

Piché, C. (2001). Kant et les organismes non vivants. In L. Cournarie & P. Dupond (Eds.), *Préparer l'agrégation de philosophie. La nature* (pp. 83–93). Ellipses.

Pittendrigh, C. S. (1993). Temporal organization: Reflections of a Darwinian clock-watcher. *Annual Review of Physiology, 55*, 16–54.

Plessner, H. (1928). *Die Stufen des Organischen und der Mensch*. de Gruyter 1975.

Polanyi, M. (1968). Life's irreducible structure. *Science, 160*, 1308–1312.

Porphyrios, De abstinentia. In *Opuscula selecta* (A. Nauck, Ed.). Leipzig 1886, Engl.: On abstinence from killing animals, transl. by G. Clark. Ithaca. Cornell University Press 2000.

Quarantotto, D. (2010). Aristotle on the soul as a principle of biological unity. In S. Föllinger (Ed.), *Was ist ,Leben'? Aristoteles' Anschauungen zur Entstehung und Funktionsweise von Leben* (pp. 35–53). Steiner.

Reuleaux, F. (1875). *Theoretische Kinematik. Grundzüge einer Theorie des Maschinenwesens*. Vieweg.

Roux, W. (1895). Ziele und Wege der Entwickelungsmechanik. In *Gesammelte Abhandlungen über Entwickelungsmechanik der Organismen* (Vol. 2, pp. 55–94). Engelmann.

Ruiz-Mirazo, K., & Moreno, A. (this volume). On the evolutionary development of biological organization from complex prebiotic chemistry. In M. Mossio (Ed.), *Organization in biology*. Springer.

Ruiz-Mirazo, K., Peretó, J., & Moreno, A. (2004). A universal definition of life: Autonomy and open-ended evolution. *Origins of Life and Evolution of Biospheres, 34*, 323–346.

Ruiz-Mirazo, K., Briones, C., & Escoura, A. (2017). Chemical roots of biological evolution: The origins of life as a process of development of autonomous functional systems. *Open Biology, 7*, 170050.

Schiller, J. (1978). *La notion d'organisation dans l'histoire de la biologie*. Maloine.

Schmid, C. C. E. (1799). *Physiologie philosophisch bearbeitet* (Vol. 2). Akademische Buchhandlung.

Sextus Empiricus, Adversus dogmaticos libros quinque (= Adversus mathematicus VII–XI). Against the Logicians, transl. by R. Bett. Cambridge University Press 2005.

Siegel, R. E. (1973). *Galen on psychology, psychopathology, and function and diseases of the nervous system*. Karger.

Simmel, G. (1918). *The view of life. Four metaphysical essays with journal aphorisms* (J. A. Y. Andrews, & D. N. Levine, Trans.). The University of Chicago Press 2010.

Sombart, W. (1925). *Die Ordnung des Wirtschaftslebens*. Springer.

Stahl, G. E. (1707). *De vera diversitate corporis mixti et vivi*. Orphanotropheum.

Toepfer, G. (2011a). *Historisches Wörterbuch der Biologie* (3 vols.). Metzler.

Toepfer, G. (2011b). Kant's teleology, the concept of the organism, and the context of contemporary biology. *Logical Analysis and History of Philosophy, 14*, 107–124.

Valéry, P. (1933). Bios. In Cahiers, Vol. 2. Gallimard, 1974.

Varela, F. J. (1979). *Principles of biological autonomy*. North-Holland.

Voltaire (1772). Vie. In Questions sur l'encyclopédie, vol. 9. pp. 55–58.

von Bertalanffy, L. (1928). *Kritische Theorie der Formbildung*. Borntraeger.

von Bertalanffy, L. (1951). Towards a physical theory of organic teleology. *Human Biology, 23*, 346–361.

von Uexküll, J. (1903). Studien über den Tonus, I. Der biologische Bauplan von *Sipunculus nudus*. *Zeitschrift für Biologie, 44*, 269–344.

Walsh, D. (2006). Organisms as natural purposes: The contemporary evolutionary perspective. *Studies in History and Philosophy of Biological and Biomedical Sciences, 37*, 771–791.

Wicken, J. S. (1987). *Evolution, thermodynamics, and information: Extending the Darwinian program*. Oxford University Press.

Wolf, J. H. (1971). Der Begriff »Organ« in der Medizin. Grundzüge der Geschichte seiner Entwicklung. Fritsch.

Wolfe, C. T. (this volume). Varieties of organicism: A critical analysis. In M. Mossio (Ed.), *Organization in biology*. Springer.

Chapter 3
Varieties of Organicism: A Critical Analysis

Charles T. Wolfe

Abstract In earlier work I wrestled with the question of the "ontological status" of organisms. It proved difficult to come to a clear decision, because there are many candidates for what such a status is or would be and of course many definitions of what organisms are. But what happens when we turn to theoretical projects "about" organisms that fall under the heading "organicist"? I first suggest that organicist projects have a problem: a combination of invoking Kant, or at least a Kantian "regulative ideal," usually presented as the *epistemological* component (or alternately, the complete overall vision) of a vision of organism – as instantiating natural purposes, as a type of "whole" distinct from a merely mechanistically specifiable set of parts, etc. – and a more ontological statement about the inherent or essential features of organisms, typically presented according to a combination of a "list of heroes" or "laundry list" of properties of organisms. This amounts to a category mistake. Other problems concern the too-strict oppositions between mechanism and organi(ci)sm, and symmetrical tendencies to "ontologize" (thus objectifying) properties of organisms and to "subjectify" them (turning them into philosophies of subjectivity). I don't mean to suggest that no one should be an organicist or that Kant is a name that should be banished from civilized society. Rather, to borrow awkwardly from Sade, "organicists, one more effort!" if one wants a naturalistic, non-foundationalist concept of organicism, which is indeed quite active in recent theoretical biology, and which arguably was already alive in the organismic and even vitalist theories of thinkers like Goldstein and Canguilhem.

In an earlier paper (Wolfe, 2010), I wrestled with the question of the "ontological status" of organisms. It proved difficult to come to a decision on this matter, because there are many candidates for what such a status is or would be and of course many definitions of what organisms are (see also Pepper and Herron (2008), Pradeu (2010a), and Pradeu (2016), Bouchard & Huneman, 2013) – the "fungus problem," as it were. But I did not focus in any detail on actual contemporary projects that rely on, or even assert, a certain concept of organism – namely, antireductionist,

C. T. Wolfe (✉)
Université de Toulouse 2 Jean-Jaurès, Toulouse, France

M. Mossio (ed.), *Organization in Biology*, History, Philosophy and Theory of the Life Sciences 33, https://doi.org/10.1007/978-3-031-38968-9_3

organicist projects (I'm fully aware that I haven't defined organicism yet). And I wish to show that these projects have an "issue," a sort of conceptual confusion, at least in part. My concern is thus not to argue for or against organicism, or to reconstruct a variant of organicism that is immune to some recent objections (a common enough practice), but rather, echoing two very different authors, to say "One more effort to achieve an organicism worth wanting."[1]

3.1   Organicism as Ontology, as Epistemology or as a Blurry Mix

A curious consensus without examination of presuppositions reigns today in the various subfields concerned with the properties (organizational, systemic, etc.) of living beings, at least those which define themselves as more or less "organicist." There is no strictly defined group of organicists, but attempts are regularly made to group various antireductionist trends in biology under this heading (e.g., Gilbert & Sarkar, 2000; Nicholson, 2014). Descriptions of how "the whole is greater than the sum of the parts" or, slightly less generically, how specific "system-level" or "higher-level" properties need a set of explanations (and perhaps an ontology) appropriate to them abound and have been repeated almost identically since the early twentieth century, with predecessors such as Aristotle and Kant invoked almost as often. At times, the concept of organism at work in these discussions is mostly a placeholder, such as when Lewontin, and developmental systems theory more generally, stresses that the focus is on organism-environment relations ("Just as there is no organism without an environment, so there is no environment without an organism"[2]). But more current organicism included various recent proclaimed "returns" of the organism concept[3] and does not just treat the concept as a placeholder, a functional or a heuristic term. They move toward what I'll call an *ontologization* of the notion: at that point, organicism has an ontology, whether it is clearly stated as such or not – it is then an *objectified* account of the nature of organisms.

---

[1] I am thinking both of Sade's "Français, encore un effort si vous voulez être républicains" (a pamphlet read out in the fifth dialogue of *La philosophie dans le boudoir*: Sade 1795/1998) and Dennett's elegant notion (the subtitle of his *Elbow Room*) of a notion of free will "worth wanting" (e.g., a pure unconditioned free will might be a lovely notion, but not accessible in our physical, causal universe, so which notions are *worth wanting*? Dennett, 1984).

[2] Lewontin, 1983/1985, 99. Pradeu has argued that in certain formulations of DST, the notion of "organism" is lost when it is posited that what evolves is actually the "organism-environment system"; he speaks of the "loss of the uniqueness of the organism" (Pradeu, 2010b, 219). Thanks to Alejandro Fábregas Tejeda for this reference.

[3] See Laubichler, 2000a, b and Baedke, 2019 (as well as Baedke's project at Bochum titled "The Return of the Organism in the Biosciences: Theoretical, Historical and Social Dimensions" (https://rotorub.wordpress.com/ although this aspect is primarily an effect of the 'project description' language), and the less commemorative or exhortatory special issue I coedited (Huneman and Wolfe, eds., 2010).

Treatments of organisms as more than just placeholders in a theoretical biological system, indeed, as entities with objective properties that should be the object of an ontology do not pose a problem as such – except perhaps to the instrumentalist, for whom it is a mistake to attribute any kind of innate meaning or definition to biological entities, as these are merely instrumental (Wolfe, 2010). What seems more curious to me is the way these treatments variously blend epistemological and ontological versions of organicism, sometimes in awkward ways. Generally, these accounts involve a combination of:

(a) an invocation of Kant, or at least a Kantian "regulative ideal," usually presented as the *epistemological* component (e.g., organicism as a theory about what constitutes "knowledge of life") or alternately, as the complete account of a vision of organism – as instantiating natural purposes, as a type of "whole" distinct from a merely mechanistically specifiable set of parts, interdependency, self-organization, and so on (Weber & Varela, 2002; Longo et al., 2012).

(b) a more *ontological* statement about the inherent or essential features of organisms (organicism as a theory about the actual properties of living systems) – essentially a "laundry list" of core features, e.g., "reproduction, life-cycles, genetics, sex, developmental bottlenecks, germ-soma separation, policing mechanisms, spatial boundaries or contiguity, immune response, fitness maximization, cooperation and/or conflict, codispersal, adaptations, metabolic autonomy, and functional integration" (Clarke, 2013, 415), itself usually tied to a kind of special tradition of thinkers who had a "feeling for the organism": typically, Claude Bernard,[4] Sherrington and Waddington, and more recently Maturana and Varela with the notion of autopoiesis, Rosen's (M,R) systems, Gánti's chemoton, or Luisi's minimal cell, all of which at one time or another are presented as the defining "characteristics of life," or "life criteria" (Griesemer & Szathmáry, 2008). These authors, or perhaps their theoretical constructs, are "heroes of homeostasis," as it were, or "masters of metabolism," given the latter concept's close association with definitions of life, e.g., when it is defined as "the collection of chemical reactions that define a living organism and allow it to make its components and obtain the energy required for staying alive" (Cardenas & Cornish-Bowden, 2011). To be precise, Bernard and others named above do not all perfectly match Clarke's "laundry list" of core features of life, e.g., fitness maximization. I used her list because it is one of the more expansive yet precise such lists I've seen.

The combination of (a) and (b) poses several problems. Note that I am not suggesting that a theory of organism has to be either epistemological or ontological. It can be both, or neither (what one might term constructivism, i.e., the view that organisms are constructs, and furthermore inasmuch as they are "sense-making"

---

[4] See Bechtel and Turner's contributions to Normandin and Wolfe eds. (2013) (which strongly present Bernard as a founder of the tradition they work in), as well as Denis Noble's influential lecture (Noble, 2008) which literally speaks in its title of Bernard as the "father of systems biology."

entities, they need to be understood via acts of cognitive construction).[5] What seems problematic to me is to combine epistemological and ontological claims in an "ungrounded" way, not always clearly presented as such (but, conversely, presented as straightforwardly empirical).

Perhaps the most obvious difficulty created by the unreflexive combination of (a) and (b) is something like a category mistake, in this case treating an epistemological property as an ontological property;[6] it is not quite right to invoke the authority of the Kantian "projective" approach to organisms (Huneman, 2007) in order to assert a set of ontological specificities about organisms. Because this is precisely what the Kantian regulative ideal concept was designed to avoid, in explicit contrast to what he would have called "rational metaphysics." That is to say, to provide an empirical set of criteria for why living beings are special and to claim that this supports or is supported by a Kantian framework is not a good idea given that the latter framework explicitly rejects the idea of giving empirical definitions of organism, inasmuch as Kant's organism concept is explicitly built around his notion of regulative ideal.[7] For Kant, organism is a "reflective" construct rather than a "constitutive" feature of reality, and reflective judgments are "incapable of justifying any objective assertions."[8] The purposive character of organisms, on this view, imposes itself upon on us by empirical experience, forcing us to (cognitively) recognize such entities as "organisms," rather than as machines or mere bundles of matter. Note that this more "constructivist" approach is by no means ruled out in biology, where such regulative concepts can serve as heuristics, e.g., regulative teleology;[9] the problem lies more in claiming to have an "empirical Kantian" foundation for organicism, because of the inherent tension therein – as when, e.g., Weber and Varela state that Kantian teleology is somehow "true" of organisms; "Our immodest conclusion is that Kant, although foreseeing the impossibility of a purely mechanical, Newtonian account of life, nonetheless was wrong in denying the possibility of a coherent explanation of the organism" (Weber & Varela, 2002, 120). As noted earlier, I am not suggesting that all organicists are Kantians or (a narrower claim) that all organicists make a category mistake with respect to Kantian notions; indeed, some contemporary organicists call for a return to Aristotle, while earlier organicists could also place themselves under the patronage of Hegel (thereby, by the way, eliminating

---

[5] I have developed this idea of a constructivist standpoint on organisms further with reference to Canguilhem and Goldstein in Wolfe (2015b, 2017) and more broadly in Wolfe (2014a).

[6] I am using this term to simply mean "The error of assigning to something a quality or action which can only properly be assigned to things of another category, for example treating abstract concepts as though they had a physical location" (Stevenson, 2010), rather than the more specific sense of a semantic error in a syntactically valid statement, like Chomsky's famous "colorless green ideas sleep furiously."

[7] Kant, 1987, § 73, p. 276.

[8] Kant, 1987, § 67, p. 259; § 73, p. 277.

[9] On regulative concepts as heuristics, see Sloan (2007) (discussing Grene) and Grene (1968, 235 f.); for a more critical view of the utility of Kantian regulative concepts in biology, see Zammito (2006); and for a more "objectivist" view, see Mossio and Bich (2017).

the category mistake in one stroke, as that is a full-blown ontological realism about organisms) (for a rare recent Hegelian move in organicism, see Gambarotto and Illetterati (2020)).

The second difficulty in certain organicist arguments lies in their empirical stance: it is not clear in any case why it counts as an argument against "mechanism" or "reductionism" to present a list of key features (such as the "generative" circular causality in organisms: Mossio, 2020, 55, 57), especially from a standpoint that curiously combines Kantianism, or at least Kantian elements, and the ontologization of organisms. I called this the "laundry list" problem above;[10] notice that the problem with this form of organicism – a kind of literal, empirical claim listing key definitory features of organisms – is very close to that found in defenses of "definitions of life," particularly "criterial" definitions of life (Malaterre, 2010; see Bich and Green (2017) for a defense of definitions of life).

A different reason why the laundry list of the organic is not the argument-stopper some treat it as is the flexibility of mechanism: mechanism is not reducible to the strictures of say early modern mechanistic ontology (in which what is real is what is specifiable in terms of size, shape, and motion) or to a "toy model" ontology in which what is real in a system is its decomposition into parts (indeed, decomposition is typically presented as an *explanatory* strategy in contemporary mechanism, e.g., in Bechtel & Richardson, 1993, 30; here I primarily mean decomposition into parts). Organicism in this sense neglects, or almost forces itself to neglect, the reality of "expanded mechanism," that is, the existence of nondogmatic, or nonfoundationalist, or ontologically open mechanistic programs. In this sense organicism can be overly polarized (this is my disagreement with Nicholson, specifically Nicholson (2013); for a more open-ended view, reflecting its historical focus, see Toepfer, this volume). By extension, organicism can at times run the risk of verging on antinaturalism, although antimechanism is obviously not synonymous with antinaturalism (I'll return to this below). Either organicism neglects the possibility of the reducibility of the laundry list (i.e., that it is potentially reducible to more basic features), or conversely, of the existence of expanded mechanism: the "remit," the "span" of mechanism is broader than we are accustomed to think when we repeat the mechanism-vitalism or mechanism-teleology opposition. One could adduce the sheer historical complexity of mechanism and its various "expansions," or the concept of teleomechanism – not in its historically "restricted" sense as in the Lenoir Thesis (Lenoir, 1982), but in a broader sense (Wolfe, 2014c), which stresses that even paradigm cases of "mechanism" or "the mechanical philosophy" (from Descartes and iatromechanism onward) are filled with functional language, including "function," "use," and the "office" of an organ. And they are concerned with properties of the organism such as health and survival, that is, not just with

---

[10] Within the autopoiesis discussions, I note that E. Di Paolo diagnosed this problem (see Di Paolo, 2018). But he also suggests that Varela himself became critical of the autopoietic theory for this "lazily empirical" character and refers to the 1996 preface Varela added to the second Spanish edition of the canonical text *De Máquinas y Seres Vivos*.

microstructure or size, shape, and motion, if indeed, "medicine is the most useful of the sciences" for Descartes (*Discourse on method*[11]).

3.2   A Current Organicist Consensus?

Of course, the theoretical biologist interested in articulating a concept of organizational closure[12] might say: none of this matters, because what these "positions" are, are simply theoretical constructs, *bricolages* intended to facilitate the articulation of an ongoing research project. This can be seen in an example from the earlier part of the twentieth century, Kurt Goldstein's *The Structure of the Organism* [Goldstein, 1939/1995], a work which combines a "bottom-up" study of brain-damaged soldiers from World War I with a "top-down" set of ontological, not just tacit commitments but explicit *claims* about the "whole": if we treat Goldstein's references to Goethe and *Naturphilosophie* seriously, we can charge him with "romanticism," with being "antimodern science," and so on. But if we treat these references as simply a typical case of an educated German scientist of the early twentieth century displaying his "humanistic" breadth, then we can study the overall argument in much more naturalistic, or naturalism-friendly terms.

But the goal of the present remarks is to clarify the possible or actual commitments of these organicist projects in current research, to ask if they hang together successfully or not, and to distinguish between them in a way that has not previously been attempted. This is justified also by the fact that many of the actors involved go out of their way to claim a Kantian (or in some cases Aristotelian[13]) pedigree for their experimental, modeling, or otherwise "empirical" projects.

Now, it is not easy to produce a typology of current organicist views, partly because the actors involved waver between positions that we might consider clearly demarcated (notably, epistemological vs ontological positions on organisms and their "reality," as in the (a) and (b) positions above). Varela, for one, sometimes stresses an "epistemological" standpoint but sometimes also speaks the language of

---

[11] Descartes writes that the preservation of health is the most important value and the foundation of all other values; that states of mind are dependent on the organization of bodily organs; and that (hence) medicine is the most useful and necessary science (*Discourse*, VI, AT VI, 62–63).

[12] Mossio et al., 2009; Mossio and Moreno 2010; Ruiz-Mirazo et al., 2000; and discussion in Bechtel, 2007; Moreno & Mossio, 2015.

[13] The work of Marjorie Grene (who called for an Aristotelian perspective in biology and 'philosophical biology" in the 1960s), James Lennox and Denis Walsh is influential here: see notably Walsh, 2021. In the context of the present paper, Aristotelian perspectives are squarely ontological. As Walsh puts it, Aristotelian biology (indeed, any "comparative biology") "seeks to account for the fit and diversity of organismal form" (Walsh, 2021, 281). While Walsh also speaks of explanations, I would understand his insistence on the need for an 'agent theory" with which to understand organisms (because it is able to accommodate teleological explanations), rather than an 'object theory", as ontologically revisionary (in Strawson's sense).

the *key features of organisms*. Rosen (1991) says he is speaking about "models" but then calls his book *Life Itself* and indeed complains that mechanistic science has not grasped "life itself." He also holds that mechanistic laws are the special cases, the outliers, whereas biology gives us a truer picture of nature; but that is not our concern here.

3.3   Organicism Strong and Weak, and the Ghost of Vitalism

Interestingly, as regards the "autonomy and organizational closure," researchers never seem to make this particular move (different from what I called a category mistake above) of reifying "life" or "organization" in a foundationalist way. The concept that emerges here which seems to overcome the epistemological/ontological divide is *organization*. Because, as can be seen with the case of Bechtel, both "organicists" and "mechanists" can help themselves to the concept of organization; differently put, organization can be a key concept for both of these approaches, and it can end up being more a matter of perspective than of some purported deep ontology that one specifies this concept in mechanistic terms (especially, expanded-mechanist terms) or in organicist terms (especially, non- or weakly ontologized organicist terms: Wolfe, 2010) – the same can be said of the notion of "structure." I use the term "perspective" because the choice of studying a system as mechanistic and/or as organizational is also a choice of perspective, or "standpoint" as Needham put it: "all things are organisms and all things are atomic systems also. You choose your standpoint" (Needham, 1930, 85, longer citation in Wolfe, 2014a). However, it is nevertheless possible that any robust notion of organization, unlike older notions of systems à la von Bertalanffy (1933), does rely on a concept of life (and thus is not tantamount to mechanism).[14]

Further, the Kantian/epistemological view can also be transfigured into something "ontological," not as a claim about the world (i.e., entities such as livers and hearts or microbes or finches or coral reefs) but as a claim about the unique nature of the subject (interiority, subjectivity, first-person knowledge, etc.,[15] in which these are distinguished from the physical world as a whole). This is not so surprising if we consider Kant's famous proclamation that "there will never be a Newton of a blade of grass": a point which is both about our cognitive capacities *and* – at least the ambiguity is unresolved – about the uniqueness of organic entities in comparison with the entities studied by physics and the physicomechanical sciences more broadly. In the view of some notable interpreters, "the third *Critique* essentially proposed the reduction of life science to a kind of pre-scientific descriptivism, doomed never to attain authentic scientificity, never to have its 'Newton of the blade

---

[14] I am grateful to Bohang Chen for this suggestion.
[15] Varela, 1996; Varela & Shear, 1999; Thompson, 2007.

of grass'" (Zammito, 2006, 755). Actually, Newtonianism was very fruitful as a methodology and an analogy in eighteenth-century life science, both in obvious cases like Albrecht von Haller's physiology— the first sentence of his textbook on physiology famously proclaims that "the fibre is to the physiologist what the line is to the geometrician" – and in less obvious cases like Montpellier vitalism, which is also filled with Newtonian methodological invocations (Wolfe, 2014b).

This slippage from the epistemological to the ontological is exemplified in earlier twentieth-century theoretical biology in work like Uexküll's, in which the "epistemic" (the states of knowledge of the world of given biological entities such as the tick: Uexküll, 1934/2010) becomes a feature of the organic as such. Similarly, Goldstein's theory of organism (Goldstein, 1939/1995), which is a really a theory of the organismic (actually, holistic) capacities of meaning-making and world-making by agents such as humans; it is really more about holism and personhood than about biological organisms per se, but in any case Goldstein's theory actively moves between these two positions, but that should not prevent us from trying to articulate the difference for the sake of clarification. When Grene discusses phenomenological and philosophico-anthropological approaches to life, including Goldstein's (esp. Grene, 1966), like Goldstein himself, she elides the difference between states and processes of "the knower" (the subject) and objective properties of "the known" (the object). While most phenomenologists would claim to only be describing the former, Grene shifts the ground to describing the latter: matters of "ontological import" (Grene, 1968, 239).

When organicism is a strong ontology (e.g., Grene, 1966, 1968), which I have also discussed under the heading of "ontologization," is it so different from its yet-stronger cousin, vitalism? That is to say, organicism recurrently tries to present itself as a "safe word," the reasonable version of antireductionism (Gilbert & Sarkar, 2000: vitalism is the weird version of our view, so let's call it organicism), while some biologists will playfully invoke "molecular vitalism" in mainly provocative ways (Kirschner et al., 2000). Indeed, this attempt to present organicism as, typically, the less metaphysical version of vitalism – today one would tend to say the "naturalized" version is an old, perhaps defining feature of this theory (or family of theories, since early twentieth-century organicism is not exactly cohesive in those terms). For instance, Needham wrote that organicism combined "the insistence of vitalism on the real complexity of life with the heuristic virtues of the mechanistic practical attack."[16]

---

[16] Needham (1936), 9; cf. also Needham (1928), discussed in Nicholson and Gawne (2015); see also Peterson (2017). Autopoietic theory also will state that its novelty lies in its being neither reductionism nor vitalism (Varela, 1991). Interestingly, Beckner (1974) thought almost the perfect opposite of Needham, in the sense that for him, organicism only really was defensible (or could survive) if supplemented by vitalism. Beckner also uses the language of "weak" and "strong" organicism, but in his case these terms are indexed to the concept of emergence, that is, they mean the same as "weak emergence" and "strong emergence." O. Sartenaer (discussion) notes nicely that anything in between reductionism and (strong) vitalism is emergentism.

Now, it would be unfortunate if organicism operated with as caricatural a presentation of vitalism as the Cricks and Monods do, for vitalism itself lives in serious tension between different meanings, which indifferent history of science or philosophy often neglects. An eighteenth-century Montpellier vitalist, Bichat, Bernard (both of whom combine denunciations of earlier forms of vitalism with insistence on the existence of uniquely vital properties, Wolfe (2022)), or Hans Driesch – not to mention Bergson or Canguilhem, moving into the twentieth century – are defending very different and at times incompatible programs (Wolfe, 2015a). Indeed, vitalism itself exhibits a tension or plurality between more metaphysical and less metaphysical views and, contrary to a view still found, does not always maintain the existence in an ontologically specific sense of a vital principle, force, entelechy, élan vital, etc. In that sense, when organicism states "there is legitimate non-reductionist investigation of the core properties of life/organization/organism, and there is illegitimate, metaphysical speculation about or hypostatization of these properties," it is just replaying the standard exclusionary game like mechanism in earlier times; the organicist who treats vitalism as the absolute cardinal error, in a self-congratulatory act of exclusion, should remember the warning "Don't throw stones in glass houses" (see also Oyama (2010), and for a different version of this point on how organicism may have difficulty demarcating itself from vitalism, especially the less metaphysical variants of vitalism, see Chen (2019)).

In a standard case of organicism defining itself over and against a "bad version of itself" called vitalism, Cardenas and Cornish-Bowden insist that the recognition of metabolism is tantamount to a decline in vitalism, which they see in the nineteenth century: "Although a living organism is identified and recognized by its physical appearance, and hence by its structure, its status as living is defined by its chemistry, and thus by its metabolism" (Cardenas & Cornish-Bowden, 2011, 1016). This approach to metabolism emphasizes that the processes occurring in a living organism are fundamentally chemical reactions and implicitly echo nineteenth-century affirmations (Bernard, Wöhler) according to which chemistry rules out vitalism.

One need only contrast Jonas (see Jonas, 1966/2011) and Cornish-Bowden and Cardenas on metabolism to see the way organicism can be more or less "ontologized," more or less subjectivist. That is to say, metabolism and homeostasis are good examples of important physiological properties or processes which allow of a kind of philosophical "overdetermination" or "overinvestment" (as happens with Jonas, when a notion like metabolism cannot just mean what it means to a biologist but has to mean *more*). Notably, metabolism is taken to raise the problem of "subjectification," i.e., when an organic property becomes construed as requiring a "self" or "subject." Thus, Jonas presents his work as "a new reading of the biological record [to] recover the inner dimension" and describes metabolism as constituting the organization of organisms for "inwardness, for internal identity, for individuality" while also turning the organism outward "toward the world in a peculiar relatedness of dependence and possibility"; life is "self-centered individuality, being for itself and in contraposition to all the rest of the world, with an essential boundary dividing 'inside' and 'outside' –notwithstanding, nay, on the very basis of the actual exchange."[17]

One imagines that most physiologists from Bernard and Sherrington onward would find it quite strange to have metabolism equated with "inwardness"; doubtlessly, Jonas is (i) stressing that metabolic entities have an inner/outer relation and (ii) hypostasizing this relation as a philosophical justification of interiority. Organicism is not, then, automatically vaccinated against the excesses of metaphysical vitalism (whether it is Jonas on metabolism, Merleau-Ponty on the flesh, writing "Just as the sacrament not only symbolizes ... an operation of Grace, but is also the real presence of God ... in the same way the sensible has not only a motor and vital significance but is nothing other than a way of being in the world that our body takes over [...] sensation is literally a communion" (Merleau-Ponty, 1945, 245),[18] or Evan Thompson's reworking of Merleau-Ponty in an even more explicitly dualist direction: "Life is not physical in the standard materialist sense of purely external structure and function. Life realizes a kind of interiority, the interiority of selfhood and sense-making. We accordingly need an expanded notion of the physical to account for the organism or living being" (Thompson, 2007, 238)[19]), and because its reasonable variants themselves vary on issues like holism (itself just a word, to be sure).

Of course, organicism (or at least some forms of organicism) seems to steer clear of considerations concerning "subjectivity," "the transcendental," "embodiment," and so on, due to its resolute focus on organization (for a different perspective, see Van de Vijver & Haeck, this volume). Granted, the latter notion is not always strictly defined, but even if it is related to "closure," it does not seem to require a "controller," a "hegemon," as the notion of constraint indicates. However, it allows of more "abstract" and more "literally real" renditions.[20] But organicism comes in different strengths, different degrees of ontologization. Some contributors to the history of organicism in the twentieth century, e.g., Jonas, seem to present organisms (which in

---

[17] Jonas 1966/2011, xxiii, 84, 79. For discussion of these passages and more general reflections on metabolism as a concept that go in a direction quite different from that of this paper, see Landecker (2013).

[18] It falls outside the scope of the present paper to present a critique of the phenomenology of embodiment in terms of its revival of subjectivity, but I note that Canguilhem diagnosed this problem in what was actually a "friendly critique," in a late lecture on health, commenting on the way this sense of privacy and ineffability in Merleau-Ponty presented a kind of "wall," an end to discussion (Canguilhem (2008b); see Wolfe (2015b) for brief discussion).

[19] I realize that early twentieth-century organicism (like that of E.S. Russell or J.H. Woodger), midtwentieth century philosophy of biology with an antireductionist focus (like Jonas and some Merleau-Ponty), and contemporary organicism (like Moreno and Mossio, with Varela lying somewhere in between) are not carrying out one and the same monolithic "program" (indeed, this monolithic, somehow uncontextualized view is something I've criticized organicism for in the present paper. But I suggest that my criticism of (a) an uncritically asserted empirical-realist appropriation of the Kantian perspective and (b) an unfounded demarcation between organicism and vitalism in which the former is the legitimate, naturalism-friendly project can apply to some different moments of the organicist history.

[20] Bich and Damiano (2008) Moreno, and Mossio et al. describe organization as a "specific causal régime" (Mossio & Bich, 2017; Moreno & Mossio, 2014; DiFrisco & Mossio, 2020).

this case are never nematodes or slime molds, but typically humans or creatures resembling them) as almost outside the realm studied by natural science. But even the "weak organismic view" will sometimes rest its claims on the existence of certain irreducible, specific, *key features* of organisms (à la Claude Bernard) – what I called the "laundry list" approach above.

Now, this approach remains "biochauvinist," to use Ezequiel Di Paolo's suggestive term:[21] it is essentially a naturalized form of (substance) vitalism, that is, claims about the "uniqueness" of biological entities. Indeed, besides seeking to define itself in opposition to the purportedly metaphysical version, vitalism, organicism can also subdivide into warring *groupuscules*: to cite one example among many, Needham distinguishes between "dogmatic" and "legitimate" organicism (regarding the concept of organization).[22] One could also ask whether all forms of organicism are naturalist; how naturalist can one be here?[23] This is a delicate matter, because most or all contemporary organicists either accept naturalism in a broad sense or at least would be wary of endorsing some antinaturalist position. But at the same time, when they invoke, inter alia, Merleau-Ponty or Jonas, they are explicitly using conceptual accounts of, e.g., life and embodiment, which insist that "life," "mind," "selfhood," "subjectivity," and "interiority" are somehow not in physical space, outside of physical space, and require almost a separate ontology: on most accounts, that is antinaturalism. If naturalism is a desideratum – if it is desirable for the position argued for in the context of discourses on biology and organisms to be naturalistic or at least broadly compatible with naturalism – then organicism has two different ways of achieving it: it can opt for the epistemological view, as Goldstein does, stressing that organisms are "knowers" who construct their worlds, or it can opt for naturalizing, e.g., declaring teleology to be integrated in a naturalistic framework (e.g., Mossio & Bich, 2017).

3.4 Ontology and Ontologization

One organicist response to my raising the issue of ontologization could be that in fact, they have moved on, and the issue is really *explanations*: that is, what is unique about biological entities is the type of *explanations* they require (an idea that might remind us of the epistemological position). So, for instance, Arno Wouters writes

---

[21] Di Paolo (2009). Di Paolo does not actually define or expand on the term, but it is close, e.g., to Andy Clark's discussion of "pressing the flesh" (Clark, 2008) in terms of distinguishing different "strengths" of embodiment as a perspective on cognition (embodied cognition), what I would term different degrees of ontologization of the biological, the living body, the flesh, etc.

[22] Needham (1936), 18, cit. in Nicholson and Gawne (2015). In the earlier *The Great Amphibium* (1932), Needham had spoken of the organicists who accept the notion of organization without seeking to investigate it scientifically as "passive organicists" (thanks to Alejandro Fábregas Tejeda for this reference).

[23] On the idea of "naturalism friendly," see notably the discussions concerning "naturalized teleology" (e.g., Walsh (2007, 2013). Keijzer (2016) comments briefly, with regard to Moreno and Mossio's concept of autonomy, on the pros and cons of "naturalizing."

that "Functional explanations are, nevertheless, crucial to understand life, because they show us how the characteristics of an organism fit into the requirements for being alive."[24] Of course, such specific explanations may in turn be "generated" or "induced" by features that are claimed to be ontologically real (explanatory autonomy is not an argument-stopper). Indeed Wouters does not seem content to dwell at the level of explanations, as the dimension of organization he emphasizes is ontological: "Organization can be defined in terms of dependence on the composition, arrangement and timing of the relevant system, its parts and their activities."[25] When Dan Nicholson writes that "Organicists, in contrast, are committed to the belief that the integrated and inherently purposive nature of organisms reflects the very essence of what they actually are in reality,"[26] I think this is right about a number of forms of organicism (and this leads to some of the problems surrounding "ontologization" I have discussed above), but it does not apply, for instance, to the organicism of Kurt Goldstein (together with another important figure who I have not discussed here, Georges Canguilhem[27]), because the latter form of organicism seeks to integrate – or better, it *builds on* – the irreducible existence of a feature of organisms one might call existential or perspectival: the fact that an organism always has an existential attitude, a perspective on other organisms, and that, thereby, part of its vital activity involves acts of (cognitive) construction.

It is as if the Kantian constructive concept of organisms had been ontologized without being subsumed under a potentially refutable or reducible list of vital features. Because, as Canguilhem argues powerfully in his essay "Aspects of vitalism," vitalism (and here we could replace this term with "organicism," *if this term is not construed in the strictly empirical, ontologized, criterial, "laundry list" sense*) is not refutable like geocentrism or the phlogiston theory are. It is a different kind of theory. If organicism were like geocentrism or the phlogiston theory, it would be because it was a strictly empirical (and thus refutable) set of claims. But, on the constructivist view, according to which organisms both *require* acts of cognitive construction and *perform* such acts, organicism is not reducible to an empirical set of claims (a.k.a. definitions of life). Goldstein and Canguilhem do not give a set of empirical features of organisms (purposiveness/teleology, integration, reflexivity, etc.) and then state "Here are organisms and here is why organicism is the theory that does justice to them." They argue less in empirical, "realist" terms and more in existential, constructivist terms.

Now, reflecting on this existential or perspectival dimension (which will be familiar to students of Uexküll, of course) also helps us with our rough typology of organicisms, since it leads us to nuance, not so much the question of ontologization and weak or strong organicism, but at the other end of the spectrum, the question of

---

[24] Wouters (2013a). Compare Walsh (2015).

[25] Wouters (2013b), 463, 460; see also Wouters 2013c.

[26] Nicholson (2010), Chap. 2, 63n.

[27] For an excellent discussion of the connection between the two thinkers in this context, see Gayon (1998); I have developed this point a bit further in Wolfe (2015b, 2017).

the "subject." That is to say, if earlier I tried to diagnose a problem of "ontologiza-tion," in that sense, *objectification* (of properties of organisms), there is a kind of symmetrical problem, the other horn of the dilemma, which is *subjectification*. By this I mean the kind of "romantic" vision of organism as deeply opposed to machines, or inert matter, because it is a version of subjectivity, because it has a "self" (this is explicit in Hegel's philosophy of nature, Schelling's and I think Jonas's – as defended again in Michelini et al. (2018); it is – debatably and controversially – also present in the later Varela (Varela (1991), Varela (1996), Varela & Shear, (1999)), and definitely in Evan Thompson). Arguably, the organizational perspective on organisms allows the organicist to escape the danger of subjectification, but so does the Goldstein-Canguilhem approach, i.e., the "constructivist" vision of organism.

3.5 Conclusion

To sum up, my point is not to assert or deny the reality or pertinence of organisms or organismic theories in biology. I notice that no one seems to bother to do this anymore: Bickle (2003) argues aggressively for reductionism (of cognitive neuro-science to cellular-level neuroscience), but otherwise it has become more than rare – irrelevant – to argue "reductionism" against "organicism" (contrast Nagel in the 1960s). Present-day neomechanists need not worry, if we consider Bechtel's own "organizational" commitments. For Bechtel, mechanism and organization stand in a complex and fruitful interaction, and mechanism can provide an adequate account of organization by "placing as much emphasis on understanding the par-ticular ways in which biological mechanisms are organized as it has on discovering the component parts of the mechanisms and their operations"; "It is the fact that the system is organized (and the type of organization it has) that makes it amenable to mechanistic description and analysis" (Bechtel 2007, 270; Levy & Bechtel 2013, 244). My point is not to defend "neomechanism" in extenso (it has received a num-ber of quite different, and significant criticisms) but to observe that one of its most useful features in the present context is to not just oppose mechanistic and organi-zational approaches. As Moreno puts it, "holism and mechanistic decomposition can be combined for the purposes of biological explanation," increased complexity creates "selective functional constraints," giving rise to "levels of organization in which a mechanistic decompositional strategy might be locally applicable."[28]

Similarly, my point is not to simply put forth a "safe," reductionist critique of all or any such programs. On the contrary, I view it as extremely useful to track such articulations, whether or not we fully commit to them (in this I feel kinship with the "epistemology of Life" in Canguilhem; see Canguilhem (2008a) and Wolfe (2017)). To clarify one potential ambiguity, my point here is deflationary (about organicism,

---

[28] Moreno (2013, 901); for critical discussion of Bechtel's way of articulating mechanism and organization, see Eronen, 2015 and Mossio (2020, 73).

i.e., not reductionist in the sense of being committed to a more minimal ontology, but deflationary inasmuch as I challenge some higher-level concepts in favor of stripped-down versions thereof) but does not take a stand on organisms (their ontological status); I've taken several different stances on organism in earlier papers, e.g., in my 2010 "Do organisms have an ontological status?" arguing for a weakly ontological status – indeed rather consonant with what I'm saying here about organicism – and in my 2014 "The organism as ontological go-between" emphasizing rather the "nomadic concept" aspect, the hybridity in its fertility (Wolfe, 2010, 2014a). Organisms as conceptual hybrids are not so directly opposed to mechanism(s) as in the usual organicist party line. They are also so not ontologically solid or essentialist (that's what the word hybrid indicates).

I have tried to suggest (a) that organicism is not necessarily so far removed from the more naturalistic forms of vitalism (and this, to be clear, is not a bad thing); but (b) that when organicism relaunches itself as a fully respectable scientific theory – no one's mistress, it might say to Haldane – it creates for itself several potential conceptual problems. One is the unstable combinations of the Kantian epistemological perspective with deliberately empirical, ontologized claims – as if organicists wanted to play both sides in an unprincipled way (sometimes claiming features must really be part of organisms, other times claiming they are only regulative ideals).[29] The other is, in a sense, another kind of instability resulting from the reduction of organicism to a "merely empirical," criterial set of definitions (the "laundry list"). This instability is, ironically, the more Kantian component of the present remarks: faced with the confidence of thinkers asserting the "the soul and the world exist," Kant suggested an approach that would not be founded on directly refutable empirical claims. Founding claims for the irreducible nature of organisms on directly refutable empirical definitions create this instability.[30] Not all contemporary organicists make this Kantian move, or simply the (ontologized) empirical criteria move; I noted in addition (c) that symmetrically to this danger of the excess, the addiction to ontologization, there is the potential to reintroduce into organicism a kind of Romantic love affair with subjectivity, ineffability (as we saw with Merleau-Ponty and Canguilhem's sensitive commentary) and the first-person perspective. It would be a good idea to try and achieve some conceptual clarity so that the observer can have an idea of what is involved in the current smorgasbord of mildly or strongly organismic projects.

Acknowledgments   Earlier versions of this paper were presented in Vienna and Paris. Thanks to Sebastjan Vörös and Matteo Mossio for their invitations and comments and to Arantza Etxeberria, Alejandro Fábregas Tejeda, Phillip Honenberger, and Andrea Gambarotto for further suggestions.

---

[29] I thank Phillip Honenberger for this formulation.

[30] While this is not a reflection on the "definitions of life" debates (e.g., Machery, 2012 versus Bich & Green, 2017), one can extrapolate from my position here: the idea would then be not necessarily to side with Machery's deflationary position, but to acknowledge the Sisyphean (or Pyrrhic?) character of attempts to define life.

References

Baedke, J. (2019). O organism, where art thou? Old and new challenges for organism-centered biology. *Journal of History of Biology, 52*, 293–324.

Bechtel, W. (2007). Biological mechanisms: Organized to maintain autonomy. In F. Boogerd, F. J. Bruggeman, J.-H. S. Hofmeyr, & H. V. Westerhoff (Eds.), *Systems biology: Philosophical foundations* (pp. 269–302). Elsevier.

Bechtel, W., & Richardson, R. C. (1993). *Discovering complexity: Decomposition and localization as strategies in scientific research*. Princeton University Press.

Beckner, M. (1974). Reduction, hierarchies and organicism. In F. Ayala & T. Dobzhansky (Eds.), *Studies in the philosophy of biology: Reduction and related problems* (pp. 163–177). University of California Press.

Bich, L., & Damiano, L. (2008). Order in the nothing: Autopoiesis and the organizational characterization of the living. In I. Licata & A. J. Sakaji (Eds.), *Physics of emergence and organization* (pp. 343–373). World Scientific.

Bich, L., & Green, S. (2017). Is defining life pointless? Operational definitions at the frontiers of biology. *Synthese, 195*, 3919. https://link.springer.com/article/10.1007/s11229-017-1397-9

Bickle, J. (2003). *Philosophy and neuroscience: A ruthlessly reductive account*. Kluwer.

Bouchard, F., & Huneman, P. (Eds.). (2013). *From groups to individuals: Evolution and emerging individuality*. MIT Press, Vienna Series in Theoretical Biology.

Canguilhem, G. (2008a). *Knowledge of life* (S. Geroulanos, & D. Ginsburg, Trans.). Fordham University Press (A translation of *La connaissance de la vie*, 1965).

Canguilhem, G. (2008b). "Health: Crude concept and philosophical question" (translation of "La santé, concept vulgaire et question philosophique" (1988), by T. Meyers and S. Geroulanos). *Public Culture, 20*(3), 467–477.

Cardenas, M., & Cornish-Bowden, A. (2011). Metabolism (Biological). In M. Gargaud et al. (Eds.), *Encyclopedia of astrobiology* (pp. 1016–1017). Springer.

Chen, B. (2019). *A historico-logical study of vitalism: Life and matter*. PhD thesis, Department of Philosophy and Moral Sciences, Ghent University.

Clark, A. (2008). Pressing the flesh: A tension in the study of the embodied embedded mind? *Philosophy and Phenomenological Research, 76*(1), 37–59.

Clarke, E. (2013). The multiple realizability of biological individuals. *The Journal of Philosophy, 110*(8), 413–435.

Dennett, D. (1984). *Elbow room. The varieties of free will worth wanting*. MIT Press.

Di Paolo, E. (2009). Extended life. *Topoi, 28*, 9–21.

Di Paolo, E. A. (2018). The enactive conception of life. In A. Newen, S. Gallagher, & L. de Bruin (Eds.), *The Oxford handbook of 4E cognition: Embodied, embedded, enactive and extended* (pp. 71–94). Oxford University Press.

DiFrisco, J., & Mossio, M. (2020). Diachronic identity in complex life cycles: An organizational perspective. In A. S. Meincke & J. Dupré (Eds.), *Biological identity. Perspectives from metaphysics and the philosophy of biology* (pp. 177–199). Routledge.

Eronen, M. I. (2015). Levels of organization: A deflationary account. *Biology and Philosophy, 30*, 39–58.

Gambarotto, A., & Illetterati, L. (2020). Hegel's philosophy of biology? A programmatic overview. *Hegel Bulletin, 41*(3), 349–370.

Gayon, J. (1998). The concept of individuality in Canguilhem's philosophy of biology. *Journal of the History of Biology, 31*(3), 305–325.

Gilbert, S. F., & Sarkar, S. (2000). Embracing complexity: Organicism for the twenty-first century. *Developmental Dynamics, 219*, 1–9.

Goldstein, K. (1939/1995). *The Organism: A holistic approach to biology derived from pathological data in man*. American Book Company/Zone Books. (A translation of *Der Aufbau des Organismus*, 1934).

Grene, M. (1966). *The Knower and the Known*. London: Faber & Faber.

Grene, M. (1968). *Approaches to a philosophy of biology*. Basic Books.

Griesemer, J. R., & Szathmáry, E. (2008). Gánti's chemoton model and life criteria. In S. Rasmussen, L. Chen, N. Packard, M. Bedau, D. Deamer, P. Stadler, & D. Krakauer (Eds.), *Protocells: Bridging nonliving and living matter* (pp. 407–432). MIT Press.

Huneman, P. (2007). *Understanding purpose. Collected essays on Kant and the philosophy of biology* (North American Kant society publication series). University of Rochester Press.

Huneman, P., & Wolfe, C. T. (Eds.). (2010). *The Concept of Organism: Historical, Philosophical, Scientific Perspectives, special issue of History and Philosophy of the Life Sciences, 32*, 2–3. https://www.jstor.org/stable/i23335068

Jonas, H. (2011). *The phenomenon of life: Toward a philosophical biology* (Northwestern University Press, [1966]).

Kant, I. (1987). *Critique of Judgment* (1790) (W. Pluhar, Trans.). Hackett.

Keijzer, F. (2016). Matching concepts and phenomena: A review of *biological autonomy. Adaptive Behavior, 24*(6), 479–486.

Kirschner, M., Gerhart, J., & Mitchison, T. (2000). Molecular 'vitalism'. *Cell, 100,* 79–88.S.A.

Landecker, H. (2013). The metabolism of philosophy, in three parts. In B. Malkmus & I. Cooper (Eds.), *Dialectic and paradox: Configurations of the third in modernity* (pp. 193–224). Peter Lang.

Laubichler, M. (2000a). The organism is dead. Long live the organism! *Perspectives on Science, 8*(3), 286–315.

Laubichler, M. (2000b). Symposium 'The organism in philosophical focus' – An introduction. *Philosophy of Science,* 67, 3, supplement.

Lenoir, T. (1982). *The strategy of life. Teleology and mechanism in nineteenth century German biology*. University of Chicago Press.

Levy, A., & Bechtel, W. (2013). Abstraction and the organization of mechanisms. *Philosophy of Science, 80*(2), 241–261.

Lewontin, R. C. (1985). The organism as the subject and object of evolution (1983). Reprinted in Levins, R., Lewontin, R.C., *The dialectical biologist*, 85–106. Harvard University Press.

Longo, G., Montévil, M., & Kauffman, S. (2012). No entailing laws, but enablement in the evolution of the biosphere. In *GECCO '12: Proceedings of the 14th annual conference companion on Genetic and evolutionary computation* (pp. 1379–1392). https://doi.org/10.1145/2330784.2330946

Machery, E. (2012). Why I stopped worrying about the definition of life... and why you should as well. *Synthese, 185,* 145–164.

Malaterre, C. (2010). On what it is to fly can tell us something about what it is to live. *Origins of Life and Evolution of the Biosphere, 40*(2), 169–177.

Merleau-Ponty, M. (1945). *Phénoménologie de la perception*. Gallimard.

Michelini, F., Wunsch, M., & Stederoth, D. (2018). Philosophy of nature and organism's autonomy: On Hegel, Plessner and Jonas' theories of living beings. *History and Philosophy of Life Sciences, 40*(3), 56. https://doi.org/10.1007/s40656-018-0212-3

Moreno, A. (2013). Holism. In W. Dubitzky et al. (Eds.), *Encyclopedia of systems biology* (pp. 900–902). Springer.

Moreno, A., & Mossio, M. (2015). *Biological autonomy. A philosophical and theoretical enquiry*. Springer.

Mossio, M. (2020). *Organisation biologique et finalité naturelle,* Mémoire de synthèse, Habilitation à diriger des recherches, Université Paris 1 Panthéon – Sorbonne.

Mossio, M., & Bich, L. (2017). What makes biological organisation teleological? *Synthese, 194*(4), 1089–1114. https://doi.org/10.1007/s11229-014-0594-z

Mossio, M., & Moreno, A. (2010). Organizational closure in biological organisms. *History and Philosophy of the Life Sciences, 32*(2–3), 269–288.

Mossio, M., Saborido, C., & Moreno, A. (2009). An organizational account of biological functions. *British Journal for the Philosophy of Science, 60,* 813–841.

Needham, J. (1928). Organicism in biology. *Journal of Philosophical Studies, 3,* 29–40.

Needham, J. (1930). *The sceptical biologist.* W.W. Norton.

Needham, J. (1936). *Order and life.* Yale University Press.

Nicholson, D. (2010). Organism and mechanism. A critique of mechanistic thinking in biology. PhD, Department of Philosophy, University of Exeter.

Nicholson, D. J. (2013). Organisms≠ machines. *Studies in History and Philosophy of Science Part C: Studies in History and Philosophy of Biological and Biomedical Sciences, 44*(4), 669–678.

Nicholson, D. (2014). The return of the organism as a fundamental explanatory concept in biology. *Philosophy Compass, 9*(5), 347–359.

Nicholson, D., & Gawne, R. (2015). Neither logical empiricism nor vitalism, but organicism: What the philosophy of biology was. *HPLS, 37*(4), 345–381.

Noble, D. (2008). Claude Bernard, the first systems biologist, and the future of physiology. *Experimental Physiology, 93*(1), 16–26.

Normandin, S., & Wolfe, C. T. (Eds.). (2013). *Vitalism and the scientific image in post-enlightenment life science, 1800–2010.* Springer.

Oyama, S. (2010). Biologists behaving badly: Vitalism and the language of language. *History and Philosophy of the Life Sciences, 32*(2–3), 401–423.

Pepper, J., & Herron, M. (2008). Does biology need an organism concept? *Biological Reviews, 83,* 621–627.

Peterson, E. (2017). *The life organic: The theoretical biology Club and the roots of epigenetics.* University of Pittsburgh Press.

Pradeu, T. (2010a). What is an organism? An immunological answer. *History and Philosophy of the Life Sciences, 32*(2–3), 247–267.

Pradeu, T. (2010b). The organism in developmental systems theory. *Biological Theory, 5,* 216–222.

Pradeu, T. (Ed.). (2016). *Biological Individuality,* special issue of *Biology and Philosophy,* 31(6), 31.

Rosen, R. (1991). *Life itself: A comprehensive inquiry into the nature, origin and fabrication of life.* Columbia University Press.

Ruiz-Mirazo, K., Etxeberria, A., Moreno, A., & Ibáñez, J. (2000). Organisms and their place in biology. *Theory Bioscience, 119,* 209–233.

Sade, D. A. F. (1998). *La philosophie dans le boudoir* (1795). In M. Delon, (Ed.), *Œuvres,* III. Gallimard-Pléiade.

Sloan, P. R. (2007). Teleology and form revisited. In R. Burian & J. Gayon (Eds.), *Conceptions de la science, hier, aujourd'hui et demain. Hommage à Marjorie Grene* (pp. 343–367). Ousia.

Stevenson, A. (Ed.). (2010). *Oxford dictionary of English* (3rd ed.). Oxford University Press.

Thompson, E. (2007). *Mind in life: Biology, phenomenology, and the sciences of mind.* Harvard University Press.

Toepfer, G. (this volume). 'Organization': Its conceptual history and its relationship to other fundamental biological concepts. In M. Mossio (Ed.), *Organization in biology.* Springer.

Van de Vijver, G., & Haeck, L. (this volume). Judging organization. A plea for transcendental logic in philosophy of biology. In M. Mossio (Ed.), *Organization in biology.* Springer.

Varela, F. (1991). Organism: A meshwork of selfless selves. In A. Tauber (Ed.), *Organism and the origin of self* (pp. 79–107). Kluwer.

Varela, F. (1996). Neurophenomenology: A methodological remedy for the hard problem. *Journal of Consciousness Studies, 3*(4), 330–349.

Varela, F. J., & Shear, J. (1999). First-person methodologies: Why, when and how. *Journal of Consciousness Studies, 6*(2–3), 1–14.

von Bertallanfy, L. (1933). *Modern theories of development* (J. H. Woodger, Trans.). Oxford University Press, H. Milford.

von Uexküll, J. (2010). *A foray into the worlds of animals and humans: With a theory of meaning.* University of Minnesota Press. [1934].

Walsh, D. (2007). Teleology. In M. Ruse (Ed.), *Oxford handbook of the philosophy of biology* (pp. 113–137). Oxford University Press.

Walsh, D. (2013). Mechanism, emergence, and miscibility: The autonomy of Evo-devo. In P. Huneman (Ed.), *Functions: Selection and mechanisms* (pp. 43–65). Springer.

Walsh, D. M. (2015). *Organisms, agency, and evolution.* Cambridge University Press.

Walsh, D. M. (2021). Aristotle and contemporary biology. In S. M. Connell (Ed.), *The Cambridge companion to Aristotle's biology* (pp. 280–297). Cambridge University Press.

Weber, A., & Varela, F. J. (2002). Life after Kant: Natural purposes and the autopoietic foundations of biological individuality. *Phenomenology and the Cognitive Sciences, 1,* 97–125.

Wolfe, C. T. (2010). Do organisms have an ontological status? *History and Philosophy of the Life Sciences, 32*(2–3), 195–232.

Wolfe, C. T. (2014a). The organism as ontological go-between. Hybridity, boundaries and degrees of reality in its conceptual history. *Studies in History and Philosophy of Biological and Biomedical Sciences, 48,* 151–161.

Wolfe, C. T. (2014b). On the role of Newtonian analogies in eighteenth-century life science: Vitalism and provisionally inexplicable explicative devices. In Z. Biener & E. Schliesser (Eds.), *Newton and empiricism* (pp. 223–261). Oxford University Press.

Wolfe, C. T. (2014c). Teleomechanism redux? Functional physiology and hybrid models of life in early modern natural philosophy. *Gesnerus, 71*(2), 290–307. (Special Issue: *Teleology and Mechanism in Early Modern Medicine*).

Wolfe, C. T. (2015a). Il fascino discreto del vitalismo settecentesco e le sue riproposizioni. In P. Pecere (Ed.), *Il libro della natura* (*Scienze e filosofia da Copernico a Darwin*) (Vol. 1, pp. 273–299). Carocci.

Wolfe, C. T. (2015b). Was Canguilhem a biochauvinist? Goldstein, Canguilhem and the project of 'biophilosophy'. In D. Meacham (Ed.), *Medicine and society, new continental perspectives* (pp. 197–212). Springer.

Wolfe, C. T. (2017). La biophilosophie de Georges Canguilhem. *Scienza & Filosofia, 17,* 33–54.

Wolfe, C. T. (2022). Vitalism. In T. Kirchhoff (Ed.), *Online Encyclopedia Philosophy of Nature / Online Lexikon Naturphilosophie.* Heidelberg University. https://journals.ub.uni-heidelberg.de/index.php/oepn/article/view/87350

Wouters, A. (2013a). Explanation, functional. In W. Dubitzky et al. (Eds.), *Encyclopedia of systems biology* (pp. 717–719). Springer.

Wouters, A. (2013b). Biology's functional perspective: Roles, advantages and organization. In K. Kampourakis (Ed.), *The philosophy of biology: A companion for educators* (pp. 455–486). Springer.

Wouters, A. (2013c). Explanation in biology. In W. Dubitzky et al. (Eds.), *Encyclopedia of systems biology* (pp. 706–708). Springer.

Zammito, J. (2006). Teleology then and now: The question of Kant's relevance for contemporary controversies over function in biology. *Studies in History and Philosophy of Biological and Biomedical Sciences, 37,* 748–770.

Chapter 4
Judging Organization: A Plea for Transcendental Logic in Philosophy of Biology

Gertrudis Van de Vijver and Levi Haeck

Abstract Even if the concept of organization is increasingly recognized as crucially important to (philosophy of) biology, the fear of thereby collapsing into vitalism, understood as the metaphysical thesis that "life" involves special principles irreducible to (and that perhaps even run counter to) the principles governing the physical order, has persisted. In trying to overcome this tension, Georges Canguilhem endorsed an *attitudinal* form of vitalism. This "attitudinal stance" (a term coined by Charles Wolfe) shifts the issue of organization away from ontological commitments regarding the nature of things as they are in themselves, in favor of epistemological issues concerning the stance of the knowing subject. However, it is based on some epistemological tenets that deserve further examination. Firstly, in spite of its anti-Cartesian spirit, the attitudinal stance implicitly relies on a Cartesian perspective on the relation between subject and object. Secondly, it rests on the idea that some objects can meaningfully be identified as persisting individuals—living organisms—in a way in which others cannot, even if it denies that the capacity to be meaningfully identified as such reflects an actual property of them. This chapter outlines a possible alternative viewpoint that takes these challenges to heart by developing a co-constitutive picture of the relation between subject and object—a picture based on Georges Canguilhem's own theory of judgment, but supplemented by Immanuel Kant's transcendental logic. Most fundamentally, it is argued that the (self-)organization of living beings draws attention to and is structurally intertwined with the (self-)organization of the thinking subject's rational (i.e., logical, conceptual, judging) capacities.

G. Van de Vijver (✉)
Ghent University, BE, Ghent, Belgium
e-mail: gertrudis.vandevijver@ugent.be

L. Haeck (✉)
The Research Foundation Flanders (FWO Flanders), Ghent University, BE, Ghent, Belgium
e-mail: levi.haeck@ugent.be

59

M. Mossio (ed.), *Organization in Biology*, History, Philosophy and Theory of the Life Sciences 33, https://doi.org/10.1007/978-3-031-38968-9_4

4.1 Introduction

What might it mean for the organization of living beings to be intrinsically linked to the organization of our rational capacities as thinking subjects? In "The concept of the organism", Joseph Henry Woodger brings this traditionally philosophical question into the heart of biology. He writes that we must distinguish between 'investigation' and 'interpretation', two processes that appear to be "different in their nature, their outcome, and in the 'canons' which regulate them" (1930, p. 2). The *investigatory process* "reduces at bottom either to observing organisms or parts of organisms in their natural relations, or to altering their natural relations in a systematic way, and recording the results […]," with heuristic success as the touchstone "by which the investigator will measure all things." The *interpretative process*, on the other hand, requires "knowledge about the properties of knowledge itself, and will not be natural scientific knowledge" (1930, p. 2). Woodger assumes that people who pursue natural scientific knowledge do not pay much attention to the dimension of "knowledge about knowledge" and vice versa, that people who make knowledge into an object of investigation "do not always know much about the subject matter of natural scientific knowledge" (1930, p. 2). He regrets this mutual disregard and compares it to the relation between the cook and the baking powder: "the people who pursue natural knowledge may be said, as a rule, and from one point of view at least, not to know what they are doing in somewhat the same sense in which a cook may be said not to know what she is doing when she uses baking powder" (1930, p. 1). In relation to the investigatory process, Woodger deplores that little appreciation is left for "understanding the properties of the intellectual tools involved—concepts, propositions, principles of inference, 'working hypotheses,' postulates, etc." (1930, p. 3). He coins the latter as "the logical realm" and considers it to be an intrinsic part of the pursuit of scientific knowledge: "Natural scientific knowledge springs from a fertilizing union of two 'realms': the realm of sense experience or perception, on the one hand, and the 'logical realm' or the realm of abstract logical entities and relations, on the other" (1930, p. 4).

In line with Woodger's focus on "knowledge about knowledge" and with the help of Immanuel Kant and Georges Canguilhem, we develop the idea that the workings of the "logical realm," as Woodger calls it, become manifest first and foremost where something *resists* the investigatory procedures pertaining to natural science. The history of the modern sciences shows that this is most prominently the case in the study of organisms or living systems. This becomes clear from the many forms of vitalism that have been put forward (cf. Normandin & Wolfe, 2013), from the recurrent attention to the organism seen as more than the sum of its parts and as a complex and self-organizing entity vis-à-vis evolutionary theory (e.g., Kauffman, 1992; Salthe, 2010), as well as from the attention to developmental and emergentist dynamics and the accompanying criticisms of (genetic) reductionism (Webster & Goodwin, 1996; Kauffman, 1992; Salthe, 2010; Oyama et. al., 2000; Van de Vijver, 2009; Robert, 2004). As illustrated by the contributions to this volume, a recent outcome of these developments is the rising recognition of the value of the concept of biological organization in studying organisms (e.g., Mossio et al., 2009, 2016) and their distinguishing features (e.g., Saborido et al., 2011; Nunes-Neto et al., 2014; Montévil & Mossio, 2015; Pontarotti, 2015; Mossio & Bich, 2017). As we see it, this trend involves recurrent "moments of crisis" of the logical realm prevailing in the modern sciences.[1] In relation to the living organism, the conceptual space itself comes under pressure and cannot but change gear, moving from "knowledge about the object' to 'knowledge about knowledge." The attention to the organism appears to be the point where the conceptual space is compelled to investigate its own structural procedures and dynamics—i.e., where it is compelled to fold back onto itself.

The hypotheses taken as a guiding thread here are threefold: (i) we interpret the crisis of the logical realm to concern at heart the relation between "subjects" and "predicates" and in particular the resistance of certain things (such as organisms) to being reduced to predicative descriptions; (ii) we take it that Kant was after something that diverges from the subject/predicate structure encountered in most of our Western languages and considered our logical and conceptual capacities to *constitute* objects, rather than merely being instruments to develop (predicative) knowledge *about* objects existing independently from us;[2] thus, Kant contributed to a fundamental rearrangement of our viewpoint on what counts as an object and on the place the knowing subject can have in the process of knowing; and (iii) Canguilhem was on a similar track to Kant in relation to judgment and knowledge, even if some of his writings on living phenomena are at times at odds with it.

Our discussion of Canguilhem and Kant, which calls for a focus on logic in matters relating to organization in biology, suggests that the crisis of "knowledge about" requires a more critical viewpoint on what can be qualified as a knowing subject. Insofar as one wishes to take on board the Kantian premise that the knowable is always *for us*, never *in itself*, the most straightforward move to make is to *include*, in the heart of conceptuality, the knowing subject's participation in living dynamics. More precisely, *both* the living organism *and* the knowing subject seeking to describe the living organism appear to be characterized by an internal logic of reciprocity. Pace Brilman (2018, p. 26), then, whose analysis suggests that life must *either* be external to rationality (which she takes to be Kant's view) *or* internal to rationality (which she takes to be Canguilhem's view), we suggest that the epistemological difficulty of objectifying living organisms is in a profound sense due to their commonality with the organization of our rational capacities. This commonality, we conjecture, impacts *both* what can be called an object and what can be called a knowing subject. From this point of view, the *attitudinal* type of vitalism, as articulated and to a certain extent also upheld by Canguilhem (cf. Wolfe, 2011),[3] uncritically leaves the knowing subject intact and distinct from the things it seeks to

---

[1] This trend has, in relation to vitalism, been called a "counter-history" of biology, a term that has recently been used in the context of the FWO-research project: "Vitalism: A counter-history of biology" (2019–2021).

[2] For a technical (and compelling) analysis of the nonpredicative aspects of Kant's logic, see Codato (2008).

[3] Cf. Normandin & Wolfe (2013); Etxeberria & Wolfe (2018)

investigate, even if it acknowledges, as Canguilhem does, that a knowing subject is fundamentally a living subject.[4]

In what follows, we begin by briefly explaining Canguilhem's attitudinal vitalism (in Sect. 4.2.1). This brings us to an analysis of what we take to be his own theory of judgment and account of "knowledge about knowledge" (in Sects. 4.2.2 and 4.2.3), on the basis of which we offer a first critical assessment of attitudinal vitalism (in Sect. 4.2.4). Both the flaws and the assets of Canguilhem's thought lead the way to an in-depth analysis of Kant's philosophy (in Sect. 4.3). First, we discuss Kant's take on logic (in Sect. 4.3.1) and his transcendental theory of judgment (in Sects. 4.3.2 and 4.3.3), which is then brought in relation to Canguilhem's thought (in Sect. 4.3.3) as well as to Kant's own account of biological organization (in Sect. 4.3.4). This allows us to dissect in a renewed and more precise manner the uncritical tenets of Canguilhem's attitudinal vitalism while simultaneously shedding a more distinctive light on how the organization of our rational capacities is intertwined with the organization of living beings (in Sects. 4.4 and 4.5).

4.2   Canguilhem's Theory of Judgment vis-à-vis Life

4.2.1   Canguilhem's Attitudinal Vitalism

We focus on the work of Canguilhem, because his questions in relation to living phenomena are very often questions of knowledge and vice versa. In Thought and the Living, for instance, he writes that we "accept far too easily that there exists a fundamental conflict between knowledge and life, such that their reciprocal aversion can lead only to the destruction of life by knowledge or to the derision of knowledge by life." This presupposition leaves us "with no choice except that between a crystalline (i.e., transparent and inert) intellectualism and a foggy (at once active and muddled) mysticism" (Canguilhem, 2008c, p. xvii).[5]

---

[4] In his contribution to this volume ("Varieties of Organicism — A Critical Analysis"), Charles Wolfe (2023) also touches on the connection between the knowing subject and the known object in the context of organization. He addresses, differently from but still remarkably in line with our own analysis, that with regard to "the problem of [...] objectification (of properties of organisms), there is kind of symmetrical problem, the other horn of the dilemma, which is subjectification." He adds that "the Kantian / epistemological view can also be transfigured into something 'ontological', not as a claim about the world [...] but as a claim about the unique nature of the subject [...]," which is indeed not surprising "if we consider Kant's famous proclamation that 'there will never be a Newton of a blade of grass': a point which is both about our cognitive capacities and [...] about the uniqueness of organic entities in comparison with the entities studied by physics [...]." We agree with the idea that for Kant, claims about the world (especially in respect to organization) go hand in hand with claims about the subject, but in our view this does not for Kant necessarily involve a "slippage from the epistemological to the ontological," for, as we will try to show in this chapter, Kant's transcendental system has the merit of warning against such a slippage (see esp. Sects. 4.4 and 4.5).

[5] We refer to the English translation of Canguilhem's works, except when the English translation is unavailable, in which case we translate it ourselves. As for Kant, we refer to the *Akademieausgabe*

Canguilhem contends, in line with Woodger, that the encounter with living beings confronts us with issues concerning "knowledge about knowledge" and concerning the place of the knowing subject in relation to what is known. More specifically, he stresses that the perspectival dimension in relation to life is ineliminable and should serve as the core of epistemology at large (see 2008a, c, 1966, 1971). In this respect, he describes vitalism, an umbrella term for scientific theories that treat living organisms as *fundamentally* distinct from other natural objects, as a "permanent exigency[6] of life in the living" (2008a, p. 62). Vitalism, according to Canguilhem, is not a mere theory among others but a central "orientation of biological thought" (2008a, p. 60). He speaks in this regard of a "vitality of vitalism" (2008a, p. 61). So, even if it

> may be that vitalism appears to today's biologists, as to yesterday's, to be an illusion of thought [...] the illusion in question is not of the same order as geocentrism or phlogiston theory—that is, it has a vitality of its own—in which case, one must philosophically account for the vitality of this illusion. (2008a, p. 61)

This constitutes the core premise of what is, according to Wolfe (2011), Canguilhem's *attitudinal* vitalism. However, even if this line of thought draws attention to "knowledge about knowledge," it remains rather meagre, because it leaves both terms, the knowing subject on the one hand and the exigency or demand of life on the other, relatively unquestioned. We therefore propose to focus on Canguilhem's view of judgment, which is, we think, much more promising for dealing with the relation between knowledge and life and which remains until now largely unexplored in the secondary literature on Canguilhem.

4.2.2   Canguilhem's Theory of the "Broken Judgment"

It is in "De la science et de la contre-science" (1971) that Canguilhem puts forward the core of his theory of judgment. If we observe a wooden stick in a glass, he says, we see it as broken.[7] Such will be our viewpoint, our perspective, no matter how many times we repeat the observation. Nonetheless, we will *know* that we see the stick as

---

(*Gesammelte Schriften*) from 1900 by indicating title abbreviation, volume, and page numbers (with the exception of the first *Critique*, for which we use the customary A and B indication), but we cite from the English translation if available (cf. our list of references down below).

[6] This term and its translation ("exigence") would deserve more attention. We see it here as a "requirement," a "demand," even a "need" of life.

[7] As aptly noted by a reviewer, one could object that the stick in question is actually "bent" rather than "broken." Yet Canguilhem clearly chooses to speak of the brokenness (*brisure*) both of the stick (*le bâton brisé*) and of judgment, as will be clear below ("[...] *c'est le jugement qui s'est brisé* [...]"), even after citing La Fontaine, who says that "the water" indeed "bends" the stick (*"quand l'eau courbe un baton* [...]") (1971, p. 173). Thus, we follow Canguilhem in his usage of the term *brisé* in both contexts as it highlights a *shift* from the brokenness of the stick *to* the brokenness of judgment. We assume that this word is best translated as "broken," although the same reviewer notes that *le jugement brisé* could also be translated as "the fractured judgment," but then again it seems a bit loaded to speak of a fractured stick.

broken only in so far as the possibility of another type of knowledge has opened, a knowledge that describes things in an objective way, that describes how things are *in reality*. We will then also *know* that the perceived stick is not broken at all. From that moment on, it is impossible to maintain both viewpoints simultaneously: "one cannot be both naïve and warned, credulous and critical, ignorant and learned" (Canguilhem, 1971, pp. 173–175; our translation). The initial "seeing" of the stick is from then on broken in two: *my* judgment and the *reality* judgment. What is broken is therefore not so much the stick itself as our *judgment*: "Quoique paraissant brisé, le bâton n'est plus brisé, mais c'est le jugement qui s'est brisé" (1971, p. 173).

Canguilhem stresses, moreover, that both options are "inverse" as well as "correlated" (1971, p. 173). From the moment judgment is broken, there is an exclusive disjunction between *my* perspective and the perspective *of reality*. Their very possibility rests on a mutual exclusion: on the one hand, objective knowledge cannot but exclude the moment of *naiveté* whereby the stick is seen as broken. On the other hand, the "naïve seeing" of the stick as broken, revealed from thereon as an appearance, an ignorance, an uncritical evidence, is incompatible with the "objective seeing" of the stick (1971, pp. 173–175). As such, in their being inverse and mutually exclusive, both judgments are correlated. More in particular, Canguilhem states that *if* there is (a judgment about) a reality, it is the result of a transfer that establishes a correlation with a judgment that, by virtue of this transfer, can be called mine, a judgment that is related to something being identified as "my" perspective. To Canguilhem, appearance emerges together with reality: both are posited in relation to each other, albeit the one next to the other, the one excluding the other.

In addition, both perspectives are legitimate—the one is not "truer" than the other. Yet the issue of their legitimacy only appears when something like an observer perspective, and thus also something like an observed reality, enters the scene. As a result, appearance judgments cannot be said to be about reality just as much as reality judgments cannot be said to reveal the falsity of appearance judgments. As soon as judgment is broken, the false can no longer be seen as proportionally related to the true. The false is not a moment of the true; it is what is excluded from the true as soon as a decision has been made about the truth of the proposition. Instead of considering error as a matter of consistency—whereby the two judgments are conceived to be on par with each other, with "my judgment" becoming a false moment *of* "true judgment"—Canguilhem advocates for a categorical brokenness of judgment, a fundamental heterogeneity between the two judgmental sides which does not allow for reconciliation, for "taking over" one type of judgment from within the other (1971, pp. 173–175).

We could of course (scientifically) explain the falsity of my judgment that the stick is broken by considering it as a mere optical illusion. However, this misses Canguilhem's point, which is to indicate a *structural* condition: if a human being makes judgments, it does so *both* from within its own, singular, lived perspective (potentially and at times actually shared by others) *and* from within a general perspective (shared by others as a *norm*). The general perspective, which adheres to universal laws of nature, often presents in an unjustified way the singular perspective as an initial, but false and therefore temporary moment of scientific investigation.[8] To Canguilhem, both standpoints are correlated while being inverse to each other: they exclude each other qua content (so that no comparison can be justified at that level), but they are related qua possibility. We will see below that this *reciprocal dynamic* is the key to understanding in what sense our capacity to judge participates in the living dynamics.

4.2.3   Knowledge About Knowledge: Reflexivity and Reciprocity

Two things need to be underlined in relation to Canguilhem's account of judgment, two things that are important in addressing the issue of the concept of the organism. The first has to do with *how our viewpoint on the capacity to judge*, as it took shape from within the objectifying project of the modern sciences, *is impacted by life and its knowability*. Canguilhem's work (as well as Kant's, as we will show in a moment) weighs the knowability of living beings in the field of the modern sciences in terms of broken judgment: life brings us to the point of having to express something about the structure of judgment. Indeed, in line with the previous section, judgment emerges as broken, not between a reality out there and a knowing subject but between a reality as objectively apprehended *by us*, representing normative universal laws of nature, and an observational perspective that is *mine*. An *I* does not perceive reality as a *we* knows it. This means that if there is something like (scientific) objectivity, it is *not* because of the fact that an agreement has been found between two pre-existing terms, a knowing subject and an observed object. If there is objectivity, it is because we have succeeded, from within our sensible (i.e., sensory) and conceptual activities, to obtain distinct judgments that can be qualified as a *reality* perspective, inverse to and correlated with *my perspective*.

In this regard, Herman de Vleeschauwer's work on Kant (1937, p. 434) suggests that to succeed in making such kinds of distinctions between judgments, we need *a reflexive return (un retour réflexif)* to our sensible and conceptual activities. Without such a reflexive return, there is no distinction to be made between conceptuality and sensibility—or between "the reality perspective" and "my perspective". These capacities or activities themselves (conceptuality and sensibility) are representational products *of* the reflexive return upon them, which highlights that they cannot but remain structurally caught up in a complicit and solidary totality.[9]

---

[8] Even a phenomenological or embodied cognition-inspired approach, which might highlight the singularity of the sensorimotor procedure whereby getting the stick out of the water informs us that, in fact, it is not broken at all, is irrelevant in this sense. Such phenomenological approaches, however much they stress the initial, singular, and lived conditions of experience, seemingly consider the false to be a moment *of* the true, thus neglecting the structural condition human beings are in as knowers.

[9] This dynamic is what Kant seized under the heading of an "original acquisition" (e.g., in *On a Discovery*, ÜE, 8:221) and of the "epigenesis of pure reason" (e.g., in the first *Critique*, KrV, B166–168, and in his Inaugural Dissertation, MSI, 2:395).

This brings us to our second point. *If* we come to ask ourselves the question of the knowability of life, as has been recurrently the case since the very onset of the modern sciences, it cannot but indicate that something in our "standard" representational procedures is shaken and has obliged us to undertake such a reflexive return. This is precisely what Woodger underlines in his analysis of the concept of the organism and in relation to which he comes to highlight the importance of logic. *Precisely* the standard ways of turning things into objects, i.e., the standard ways of grasping phenomenal diversity on the basis of concepts, becomes contentious in the context of scientific research about living organisms, where phenomenal diversity *resists* being subsumed under a concept as its predicate.[10] At stake here is the impossibility of identifying objects in terms of appropriate predicates (or properties), i.e., the impossibility of finding adequate concepts—a "reality perspective"—that can be stably, but inversely, correlated with the phenomenal diversity at hand, that is, "my perspective." Instead of arriving at knowledge *about* the object, there is, in the case of the living organization, a *prevailing indecision or indeterminacy* between my perspective and the reality perspective.[11] This prevailing indecision implies that to address the question of the organism solely in terms of *predicates* or *properties* adequately characterizing it as a living organization, cannot but miss the point. Fundamentally at stake is indeed not the adequacy of properties but the structure of judgment underlying our procedures of knowing.

In sum, the identification of properties as if they are more or less adequate to capture objects that exist independently from our doings misses the core of the logic that both Woodger and Canguilhem, so it appears, intrinsically relate to an epistemological project that purports to be in line with what is implied by the living. There is no sense in assuming in this process an "originary standpoint," as there is no sense, in Canguilhem's broken judgment, in assuming the *reality* judgment to be more originary than *my* judgment. Here lies, in our opinion, the core of what can be called "reflexivity," understood in terms of reciprocity: instead of referring to a knowing subject that develops knowledge *about* something, it indicates, *from within* an organic dynamic (as will be shown below), a return, a folding back onto certain activities (sensible and conceptual ones), leading to a "representational product" about that activity. As such, the prevailing indecision between my perspective and the reality perspective guides us to the insight that in the constitution of objectivity through the identification of properties, the knowing subject is an ineliminable factor.

---

[10] We shall see in Sect. 4.3 how this implies that the distinction between what counts as subject and what as predicate becomes critical.

[11] This can remind us of Von Neumann's and Burks' idea that in relation to autonomous machines, there is an inescapable confusion between the autonomy of the model and the modeling of autonomy (Von Neumann & Burks, 1966; Dupuy, 1985). See also Pask and Von Foerster on a relational definition of self-organization (1960)

4.2.4   *Life and Logic*

In "Aspects of vitalism," where the core of Canguilhem's attitudinal vitalism is articulated, it is argued that the crucial fault of classical vitalism lies in its "insertion of the living organism into a physical milieu to whose laws it constitutes an exception." He takes this to be mistaken, because "[t]here cannot be an empire within an empire without there being no longer any empire, neither as container nor as contents. [...]" (2008a, p. 70). A justification of this enigmatic statement follows right after in the text:

> One cannot defend the originality of the biological phenomenon, and consequently the originality of biology, by demarcating within the physico-chemical territory—that is, within the milieu of inertia, of externally determined movements—enclaves of indetermination, zones of dissidence, or foyers of heresy. *If one is to assert the originality of the biological, this must be in terms of the originality of one realm over the whole of experience, and not over islets of experiences.* In the end, classical vitalism sins, paradoxically, only in its excessive modesty, in its reluctance to universalize its conception of experience. (Canguilhem, 2008a, p. 70; our italics)[12]

Canguilhem appears to claim that life serves to be the sole perspective for all science, whereby the originality of the biological phenomenon reigns over "the whole of experience" as the one and only empire. There is no islet of life within a world of inertia, nor is there a "vitalist" empire contained within a "mechanistic" empire. But it is still enigmatic that Canguilhem formulates his criticism of (this excessive modesty of) classical vitalism in terms of the relation between a "contained" or "content" and a "container" (contenu et contenant). This distinction is traceable at least to Gottfried Wilhelm Leibniz' *New Essays*, where he writes the following in relation to Aristotle's logic:

> [...] rather than saying 'A is B' he [Aristotle] usually says 'B is in A' [...]. This manner of statement deserves respect; for indeed the predicate is in the subject, or rather the idea of the predicate is included in the idea of the subject. [...] when I say *Every man is an animal* I mean that all the men are included amongst all the animals; but at the same time I mean that the idea of animal is included in the idea of man. 'Animal' comprises more individuals than 'man' does, but 'man' comprises more ideas or more attributes: one has more instances, the other more degrees of reality; one has the greater extension, the other the greater intension. So it can truthfully be said that the whole theory of syllogism could be demonstrated from the theory *de continente et contento,* of container and contained. (1996, p. 486)[13]

---

[12] He goes on: "Once one recognizes the originality of life, one must *comprehend* matter within life, and the science of matter—which is science itself—within the activity of the living. Physics and chemistry, in seeking to reduce the specificity of the living, did no more than remain faithful to their underlying intention, which is to determine the laws between objects, valid without any reference to an absolute, central point of reference. Today, this determination has led them to recognize the immanence of measuring to the measured, and to see the content of observation protocols as relative to the very act of observation. The milieu in which one looks for the emergence of life only acquires its meaning as milieu in virtue of the operation of the human living being who takes measurements of it, measurements that bear an essential relation to the technical apparatuses and procedures by which they are made" (Canguilhem, 2008a, p. 70).

[13] See Book IV, Chapter XVII, §8.

Canguilhem's suggestion is that our "standard" scientific approach to nature must to some extent adhere to this logic of predication, according to which we are to mold our observations into a structure of predicates, subjects, and their internal (logical) relation, which allows for syllogistic reasoning.[14] He also seems to suggest that we simply cannot apply (this kind of) logic to nature in order to know what *life* is. This is because in an important respect life *is* the attempt to know what life is. In *The Living and Its Milieu*, for instance, he argues that:

> [...] if science is the work of a humanity rooted in life before being enlightened by knowledge, if science is a fact in the world at the same time as it is a vision of the world, then it maintains a permanent and obligatory relation with perception. And thus *the milieu proper to men is not situated within the universal milieu as contents in a container*. (2008b, p. 120; our italics)

Thus, Canguilhem appears to defend the idea that living beings somehow escape our logical procedures—i.e., cannot be seen as a contained in a (predicative) container—because these logical procedures are *themselves* to some extent "rooted in life," i.e., that "the living" reigns not only over the whole of experience but over "rationality" as well. The logical or intellectual tools (that we use to rationally study nature) discussed by Woodger are themselves living tools (used *by* living beings).[15]

However, these and other passages might also lend themselves to a diverging interpretation. Schmidgen, for instance, observes that Canguilhem's account of rationality involves a conception of life as "predominantly [manifesting] itself in organic individuals that act and react within specific environments which, in turn, are defined by the *needs and desires* of these individuals" (2014, p. 235; our italics). This is confirmed by passages such as the conclusion of *The Living and Its Milieu*:

> From this stems the insufficiency of any biology that, in complete submission to the spirit of the physico-chemical sciences, would seek to eliminate all consideration of sense from its domain. From the biological and psychological point of view, a sense is an appreciation of values in relation to a need. And for the one who experiences and lives it, a need is an irreducible, and thereby absolute, system of reference. (2008b, p. 120)

It would take us too far to analyze Canguilhem's standpoint concerning *needs* and their satisfaction in relation to his renewed epistemology, but we are certainly doubtful about the "irreducible, and thereby absolute" status of them as a "system of reference." What is clear, however, is that a viewpoint that falls back on absolute and irreducible needs misses his insight that the living is not predicatively or logically graspable (i.e., that it is not a "content" included in a "container"). Hence, the chances are thin from thereon to arrive at a consequential viewpoint in which our *logical* capacities (as representational products) are seen as genuinely undetachable from the living dynamics.

---

[14] This structure is indeed what allows for truth, according to Leibniz: "[...] always, in every true affirmative proposition, necessary or contingent, universal or particular, the concept of the predicate is in a sense included in that of the subject; the predicate is present in the subject; or else I do not know what truth is" (1967, p. 63 [G, 56]).

[15] We developed this idea in more detail in Haeck & Van de Vijver (2023).

The way in which this point is at stake in the current discussions on vitalism would certainly also merit further discussion. Let us just say here that the mere distinction made between substantival, heuristic, and attitudinal vitalism (Wolfe, 2011) potentially testifies to a similar predicament. Attitudinal vitalism refers to the viewpoint—attributable to Canguilhem, but also Kurt Goldstein (1995)—that the knowing subject adopts a certain stance in relation to life (call it epistemological, ethical, or political) and thereby refuses to ontologize and to substantivize the idea of a "vital supplement" (*force, élan*), so as not to fall into a scientifically unacceptable ontological vitalism. It does problematize the view that life is in any standard sense objectifiable, suitable for standard scientific investigation, yet it seems to do so by (uncritically) substantivizing that which is not objectifiable in terms of a (subjective) "stance" or "attitude." In this respect, attitudinal vitalism appears to operate on par with classical or substantival vitalism, however different the former claims to be from the latter.[16] In *this* respect, it potentially misses the very notion of reciprocity that Canguilhem (be it in his theory of judgment), Kant (as will be shown below), and Woodger, in his way, are at pains to articulate.

In order to make clear how essential a *logical* viewpoint is in this discussion, we now turn to Kant's theory of judgment. It will allow us to show how difficult it can be to escape from the "Cartesian trap" of ontologizing subject and object in terms of a knowing instance that cognizes a world which is more or less (or not at all) graspable through predicative means.

4.3   Kant's Theory of Judgment vis-à-vis Life

Kant's reflections on the possibility of objectivity led him to carry out a transcendental investigation into the thinking subject's logical capacities.[17] Whereas its foundations reside in the *Critique of Pure Reason* (1781/7), Kant brings his investigation to a head in the *Critique of the Power of Judgment* (1790) by engaging precisely with issues, such as the issue of living organization, that seem to *resist* objectification. In this section, we argue that the special place Kant gives to living organisms in his "transcendental logic" (broadly construed) and theory of judgment must be addressed in terms of his overall concern with the science of logic and not so much in terms of their belonging to a supposedly unique ontological class. First, we briefly present Kant's view that logic concerns a return upon unconsciously

---

[16] A similar tendency can be found in Evan Thompson's seminal *Mind in Life* and its treatment of the organism, which (arguably) plays the card of the attitudinal stance toward organisms *while* adhering to aboutness vis-à-vis "the mind." Thompson's theory of *autopoiesis* ultimately *intends* to make self-producing organization into an object of investigation, considering that we now *have* scientific accounts of circular causation and nonlinear emergence, which means that self-organization would no longer be a matter of merely regulative but also of constitutive principles (e.g., Thompson, 2007, pp. 138–9).

[17] At this point in history, the science of logic more or less amounted to a study of "thought."

operative formal rules (in Sect. 4.3.1). This allows us to move on to his so-called transcendental logic, which involves a subtle critique of the reigning logic of his time (i.e., a logic of subjects and predicates) and which brings him to develop a new theory of judgment. We discuss how this theory emerges and is developed in both the first and third *Critiques* (resp. in Sects. 4.3.2 and 4.3.3). By picking up Canguilhem along the way, we show that the Kantian point of view on judgment is closely connected to the one upheld in "De la science et de la contre-science" (Sect. 4.3.3). On this basis, we explain why the difficulty of acquiring objective knowledge about organized beings in nature is tied to the organization of our own capacity to judge—again in line with Canguilhem (Sect. 4.3.4).

4.3.1 Logic as a Return Upon Unconsciously Operative Rules

Although Kant never got to the point of formalizing his philosophical system, it cannot go unnoticed to what a tremendous degree his entire philosophical enterprise is anchored in and fed by the issue of logical formality. Striking in this respect is his attention to *rules*. He used to inaugurate his lectures on logic, for instance, with the statement that "[e]verything in nature, both in the lifeless [*leblosen*] and in the living [*belebten*] world, takes place *according to rules*, although we are not always acquainted with these rules" (Log, 9:11). One could read this as some kind of insistent—some might even say dogmatic—conviction that the world shall be seen as a rule-governed thing, even when evidence thereof is lacking. But Kant seems to be after something else: to him, rules seem first and foremost *unconsciously operative*, such that they can only be revealed in retrospect. In Kant's works, regularity is above all a *presupposition* that allows us to understand certain results or effects (thus requiring a return upon the presupposition of regularity itself). Clearly, formal rules are not merely being put to use as methodological tools for "making sense" of the world. Instead, our scientific-methodological attention to rules is an essential feature of our condition, reflecting our own powers and capacities:

> The exercise of our powers also takes place according to certain rules that we follow, unconscious [*unbewußt*] of them at first, until we gradually arrive at cognition of them through experiments and lengthy use of our powers, indeed, until we finally become so familiar with them that it costs us much effort to think them *in abstracto*. Thus universal grammar is the form of a language in general, for example. One speaks even without being acquainted with grammar, however; and he who speaks without being acquainted with it does actually have a grammar and speaks according to rules, but ones of which he is not himself conscious. (Log, 9:11)

So, when we approach "the world" in terms of a search for "its" underlying rules, we cannot exclude the "we" from the equation: *we are indeed rule-oriented by virtue of being rule-governed*. In his lectures on logic, Kant thinks through this assumption by suggesting that if we are, in a sense, condemned to rules, then this is so because our power to think these rules must itself, in its activities, be bound by rules. Investigating into these rules is the task of logic:

> [...] as sensibility is the faculty of intuitions, so the understanding is the faculty for thinking, i.e., for bringing the representations of the senses under rules. Hence it is desirous [*begierig*] of seeking for rules and is satisfied [*befriedigt*] when it has found them. Since the understanding is the source of rules, the question is thus, according to what rules does it itself proceed? [...] Now what are these rules? (Log, 9:11–12)

These opening passages of his lectures on logic indicate that in construing scientific theories about the natural world—both in respect to what we call lifeless matter in physics and to what we call living organization in biology—it is crucial to reflect on our rational capacities, which are indeed of a *logical* nature. And, as will become clear below, this need to reflect on or "fold back onto" our rational capacities is indeed *especially* felt in relation to organisms. But first, we must consider the *Critique of Pure Reason* (the first *Critique*), in which Kant lifted to a higher level the idea that logic is not so much instrumental as it is constitutive vis-à-vis what we can call an object.

4.3.2 Transcendental Logic: The Emulsifying Function of Judgment

One of Kant's central ideas in the first *Critique* is that analysis always presupposes synthesis: if we are to *analyze* the material world, we have to presuppose that it is always already *synthesized* following certain rules (KrV, B 130).[18] It is for this reason that, by investigating the capacity to have knowledge about the material world, one immediately lays bare the conditions of possibility of the material world itself. In other words, if knowledge of an object can generally be dissected into a conceptual constituent on the one hand and a sensory constituent on the other, this means that an object is as such *made possible* by a synthesis of sensibility and conceptuality (KrV, B 137). This is the heart of Kant's transcendental logic. It presupposes the joint cooperation of two basic human faculties: the understanding and sensibility, each accompanied by representations of their own that are fundamentally heterogeneous to one another (*concepts* and *sensible intuitions*, respectively).

Besides heterogeneity, then, there is also reciprocity at play between the two faculties. Sensible representations are entirely heterogeneous to the understanding's concepts, but they are nonetheless distinguished as such from within the understanding. Sensibility is identified as distinct from conceptuality, while being anticipated by it. If we want to maintain that sensibility and conceptuality are "inverse" terms, then we must also maintain that they are "correlated" ones. In other words: we must presuppose our conceptuality to have significance precisely in view of the fact that we have sensibility. Turning the manifoldness of our sensory

---

[18] In Kant's eloquent words: "[W]e can represent nothing as combined in the object without having previously combined it ourselves [...] for where the understanding has not previously combined anything, neither can it dissolve anything, for only through it can something have been given to the power of representation as combined" (KrV, B 130).

representations into an object is only possible by making an appeal to the opposite thereof, namely, the unity of concepts. Thus, the concept is constitutive of the object on account of the fact that it unifies the manifold of intuition delivered in sensibility (KrV, B 135). Objectification requires *overcoming* heterogeneity while being *indicative* of it. It is only through objectification that heterogeneity itself is retrospectively revealed. That is why Kant defines the object as "that in the concept of which the manifold of a given intuition is united" (KrV, B 137). The condition of possibility of the object lies in the homogenization *of* our heterogeneous condition. Kant stresses quite elaborately in this respect that the act of homogenizing is a matter of judgment: judgment is the activity that *relates* concepts to sensible intuitions (KrV, B 169/A130—B 178/A 139). In that sense, judgment is not only the linchpin of Kant's logic but also of his epistemology. The apparent "tension" behind object constitution, consisting in the homogenization of what can be called heterogeneous elements, is like mayonnaise: the watery parts could never mix with the fatty ones if it were not for an *emulsifier*—judgment.

The activity of judging, then, *cannot* represent an object "out there" by attributing predicate terms to subject terms and by drawing inferences on that basis. Such a logic of predication is potentially too compliant with epistemological realism, because it presupposes that judging merely consists in reflecting *about* the world by applying formal instruments. Instead, judgment is to Kant the condition of possibility of the object *to begin with*—its formal features are *constitutive* of objectivity.[19] We can indeed see his "transcendental logic" as a radical meditation on the constitutive *impact* of our human condition as divided between sensibility and discursivity. But all this is still more obvious in Kant's writings on judgment in relation to the knowability of living organisms, as will be shown in the following sections.

4.3.3 The Life of the Reflecting Power of Judgment

In the *Critique of the Power of Judgment* (the third *Critique* for short), Kant sheds an even more distinctive light on our divided human condition. He now shows how the conditions of possibility of objectivity as described in his first *Critique*, otherwise valid for mechanical systems, see their activities fail. This failure of objectification befalls our judgment upon confrontation with the beautiful, the sublime, and living organisms. In light of this failure, Kant seizes the opportunity to develop a more distilled account of what it means to judge, considering that upon

---

[19] It is true that, in taking up this philosophical struggle, Kant can sometimes give the impression that our logical capacities are *mere tools* employed to discover something about the world (e.g., when discussing judgment in terms of a procedure of reflection, comparison, and abstraction in his lectures on logic (Log, 9:94–5)).

confrontation with the beautiful, the sublime, and living organisms, judgment is forced, as it were, to fold back onto itself.

In what is conventionally referred to as the *First Introduction* to this work, Kant writes that the capacity to judge "is not merely a faculty for subsuming the particular under the general (whose concept is given), but that it is also, conversely, one for *finding* the general *for* the particular" (EEKU, 20: 209–210; our italics). In the first *Critique*, it was indeed argued that, by way of a *determinative* (or determining) kind of judgment, sensible particulars are subsumed under general concepts of the understanding, thus constituting objects in conformity with mechanical laws of nature. But what if there is no general concept available to us? If this is the case, we read in the third *Critique*, the capacity to judge becomes a faculty for *finding the general for the particular*, that is, for finding an adequate concept for what is presented to us by the senses. In this situation, we are dealing with a *reflective* (or reflecting) kind of judgment, which, as it *tends toward* the generality of the concept without necessarily attaining it in a final manner, stresses above all the reciprocity of our heterogeneous condition. In that sense, we must presume that the reflective kind of judgment grounds the determinative kind, because "the latter is, as it were, a dressed-up version of the former" (Haeck, 2020). It seems indeed that subsuming particulars under general concepts is nothing but an instantiation of the structural tendency to find the general for the particular.

It is at this point that the dynamics of living organization enter the scheme. Why would a human being want to know, want to generalize? The answer to this question must be found in the structure of judgment itself. The third *Critique* teaches us that there is no life for the cognizing subject when we assume that we always already have rational capacities on the one hand and, separately, a world to investigate on the other, whereby the former is simply to be "put to use" in order to teach us something about the latter. It seems rather that these two poles are *produced* from within the seemingly *purposive* structure of the power of judgment itself. The living human being is set to generalize and to know by virtue of the fact that it is condemned, on the grounds of its heterogeneous condition, to judge. Meanwhile, it acknowledges its heterogeneous condition *by* judging. That the human subject is judging is therefore indicative of the fact that it is inscribed in the dynamics of the general and the particular and that it attempts to orient itself on this basis.

The drive to find the general for the particular is related to the fact that the legislative power of the understanding (which is set to subsume the particular under the general, thus constituting it as an object of nature) is not always *satisfying* to the power of judgment. There is a need to reflect on the particularity of nature in its very contingency and infinite manifoldness and to seek the general *from within* the particular (KU, 5:186–7). It is important to note, in this regard, that "subsuming" (determining) and "finding" (reflecting), if they are in a sense "inverse" to each other, are certainly also "correlated" acts of judgment: determinative judgment constitutes natural phenomena adhering to natural laws, yet in doing so, it also leads the way to reflect on (those) phenomena insofar as they showcase infinite specificity, diversity, and contingent particularity.[20] This "excessive multiplicity" of nature makes the power of judgment run at full speed (KU, 5:193), pressing again and again to search for laws that explain the very contingency of those phenomena. In other words: even if particular phenomena are determined by a priori general laws of nature, they are still "determinable in so many ways" (KU, 5:183).

Consequently, Kant writes that the power of judgment "must thus assume it as an *a priori* principle for its own use that what is contingent for human insight in the particular (empirical) laws of nature *nevertheless contains a lawful unity*" (KU, 5:183; our italics). Objects and laws that are *contingent* from the point of view of the understanding (i.e., from the point of view of determinative judgment), must, in other words, appear as *purposive for* the reflective power of judgment. This principle of purposiveness "attributes nothing at all to the object (of nature), but only represents the unique way in which we must proceed in reflection on the objects of nature with the aim of a thoroughly interconnected experience." It is a "purposiveness through which nature agrees with our aim, but only as directed to cognition" (KU, 5:186). Kant adds quite tellingly in this regard that due to the fact that this principle of purposiveness is a *subjective* principle of judgment, we are "delighted (strictly speaking, *relieved of a need*) when we encounter such a systematic unity among merely empirical laws [...]" (KU, 5:184; our italics). Our human scientific rules of thumb or "stock formulae," as Kant calls them, like "nature takes the shortest route," "she does nothing in vain," "she is rich in species but sparing with genera," and so on, are "nothing other than this very same transcendental expression of the power of judgment in establishing a principle for experience as a system and hence for its own *needs*" (EEKU, 20:210; our italics). These needs instigate an endless search for satisfaction: even if it would never be truly satisfied, the power of judgment seizes every opportunity to reiterate its endeavors (KU, 5:187–8; see also Van de Vijver, 2019).[21]

Although the particular must be subsumed under the general, the act of reflection requires judgment to return its focus once again to that particular. Here, Canguilhem's theory of broken judgment is echoed, for Kant's theory of judgment involves a similar type of brokenness. Not only between subjective perception (the broken stick) and objective experience (the knowledge that the stick is in reality not broken), between sensibility and discursivity, between the ever-contingent sensible presentation and the law-like objective one, but also between judgments of reflection and

---

[20] Put differently: "in spite of all the uniformity of things in nature in accordance with the universal laws, without which the form of an experiential cognition in general would not obtain at all, the specific diversity of the empirical laws of nature together with their effects could nevertheless be so great that it would be impossible for our understanding to discover in them an order that we can grasp [...]" (KU, 5:185).

[21] In a footnote to the official introduction to the third *Critique*, in which he discusses the notion of desire, Kant seems to claim that we will try to satisfy our wishes *even if*—or rather *especially when*—we know it is impossible to do so. And indeed he explains this "tendency in our nature to consciously vain desires" precisely by stating that "we learn to know our powers only by first trying them out" (KU, 5:177).

judgments of determination. Insofar as judgment is determinative of the particular, it *opens up* an opportunity for reflection on the particular's particularity. Then again, we must at the same time assume reflecting judgment to *ground* determinative judgment. These two kinds of judgment, although inverse, must always be considered correlated to one another. The perception of the broken stick *can* go together with the knowledge that it is not broken. Even more, the reflective power of judgment *cannot* ignore the fact that we see the stick as broken, although we know from the determinative one that it is not. Kant's theory of judgment thus joins Canguilhem's conception of science as rooted in life, which also "maintains a permanent and obligatory relation with perception" (2008b, p. 120). Hence, his discovery of the reflective power of judgment in the third *Critique* serves primarily as a first hint of the idea that a "living organization" can be found not only in nature, but at least also in the power of judgment itself. We will explain below that the drive to judge *is* a drive to live, even if life becomes manifest to the extent that it escapes a perfect and neat covering of the particular by the general.[22]

4.3.4   The Organization of Judgment and/as Living Organization

It is against this backdrop that we must consider what Kant says about living organization, the so-called natural purposes or self-organizing products of nature. Kant describes the organism in diverging, yet thoroughly interconnected ways. It is, first of all, a thing that "exists as a natural end" as if it were "cause and effect of itself" (KU, 5:370–71). And, as "a thing that is to be cognized as a natural product *but yet at the same time* as possible only as a natural end," it must also "be related to itself *reciprocally* as both cause and effect" (KU, 5:372; our italics). On these grounds, Kant concludes that a "natural purpose" is to be regarded as both an *organized* and a *self-organizing* being (KU, 5:374). This concept of organization could never be objectively attributed to mechanical nature, according to Kant, because "as a concept of a **natural product** it includes natural necessity and yet at the same time a contingency of the form of the object (in relation to mere laws of nature)" (KU, 5:396). But although the concept of organization "can never be a constitutive concept of understanding or reason," it can (and must) be a *regulative* concept, suited "for guiding research into objects of this kind and thinking over their highest ground in accordance with a remote analogy with our own causality in accordance with ends" (KU, 5:375).

For Kant, living beings cannot be scientifically investigated as a phenomenon subject to general laws of nature, because they are not to be found "out there" as a special kind of matter organized in space and time. When we say they are not "out

---

[22] In this regard, it is unsurprising that Kant links the power of judgment to the feeling of pleasure and displeasure (e.g., EEKU, 20:208; KU, 5:186–7; 218).

there," we mean that "we do not actually **observe** ends in nature as intentional, but merely **add** this concept as a guideline for the power of judgment in reflection on the products of nature, [as] they are not given to us through the object" (KU, 5:399). Likewise, "it is quite certain that we can never adequately come to know the organized beings and their internal possibility in accordance with merely mechanical principles of nature, let alone explain them" (KU, 5:400). Kant's view is certainly not that an organism resists objectification *in itself*, as being qua being. His view rather implies a restriction on the *knowability* of organisms, which is to be taken seriously by the life sciences.

Assuming, moreover, on the basis of the previous section, that the knowing subject is structurally condemned to reflection rather than merely voluntarily engaged in a scientific practice, it cannot come as a surprise that organisms occupy such a prominent place in Kant's theory of judgment. The tension within judgment between particularity and generality, which seems to correspond to the heterogeneity between sensibility and conceptuality, is at its peak where the concept of a *self-organizing being* comes in. Here, the power of judgment must account for the lawfulness *of* the contingent *itself*. To regard such a lawfulness as purposive for our faculty of cognition is therefore not only an enormous *challenge* to the power of judgment. It is, on account of the structure of judgment as divided between two heterogeneous realms, also an enormous *need* of it (KU, 5:404). We have to regard "the concept of the purposiveness of nature in its products" as "a concept that is necessary for the human power of judgment in regard to nature," although it does not "pertain to the determination of the objects themselves" (KU, 5:404).

But there is more to this slightly dramatic presence of living organisms in the world. The fact that they *are* in some way dramatically present to us gives us some retrospective insights into our own (unconscious) endeavors as knowing subjects. Kant writes that when the understanding "cannot follow" the excessive multiplicity of nature (especially, we would say, in relation to organisms), it is reason itself that becomes excessive (KU, 5:401). Therefore, in order to get a conceptual hold on living organisms, the power of judgment finds itself in need of supplementation and takes recourse to supersensible ideas proper to the faculty of reason. But in contrast to Onnasch (2014), we do not believe this recourse has to be taken intentionally.[23] Instead, Kant invites us to consider the knowing subject's concepts and ideas and their logical organization as a "source of supplementation" with regard to a realm heterogeneous to them, namely, the realm of sensibility in which they take part— and certainly when, from within this realm, something like the lawfulness of the contingent, like natural purposes, becomes manifest. This organization and its dynamics are precisely what Kant's transcendental logic is about. From the first *Critique*, we know that transcendental logic is the rational organization of a discourse providing the conditions of possibility of objectivity. In the third *Critique*, through what escapes conceptualization in a principled manner, it appears most

---

[23] In that sense, we also disagree with Philonenko that the third *Critique* would be about direct communication between humans (1984, pp. 10–11).

vividly that the knowing subject, instead of being a "central directing agency," a Cartesian cogito, *participates* in this conceptual and "vital" dynamics—is an effect of it, rather than its director.[24] The human being is subject to, rather than a subject over and above, its heterogeneous condition. This conjoins our suggestion, developed in Sect. 4.3.1., that logic is able to expose the *unconscious* dynamics of rationality. Rationality, as Kant wrote about it, is fundamentally alive: insofar as it involves heterogeneity between conceptuality and sensibility, it involves *reciprocity*. Whereas the first *Critique* shows, in relation to this reciprocity, that our rational capacities amount to an organized system, the third *Critique* informs us in hindsight that it has in fact always been *self*-organizing (see also Van de Vijver, 2006).

In this regard, it should be noted that for Kant, purposive self-organization, which we ascribe to "organisms," is not necessarily identical to what he calls "life," which is first and foremost to be ascribed to the faculties of our mind (*Gemüt*).[25] Yet, as Kant suggests in his *Metaphysical Foundations of Natural Science*, both life in the biological sense and life in the "facultative" sense "in no way belong to representations of the outer senses, and so neither [...] to the determinations of matter as matter" (MAN, 4:544). This is because life is defined by Kant as "the faculty of a substance to determine itself to act from an internal principle, of a finite substance to change, and of a material substance [to determine itself] to motion or rest, as change of its state." However, we "know no other internal principle in a substance for changing its state except *desiring*, and no other internal activity at all except *thinking*, together with that which depends on it, the *feeling* of pleasure or displeasure, and *desire* or willing" (MAN, 4:544).[26]

Here, it is interesting to note that Kant describes what he calls *Gemüt* as "the principle of life itself" (KU, 5:278). This *Gemüt* is not so much "the mind" in the contemporary sense, as it is "the position or place of the *Gemütskrafte* (the *Gemüt*'s powers) of sensibility, imagination, understanding and reason" (Caygill, 1997, p. 210). It is the place where different parts of "rationality" come together in a systematic whole, being in a sense *inverse* to one another (i.e., heterogeneous), but nonetheless *correlated*. Insofar as reciprocity is concerned, it seems indeed to involve an organized reciprocity that is at the same time *self*-organizing, since it is due to the very structure of judgment—i.e., internal to the organization—that the activities of determination and reflection take place. This self-organization is also clear from Kant's *Opus Postumum*, where he cites from Friedrich Schiller's *On the Aesthetic Education of Man* (1794), according to which the human *Geist* is like a system of reason that "becomes active [*thätig*] only through suffering [*Leiden*], that

---

[24] Cf. Lu-Adler, who writes that "[r]egulative principles, which concern nature as a whole, are not just optional heuristics for the investigation of nature" (2023, p. 141). She rightly notes that for Kant, a regulative principle of reason is not "merely a device of reason for achieving economy" (KrV, A653/B681).

[25] On account of this distinction, not all organisms are necessarily "alive," such as plants. However, when we use words like "life," "alive," and "living" in this text, we mostly use it in the broader sense of "self-organization."

[26] See also Kant's *Critique of Practical Reason*, KpV 5:9n.

reaches absoluteness only through boundaries [*Schranken*]; that acts and forms only in so far as it receives matter [*Stoff*]" (Op, 21:77; our translation). More importantly, it is said here that:

> [s]uch a mind will, therefore, combine the drive for form or absoluteness [*Triebe nach Form oder nach dem Absoluten*] with a drive for matter or boundaries [*Trieb nach Stoff oder nach Schranken*], as these are the conditions without which it could neither have nor satisfy [befriedigen] the first drive. (Op, 21:76; our translation)[27]

This is yet another way to illustrate that our rational capacities are indeed fundamentally "alive" themselves: they structurally involve the reciprocal distinction and oscillation between matter and form, between receptivity and spontaneity, between sensibility and conceptuality. The need to continuously oppose yet relate sensibility and conceptuality, as inverse but correlated, is revealed from within a system of judgment. For this reason, it should not surprise us that this reciprocity is connected, as mentioned above, to a search for *satisfaction* (KU, 5:187–8; see also Van de Vijver, 2019). What is striking, however, is that we come to this conclusion by investigating why living beings—and their self-organization—resist objectification. This can show that, even if the laws of nature run counter to the order of the living, this is not to say that the order of the living runs counter to the logical organization *behind* the knowing subject's attempt to investigate the world in line with these laws of nature.

4.4   Back to Attitudinal Vitalism

In honoring the epistemological restrictions with regard to knowledge about living organisms, one might nonetheless be committed to a kind of ontological dualism between subject and object, such that there is a capacity to judge on the one hand and a world on the other. It should be clear now that this Cartesian view is foreign to Kant's theory of judgment, which is rooted in an overall rejection of the assumption of an independent world that can be adequately (or inadequately, for that matter) described and investigated. But is this Cartesian view also foreign to Canguilhem's so-called attitudinal vitalism? In spite of its obviously anti-Cartesian spirit, there seems to be a Cartesian side to it.

What is it, after all, that grounds the requirement of an *attitude*? Although this remains an open question, Canguilhem seems to be convinced that it is life itself that incites a certain attitude toward it. Recall the "vitality of vitalism" (2008a, p. 60), the idea that:

> [i]ntelligence can apply itself to life only if it recognizes the originality of life. The thought of the living must take from the living the idea of the living [...]. [T]o do mathematics, it would suffice that we be angels. But to do biology [...] we sometimes need to feel like beasts ourselves. (2008c, p. xx)

---

[27] This statement appears on pp. 370–371 of (the original edition) of Schiller's (1794/2019) work. It is quite telling that Kant wrote it down in his notes in what seems to be an approving vein.

But is this not to suggest, be it implicitly, that if life resists objectification, it does so in the capacity of a *being qua being*? This idea, in turn, seems to be contingent on the assumption that there is a world consisting at once of mechanical nature *and* of living entities that transgress the mechanical order, thus necessitating the very attitude toward them. But we know from Sect. 4.2.4 that Canguilhem would certainly deny that: "[t]here cannot be an empire within an empire without there being no longer any empire [...]" (2008a, p. 70). This means that (explicit) hints of a form of Cartesianism, upholding a conception of the world or the object as independent from our subjective doings, are not to be sought for in Canguilhem's writings.[28] The hints are rather to be found on the (implicit) flipside of the Cartesian point of view on the object, namely, in its point of view on the *subject*—or rather the *absence* thereof. As explained in Sect. 4.2.4, the position of the knowing subject endorsed by Canguilhem remains to a certain extent unquestioned precisely because it is ultimately reduced to "organic individuals that act and react within specific environments which, in turn, are defined by the needs and desires of these individuals" (Schmidgen, 2014, p. 235). This, of course, can be seen as an anti-Cartesian move. So where does the attitudinal stance's Cartesianism reside? It resides precisely in its relative silence on the topic of our rational capacities and their autonomous logical organization in which the knowing, *living* subject cannot but take part. We saw in Sects. 4.2.2 and 4.2.3 that Canguilhem himself had already come to the point of raising this issue, yet in this regard it was Kant who did justice to the arguably *attitudinal* idea that the impossibility of objectifying living organisms is fundamentally tied, not to a material state of affairs, but to the living dynamics proper to our *own* capacity to judge. When Kant employs the distinction between the nonliving (the mechanical) and the living (the organic), as two "kinds" of natural objects, his concerns are not only epistemological but also involve a circumscription of the living dynamics of the *knowing* subject itself (see also Van de Vijver & Demarest, 2013). The problem with Canguilhem's attitudinal vitalism is indeed that it fails to give due consideration to what it means for a capacity to *judge* to take part in the very living dynamics that it seeks to investigate. In rightly suggesting that we must take an attitudinal stance with regard to the living organization, thus denouncing any substantivism with regard to its real properties, this stance inadequately takes on board the significance of the *relation* between the knowing subject and the observed object, even if it would admit, as Canguilhem arguably does (cf. Wolfe, 2011), that both the organism and our capacity to judge must be treated as if they were self-organizing, "alive". To presuppose that we humans are alive *in* our rational endeavors, however, should not be a trivial fact—just *saying* it is not enough. As a presupposition, it has implications for what rationality *means*. The point is the following: by foreclosing oneself from examining these implications, one implicitly treats the knowing instance as a neutral epistemic agent.[29]

---

[28] For a recent account of Canguilhem's own reading of Descartes, most notably in his "Descartes et la technique" essay from 1937, see Sfara (2023).

[29] In our view, then, Canguilhem's implicit Cartesianism is also recognizable in his transition (in the late 1930s) from focusing on "the mental" (judgment, the intellect, etc.) to focusing on "the practical." For a very helpful historical account (and defense) of this transition, see Sfara (2023).

Although the concepts of life and organization do, in fact, go hand in hand with the almost compulsory need to assume that living organisms are to be encountered in nature, this is not essential.[30] What is essential is that insofar as we encounter life in organisms "out there" by reflectively using our capacity to judge, we retrospectively bump into (fold back onto) our capacity to judge as participating in the dynamics of life itself. In thus going beyond a mere attitudinal vitalism, we agree with Jennifer Mensch that "[w]hen reason saw organic activity in nature, according to Kant, what it was really looking at was itself" (2013, p. 144). But this, we submit, seems to hold not only for the faculty of reason, but also for the other Kantian faculties of the intellect (understanding, judgment, etc.).

However, if the living organism is epistemologically challenging, this is not just in relation to the object that escapes, but perhaps also in relation to the subject itself: there is no clear division between the knowing subject and the observed object. There seems indeed to be something fundamentally symmetrical to the relation between subject and object, whenever the former is considered to be a self-organizing system and the latter is taken to be a self-organizing being. They are symmetrical to one another in the sense that what aims to be a judgment about the so-called living organism is revealed at once to be a judgment about judgment, for in both cases we attempt to grasp something of which we must presume that it is, in a way, organized and self-organizing. Quite strikingly, this means that knowledge about our capacity to judge might very well be subject to the same epistemological challenge that pertains to knowledge about living organisms.[31] Paying attention to "knowledge about knowledge" should therefore not be seen as simply summoning some kind of a meta-perspective. This would be out of place here, because in dealing with living organisms, we are ultimately dealing with a condition to which we are subject ourselves—which we cannot investigate from "a view from nowhere."

In this regard, we have argued against the interpretations according to which Kant's system of thought is seen as distant, theoretical, or instrumental, as such too external to be able to capture the living, whereas Canguilhem's conception of our rationality would, on the contrary, be closer to the living in its creative "theoretical polyvalence" (Brilman, 2018). According to Brilman (2018, p. 26), for instance, it seems that if life is the condition of possibility of rationality, then it should not be its blind spot, such that life must *either* be external to rationality (which she takes to be Kant's view) *or* internal to it (which she takes to be Canguilhem's view). She concludes that life is the condition of possibility of rationality, "rather than" rationality's 'blind spot.'" But this is a misleading opposition, because life can be the condition of possibility of rationality *while* being its blind spot. If Canguilhem

---

[30] Phenomenologically speaking, the presupposition of life is indeed both the horizon and the origin of a knowing subject.

[31] For an interesting account of Kant's "methodological skepticism" in regard to a related issue, see Lu-Adler (2022). For an equally fascinating account of how the biological notion of epigenesis can be made relevant for the science of logic and logical cognition, see Lu-Adler (2018).

deplores that rationality is too often seen as a "crystalline (i.e., transparent and inert) intellectualism" (2008c, p. xvii), thus hiding rationality's deep connection to life, then his own and Kant's theory of judgment should come to the rescue.[32]

4.5 Conclusion

The conclusion is *not* that there exists, between organization "in nature" on the one hand and logical organization "in rationality" or "in our minds" on the other, an absolute kind of isomorphism. A precise "mapping" of the elements of the one to the elements of the other would be the answer to a question we did not pose. It would also dishonor one of our core convictions, namely, that there is no sense in attempting to develop a meta-perspective overviewing what is "knowledge about knowledge" and what is the world at large. In that sense, Woodger's programmatic essay might be a bit misleading at first. Instead, we tried to convey that the process of acquiring knowledge about living organisms *forces* us to fold back onto our own rational capacities ("knowledge about knowledge"), a process that ultimately reveals that the latter is in fact also to be understood in terms of a living organization. This revelatory moment, which involves a return upon certain (sensible and conceptual) activities, is essentially retrospective.

Through a survey of the theories of judgment put forward by both Kant and Canguilhem in connection to the problem of knowledge about living organisms, we articulated our dissatisfaction with Canguilhem's attitudinal form of vitalism. In so doing, we shed a light on what it means to form judgments about living organisms but also about what it *generally* means to judge. While attempting to formulate objective judgments about living organization, we must not trivialize the assumption that our capacity to judge is (self-)organized too. It involves an internal reciprocity between heterogeneous elements, which highlights a certain purposefulness in its tendencies. A structural *drive* to connect our sensible representations with conceptual ones in judgments functions as the motor behind our rational endeavors and intentions. In this regard, Canguilhem's theory of the "broken judgment" eloquently captures to what extent judgment is *structurally* torn between a conceptual, universal realm (the reality judgment) and a sensible, singular one (*my* judgment). Confronted with nature's infinite specificity and particularity (not excluding self-organizing beings), the attempt to unify sensible representations according to concepts or universal laws ultimately fails. This then breaks judgment in two and leaves us, as Kant would have it, with the distinction between "reflective" and "determinative" judgements.

In this regard, our take-home message is that the judging, knowing human being—according to the philosophical tradition, the conscious holder of all sorts of

---

[32] See Sfara (2023), for a very different but thought-provoking comment on Brilman's characterization of the Kant-Canguilhem relation.

intentions—is subject to, rather than a subject over and above, its rational capacities. The subject is perhaps not so much an agent that simply makes use of these capacities in view of acquiring knowledge about the world—for instance, while judging organization in nature. Rather, it is condemned to use these capacities on the grounds that it *is* a judging organization.

Acknowledgments Special thanks are due, firstly, to the "Georges Canguilhem Seminar" organized by the Centre for the History of Philosophy and Continental Philosophy (Ghent University, Belgium) in cooperation with Charles Wolfe (Université Toulouse II Jean Jaurès, France). This seminar was the birthplace of most of our basic ideas for this chapter. Secondly, to Andrea Gambarotto, for having invited us to present an early version of this chapter to the "Life and Cognition Seminar" organized by the Institut supérieur de philosophie (Université Catholique de Louvain, Belgium). Thirdly, we would like to express our gratitude to Xuansong Liu, Alexandre Métraux, and August Buholzer for their thought-provoking and sharp comments, as well as to the three reviewers for their useful criticism and questions. And, finally, to Matteo Mossio, for his always constructive feedback.

References

Brilman, M. (2018). Canguilhem's critique of Kant: Bringing rationality back to life. *Theory, Culture & Society, 35*(2), 25–46. https://doi.org/10.1177/0263276417741674.

Canguilhem, G. (1966). Le concept et la vie. *Revue Philosophique de Louvain, 64*(82), 193–223. https://doi.org/10.3406/phlou.1966.5347

Canguilhem, G. (1971). De la science et de la contre-science. In *Hommage à Jean Hyppolite* (pp. 173–180). PUF.

Canguilhem, G. (2008a). Aspects of vitalism. In P. Marrati & T. Meyers (Eds.), & S. Geroulanos & D. Ginsburg (Trans.), *Knowledge of life* (pp. 59–74). Fordham University Press.

Canguilhem, G. (2008b). The living and its milieu. In P. Marrati & T. Meyers (Eds.), & S. Geroulanos & D. Ginsburg (Trans.), *Knowledge of life* (pp. 98–120). Fordham University Press.

Canguilhem, G. (2008c). Thought and the living. In P. Marrati & T. Meyers (Ed.), & S. Geroulanos & D. Ginsburg (Trans.), *Knowledge of life* (pp. xvii–xx). Fordham University Press.

Caygill, H. (1997). A Kant dictionary. Wiley-Blackwell.

Codato, L. (2008). Judgment, extension, logical form. In V. Rohden, R. Terra, G. de Almeida, & M. Ruffing (Eds.), *Recht und Frieden in der Philosophie Kants. Akten des X. Internationalen Kant-Kongresses* (Vol. 5). De Gruyter. https://doi.org/10.1515/9783110210347.5.139

De Vleeschauwer, H. (1937). *La déduction transcendantale de 1787 jusqu'à l'Opus Postumum* (Vol. 3). De Sikkel.

Dupuy, J.-P. (1985). L'essor de la première cybernétique. *Cahiers du CREA, 7,* 7–139.

Etxeberria, A., & Wolfe, C. T. (2018). Canguilhem and the logic of life. *Transversal: International Journal for the Historiography of Science, 4,* 47. https://doi.org/10.24117/2526-2270.2018.i4.06

Goldstein, K. (1995). *The organism: A holistic approach to biology derived from pathological data in man* (p. 422). Zone Books.

Haeck, L. (2020). Exploring the deduction of the category of totality from within the analytic of the sublime. *Con-Textos Kantianos, 1*(12), 381–401. https://doi.org/10.5281/zenodo.4304113.

Haeck, L., & Van de Vijver, G. (2023). Canguilhem's divided subject: A Kantian perspective on the intertwinement of logic and life. In G. Bianco, C. T. Wolfe, & G. Van de Vijver (Eds.), *Canguilhem and continental philosophy of biology* (pp. 123–146). https://doi.org/10.1007/978-3-031-20529-3_7.

Kant, I. (1900). *Gesammelte Schriften* (Vols. 1–29). De Gruyter.

Kant, I. (1992). On the form and principles of the sensible and the intelligible world [Inaugural dissertation] (1770). In D. Walford (Ed. & Trans.), *Theoretical philosophy, 1755–1770* (pp. 373–376). Cambridge University Press.

Kant, I. (1998). *Critique of pure reason* (P. Guyer & A. W. Wood, Eds.). Cambridge University Press. https://doi.org/10.1017/CBO9780511804649.

Kant, I. (2000). *Critique of the power of judgment* (P. Guyer & E. Matthews, Trans.). Cambridge University Press. https://doi.org/10.1017/CBO9780511804656.

Kant, I. (2002). On a discovery whereby any new critique of pure reason is to be made superfluous by an older one (1790). In H. Allison & P. Heath (Eds.), & H. Allison (Trans.), *Theoretical philosophy after 1781* (pp. 271–336). Cambridge University Press. https://doi.org/10.1017/CBO9780511498015.005.

Kant, I. (2004a). *Lectures on logic* (J.M. Young, Trans.). Cambridge University Press.

Kant, I. (2004b). *Kant: Metaphysical foundations of natural science* (M. Friedman, Ed. & Trans.). Cambridge University Press. https://doi.org/10.1017/CBO9780511809613.

Kant, I. (2015). *Critique of practical reason* (M. Gregor, Trans, sec. ed.). Cambridge University Press. https://doi.org/10.1017/CBO9781316136478.

Kauffman, S. A. (1992). Origins of order in evolution: Self-organization and selection. In F. J. Varela & J.-P. Dupuy (Eds.), *Understanding origins: Contemporary views on the origin of life, mind and society* (pp. 153–181). Springer Netherlands. https://doi.org/10.1007/978-94-015-8054-0_8

Leibniz, G. W. (1996). *New essays on human understanding* (P. Remnant & J. Bennett, Eds.). Cambridge University Press. https://doi.org/10.1017/CBO9781139166874.

Leibniz, G., & Arnauld, A. (1967). *The Leibniz-Arnauld correspondence* (H. T. Mason, Ed.). Manchester University Press.

Lu-Adler, H. (2018). Epigenesis of pure reason and the source of pure cognitions—How Kant is no nativist about logical cognition. In P. Muchnik & O. Thorndike (Eds.), *Rethinking Kant.* Cambridge Scholars Publishing.

Lu-Adler, H. (2022). The subjective deduction and Kant's methodological skepticism. In G. Motta, D. Schulting, & U. Thiel (Eds.), *Kant's transcendental deduction and the theory of apperception* (pp. 341–360). De Gruyter.

Lu-Adler, H. (2023). *Kant, race, and racism: Views from somewhere.* Oxford University Press (forthcoming).

Mensch, J. (2013). *Kant's organicism.* The University of Chicago Press.

Montévil, M., & Mossio, M. (2015). Biological organisation as closure of constraints. *Journal of Theoretical Biology, 372*, 179–191. https://doi.org/10.1016/j.jtbi.2015.02.029

Mossio, M., & Bich, L. (2017). What makes biological organisation teleological? *Synthese, 194*(4), 1089–1114. https://doi.org/10.1007/s11229-014-0594-z

Mossio, M., Saborido, C., & Moreno, A. (2009). An organizational account of biological functions. *The British Journal for the Philosophy of Science, 60*(4), 813–841. https://doi.org/10.1093/bjps/axp036

Mossio, M., Montévil, M., & Longo, G. (2016). Theoretical principles for biology: Organization. *Progress in Biophysics and Molecular Biology, 122*(1), 24–35. https://doi.org/10.1016/j.pbiomolbio.2016.07.005

Neumann, J. V., & Burks, A. W. (1966). *Theory of self-reproducing automata.* University of Illinois Press.

Normandin, S., & Wolfe, C. T. (Eds.). (2013). Vitalism and the scientific image in post-enlightenment life science, 1800-2010. *Springer Netherlands.* https://doi.org/10.1007/978-94-007-2445-7

Nunes-Neto, N., Moreno, A., & El-Hani, C. N. (2014). Function in ecology: An organizational approach. *Biology and Philosophy, 29*(1), 123–141. https://doi.org/10.1007/s10539-013-9398-7

Onnasch, E.-O. (2014). The role of the organism in the transcendental philosophy of Kant's Opus Postumum. In E. Watkins & I. Goy (Eds.), *Kant's theory of biology* (pp. 239–256). De Gruyter.

Oyama, S., Taylor, P., Fogel, A., Lickliter, R., Sterelny, P. K., Smith, K. C., & van der Weele, C. (2000). *The ontogeny of information: Developmental systems and evolution.*

Philonenko, A. (1984). Introduction. In A. Philonenko (Trans.), *Critique de la faculté de juger.* Librairie philosophique J. Vrin.

Pontarotti, G. (2015). Extended inheritance from an organizational point of view. *History and Philosophy of the Life Sciences, 37*(4), 430–448. https://doi.org/10.1007/s40656-015-0088-4

Robert, J. (2004). Embryology, Epigenesis and evolution: Taking development seriously. *Embryology, Epigenesis, and Evolution: Taking Development Seriously, 1*–158. https://doi.org/10.1017/CBO9780511498541

Saborido, C., Mossio, M., & Moreno, A. (2011). Biological organization and cross-generation functions. *The British Journal for the Philosophy of Science, 62*(3), 583–606. https://doi.org/10.1093/bjps/axq034

Salthe, S. N. (2010). Development (and evolution) of the universe. *Foundations of Science, 15*(4), 357–367. https://doi.org/10.1007/s10699-010-9181-z

Schiller, F. (2019). *Über die Ästhetische Erziehung des Menschen in einer Reihe von Briefen*. In G. Stiening (Ed.). De Gruyter. https://doi.org/10.1515/9783110415254.

Schmidgen, H. (2014). The life of concepts. *History and Philosophy of the Life Sciences, 36*(2), 232–253. https://doi.org/10.1007/s40656-014-0030-1

Sfara, E. (2023). From technique to normativity: The influence of Kant on Georges Canguilhem's philosophy of life. *History and Philosophy of the Life Sciences, 45*(2), 16. https://doi.org/10.1007/s40656-023-00573-8

Thompson, E. (2007). *Mind in Life*. Harvard University Press.

Van de Vijver, G. (2006). Kant and the intuitions of self-organization. In B. Feltz, M. Crommelinck, & P. Goujon (Eds.), *Self-organization and emergence in life sciences* (pp. 143–161). Springer Netherlands. https://doi.org/10.1007/1-4020-3917-4_9

Van de Vijver, G. (2009). No genetics without epigenetics? No biology without systems biology? *Annals of the New York Academy of Sciences, 1178*, 305–317. https://doi.org/10.1111/j.1749-6632.2009.05010.x

Van de Vijver, G. (2019). Objectivity, repetition, and the search for satisfaction. In *Law, labour and the humanities: Contemporary European perspectives* (pp. 158–168). Routledge. https://doi.org/10.4324/9780429022302-11.

Van de Vijver, G., & Demarest, B. (Eds.). (2013). *Objectivity after Kant: Its meaning, its limitations, its fateful omissions*. Georg Olms Verlag. http://hdl.handle.net/1854/LU-3166411

Pask, G. & Von Foerster, H. (1960). A predictive model for self-organizing systems (I). *Cybernetica, 3*(4), 258–301.

Webster, G., & Goodwin, B. (1996). *Form and transformation: Generative and relational principles in biology*. Cambridge University Press.

Wolfe, C. T. (2011). From substantial to functional vitalism and beyond: Animas, organisms and attitudes. *Eidos, 14*, 212–235.

Wolfe, C. T. (2023). Varieties of Organicism: A critical analysis. In M. Mossio (Ed.), *Organisation in biology: Foundational enquiries into a scientific Blindspot*. Springer.

Woodger, J. H. (1930). The 'concept of organism' and the relation between embryology and genetics. Part I. *The Quarterly Review of Biology, 5*(1), 1–22. https://doi.org/10.1086/394349.

Chapter 5
On the Organizational Roots of Bio-cognition

Cliff Hooker

5.1 Introduction

The theme of this book is the place of organization in the life sciences, especially biology. In that context, this essay is concerned with the place of organization within mind and the place of mind within the life sciences, especially biology. There are many possibilities for theories of mind, ranging from noumenal to neural to nihilist (behaviorist), and for most of these, the question of the role for organization therein makes no sense; further, they escape, or are opposed to, any deep tie to biology. Even when some link to biology is acknowledged, as for physicalisms, no inherent notion of organization appears in their development. But this chapter will present a thoroughly organizational conception of mind-as-cognition, anchored in a supportive conception of biology.

There are three versions of how something – here, cognition – is bio-organizational, each more stringent than its predecessor. (I) Cognition is best understood from within a bio-cognitive organizational framework. (II) There is a key high-level organizational characterization of cognition. (III) At the core of cognitive function is organization. Here explanation is ultimately dynamical explanation, and these three characterizations of cognition are to be considered as three degrees of explanatory centrality for organization, rather than, for example, as three distinct conceptual kinds (see below).

Consider, in illustration, an unheated pot of fluid on a stove, its liquid molecules moving at random. There is neither ordering nor organizing. Then the stove is used to gently heat the bottom liquid layer. The liquid forms ordered horizontal layers, warmest at the bottom, coolest at the top. Molecular motion remains random horizontally, but the vertical symmetry of random motion is broken, replaced by an ordering of layers by temperature (random molecular energy) that conducts the heat

C. Hooker (✉)
University of Newcastle, New Lambton, NSW, Australia
e-mail: jeanandcliff@gmail.au

© The Author(s) 2024
M. Mossio (ed.), *Organization in Biology*, History, Philosophy and Theory of the Life Sciences 33, https://doi.org/10.1007/978-3-031-38968-9_5

slowly upwards. Finally, as heating increases, rolling boil (or Benard) cells form, vertical and horizontal symmetries are broken and random motion is replaced by a pattern of cells within each of which molecules move circularly, conveying hotter liquid up to the fluid surface and cooler liquid back down to be reheated, their intra-cellular circular motions so arranged horizontally that at each pair of adjacent cell surfaces they are moving in the same direction. The whole manifests moderate order (horizontal) and moderate organization (vertical and horizontal). Then, following the three nested requirements for organization as fundamental, we have the following: (I Benard) The phenomena are indeed best understood from within the molecular organizational framework given above. (II Benard) There is a key high-level organizational characterization of the phenomena as representing a succession of molecular arrangements providing increasing heat transfer capacities. (III Benard) At the core of this succession lies breaking symmetries, partly through ordering (vertical stratification), but with the largest capacity shift achieved through coordinated horizontal and vertical reorganization. Nor is more needed for core understanding; (within limits) it does not matter what the fluid is, nor what the heat source is, or what the pot is made of, the sequence of pot states will recur.

5.2   Characterizing Organization

The intracellular Krebs cycle is a useful model of organization. Its function is to transport energy into the cell and eject waste. It is made up of several molecular steps and produces several products, each combining a specific external input with the current internal chemical to dynamically lead to (produce) an output internal chemical for the next step. It is typically diagrammed as a large cycle with several nested cycles driving off it and ordered around it.[1] Organisms are congeries of such kinds of processes, nested from the subcellular (Krebs cycle) to whole organism (e.g., respiration), all component processes appropriately space-time interrelated.

This is not so different from a motor vehicle engine where all of the many kinds of components are very different from one another (cf. fuel injectors, spark plugs, cam shafts) yet are interrelated in many distinctive ways so that together they perform the transformation of fuel into linear motion. We can think of this as an inter-related structure of sub-functions – fuel injections into cylinders, cylinder heads rising up, sparking the injected fuel, etc. – that together bring about the overall global function. In a clarified ontology, each sub-function is realized as a causal process (one driven by an energy gradient) that takes its start as the function initial condition and moves dynamically to generate the function end condition. In many situations, sub-functions and their realizing dynamical processes may come and go as part of the overall function/process. Immediately after sparking a cylinder head, a large energy gradient forms in it which forces it back along its cylinder shaft. But

---

[1] See, e.g., Bechtel (2006, 2007) and Bechtel and Abrahamson (2011).

this gradient only lasts until the fuel is "burnt." Then another cylinder takes its turn. Similarly, there are many biochemical interactions in molecular biology that only briefly exist while some momentary, but precisely located, function is realized. In the case of the engine, the constraints that structure and stabilize these processes include the entire engine frame and are much longer lasting than individual cylinder processes. And this is common for current human-made machines. But in molecular biology, it frequently occurs that the whole realizing processes, energy gradients and constraints are ephemeral, changed by equally ephemeral processes of which they are temporarily a part (cf. a seasonally eroding river bank and its flow). This should be understood as normal. The core process organization that grounds cognition relies on just such a structure (see III below).

"Organization" has a narrow and a wide usage. In its narrow usage – n-organization – it means possessing internal, nested correlations of the general sort illustrated above in the Krebs cycle and car engine. In its wide usage, "organization" means no more than "is in some respect, to some degree, systematic," as in having a well-organized work desk. In this wide sense, one may speak of hierarchical organization though only an ordering by parts and composition is intended, and whether or not internal nested correlations are involved. "Self-organization" as commonly used includes molten iron cooling down to a solid bar (the ion lattice is "well organized"), and ordering coins by size through random vibration against varying mesh-widths. In neither case is n-organization part of the output. And in neither case is there any more than the faintest suggestion of a "self-active" process. But both of these examples have new constraints as output. This feature generalizes: self-organization is best conceived as a process leading to the emergence of new constraints, whether or not they produce n-organization and whether or not there is an active self involved (see Hooker, 2011c). Here we are concerned only with n-organization where, as we shall see, it forms a distinctive class of biological conditions.

Ultimately, all n-organization is grounded in dynamical processes, as are all non-organized (a-organizational) states and behavior. N-organized dynamics grades into a-organized dynamics (i.e., plain old dynamics) as the internal processes show decreasing variety, decreasing uniqueness and complexity of collective functions realized and increasing dependence on specific dynamical conditions. But the universality of dynamics is the same in all cases. Two billiard balls colliding show no n-organization but are fully dynamical; the Krebs cycle is strongly n-organizational but each transformation is fully chemo-dynamical. N-organization carries only relationship or form, not quality; quality is carried by dynamics, including the dynamics that grounds relationships or form. This applies to cognitive accounts (e.g., Russell's electrical charge, Penrose's intracellular coherent quantum states – Russell, 1927; Penrose, 1989). Here only n-organizational character will be considered.[2]

---

[2] An instructive case of dynamics in organization in this setting is the notion of levels of organization. See, e.g., Eronen and Brooks (2018). These, like emergents (see text above), can be made up of dynamical constraints, within which the dynamics takes place (e.g., systems of double pendulums for constrained chaos), but they can also be externally unconstrained, their system-wide sta-

N-organization and Order   In terms of interrelations between components, n-organization lies between complete disorder and complete order. Complete disorder is where all component interrelations are random, so that there is no simplifying multi-component pattern which constrains their interrelated behaviors. With completely ordered components, there is a governing pattern, illustrated in soldiers marching in tight formation, or crystals in a uniform lattice, and also distinct from the random collection of its components. With n-organized components, there are also governing patterns, but these can be much more complicated and subtle than the simplicities of complete randomness or complete orderedness (cf. Krebs, engine).[3] Wholly random and wholly ordered are poles of zero internal n-organization, all n-organized systems falling somewhere between them. Bennett proposed the notion of logical depth to capture a formal notion of n-organization located along this continuum (Bennett, 1985). Roughly, logical depth is the number of nested correlations within correlations in an entity. This is certainly an important step in the right direction because it places distinctive correlations at the heart of n-organization. But obtaining a satisfactory measure for degree of n-organization is not easily done.[4] Further exploration lies beyond the scope of this paper.[5]

---

bilities an outcome of their internal interactions (e.g., gravitational solar systems with planetary moons). Conditionalization within systems can also be by dynamical switching, like fast constraint formation, or as slower dynamical transitions to new interaction basins. (Cf. SDAL below.) The assumption that all these differences must instead be conceived logically has great difficulty in understanding any of them. As with weak organization (above), there is also a weak notion of level of organization where it names only commonalities of spatial scale, e.g., in the common "hierarchy of life," representation (cells, multicellular organisms, populations, etc.) Dynamical systems may have scales of statistical aggregation of various dynamical kinds, all consistently with also having cross-scale dynamical interactions. Such dynamical distinctions are likely to play important roles in accounts of brain function underlying cognition and other mental capacities, but not when confined to logical models of brain function where dynamics is neglected. (The otherwise useful review of the conventional literature by Eronen and Brooks, e.g., makes only occasional mention of dynamical levels and does not explore the consequences of a systematic dynamical approach. See further, e.g., Hooker, 2004, Sect. 5, 2011d, Sects. 4–5, for expositions of agency and cognition in dynamical terms, as in Moreno & Mossio, 2015; cf. Hooker, 2011b; Hooker & Hooker, 2018.)

[3] See, e.g., Bennett (1985), Li and Vitànyi (1990), and earlier discussion in Collier and Hooker (1999).

[4] For instance, correlations can be used to specify both ordered and n-organized states, so when is each supported? How are cycles that stay within an order (e.g., a cycle within a device) compared with those that move across functional orders (e.g., a cycle that includes both machine and regulatory administrative states)? How are these to be compared when system n-organization is vertically modular versus horizontally modular? And so on.

[5] As note 4 illustrates, there is at present little to be gained from pursuit of precise definitions, formal or otherwise, for the foregoing distinctions, or for similar notions to come concerning agency and cognition. Rather, there are good examples on which to rest creative conceptions, as a way of moving forwards constructively. This approached is buoyed by noting that even in the most developed domains, like physics, definitions, if they come at all, come after the domain has been thoroughly understood, not beforehand, while pursued too early, they can stifle deeper explorations.

5.3   N-organization and Bio-cognition

Briefly, Looking Ahead First, it is argued that a specific kind of n-organization, called autonomy, characterizes all and only living organisms. Autonomy is shown to ground all the major n-organizational aspects of agency. Second, cognitive agency, the main objective here, is in turn shown to be a sub-class of autonomous agents and so ultimately a specific class of n-organized systems. Third, cognitive agency spans a range from elementary to deep problem-solving powers, a range that can be characterized n-organizationally. In sum, autonomy > agency > cognition > deep cognition, with each step along the way, distinctively and strongly characterized n-organizationally. With this framework in mind, let us proceed.

5.3.1   Autonomy, Agency (and Robotics), Auto(self)-directedness, and Anticipation

Autonomy Our concern in this paper is with the place of n-organization in a biologically centered account of mind. Even so, it is essential to begin with at least one aspect of the wider issue of the place of n-organization within biology generally. N-organization lies at the heart of what an organism is and when we properly understand how that is, we shall have constructed the basis for an n-organizational account of organism minds.

At their most basic, all living things are thermodynamic engines, existing in a far-from-equilibrium condition only maintained by conversion of an input flow of negative entropy (food) to do work and by the export of unutilized material to the environment (wastes). Essential work is of three kinds: (i) the repair or replacement of internal infrastructure, including of any enclosing membrane, and of the capacity for suitable work, (ii) the support of action in the environment, and (iii) the export (elimination) of wastes. This is already an n-organizational arrangement, focused around two cycles, an external interaction cycle with the environment comprising resource extraction and waste elimination and an internal action cycle comprising repair and replacement.

There are various obvious constraints on successful versions of this n-organizational arrangement: (C1) the negative entropy input flows have to arrive in a timely manner, at appropriate places and in appropriate quantities, to sustain all the organism's processes; (C2) the internal work doable on these flows by the organism must produce sufficient components to fully support the internal repair work, including reproduction of the repair capacities; and (C3) at the same time, their consequent resource exploitation and waste accumulation must be extractable and exportable by the organism at sufficient rates and volumes as to avoid both direct damage to the organism internally and indirect damage via environmental damage. Despite their apparent particularity, these constraints are in fact permissive in form.

For instance, it does not matter whether the food-gathering action is largely passive (e.g., a pitcher plant trapping insects) or active (e.g., a dragon fly hunting insects), discriminating (e.g., a koala's taste for eucalypt leaves) or indiscriminate (e.g., the pitcher plant); it matters only that it satisfies at least the constraints C1–3.[6]

The condition for organism viability is that each cycle is supported and the two cycles so interact as to meet constraints C1–3 above. This n-organized dynamical viability condition is called *autonomy*.[7] It picks out all, and plausibly only, living individuals – from cells to multicellular organisms to various multi-organism communities, including many (but by no means all) business firms, cities and nations. There is an issue of how sub-function processes might exactly fit together, each helping to canalize others (e.g., Kauffman's work-constraint cycles, Kauffman, 2000), to achieve self-reproduction on a sufficiently small scale (contrast engine repair and the whole economy), but in principle some combination of longer-lasting and ephemeral process constraint formation should do it.[8]

The name is appropriate: in autonomous systems, the locus of living process regulation lies more wholly within them than in their environment. Birds use twigs to make nests, but twigs themselves have no tendency to use nests or birds to any purpose. Hence the root sense of autonomy in the traditional social sense. Moreover, there is a richness to the notion that escapes the bare appearance of inter-locked cycles. Autonomous entities have a distinctive wholeness, individuality and perspective in the world derived from the global, interconnected nature of their cycles and the regenerative condition they sustain. This gives rise to achieving (or not) an integrated condition of satisfying autonomy (or not).[9] Further, when this satisfaction

---

[6] This example makes it obvious that there will be a raft of particularities characterizing the many different ways to satisfy these constraints. In addition, further rafts of particularities will characterize nearer satisfactions that strictly do not fully satisfy all the constraints but do so nearly enough, long enough for organisms to replicate before dying, and so on. Again, there is at present little fruitfulness in attempting to explore these n-organizational byways.

[7] On autonomy, see further Hooker (2011a, Sect. 4.1.1), Christensen and Hooker (2000a, 2002), Christensen (2004), Bechtel (2007), Moreno (2007), Ruiz-Mirazo et al. (2008), Moreno et al. (2011), Ruiz-Mirazo and Moreno (2012), Arnellos et al. (2014), Moreno and Mossio (2015) and references.

[8] See Moreno and Mossio (2015, Chap. 1) and its Foreword by Hooker (pp. x, xi). This remark covers a complex issue: how is autonomy to be understood? The origins of the notion of autonomy lie with the biological ideas concerning the nature of cellular life by Maturana and Varela (Varela et al., 1974; Varela, 1979; Maturana & Varela, 1980, among others) and attempts to construct formal principles that distinguish living forms (Rosen, 1991; cf. Smithers, 1995). Here the notion of a closed set of states, e.g., one that regenerates metabolism, plays a central role. Such closures were then seen as the mark of the living and sought everywhere, e.g., among information states as the mark of the cognitive. Every organism was ipso facto a cognitive entity (Maturana & Varela, 1980). Some reflected this position back on to the constructive idea that every closure loop of states would give rise to a semantic system of symbols so that autonomous entities were ipso facto internally meaningful cognisors (e.g., Stewart, 1996; Pattee, 1993, 1995, 2007).

[9] As an alternative to the cognisor approach, others distinguished between dynamical (energetic, material) closure and functional closures (see, e.g., Barandiaran & Moreno, 2006; Barandiaran et al., 2009). It is clear that organisms cannot be energetically closed because the laws of thermodynamics require that they replace higher entropic degraded states with lower entropy (more

condition is available to the organism itself as some kind of associated signal (e.g., absence of enclosing membrane stress), then the autonomous system has a basic sense of normative requirement. A situation will come to be identified as injurious (reduced integrity), healthy (increased integrity), or neutral, an evaluation that amounts to a distinctive normative perspective. In this manner, autonomous system activities are also willful, anticipative, deliberate, normatively self-evaluated, and adaptive. Such entities are properly treated as genuine *agents*. Autonomous systems are inherently all of those things.

Agency and Robotics   Meanwhile, let us pause to briefly consider technologies in relation to autonomy and the possibilities of autonomy-based robotics. The dominant difference between biology and technology is, as the petrol vehicle illustrates, that organisms are much more active, responsive and integrated entities than technological systems are, or are often capable of being. A primary difference lies in the inner loop. Vehicles are not self-repairing. Their metabolism is outsourced to repair specialists (mechanics), and from there – via manufacture of spare parts and tools for repairing, the process strengthened by human n-organizational technologies such as pacemakers – to the rest of the economy. Plants do reproduce branches and roots, and both they and animals adaptively alter their bodies in response to environmental interactions, but in animals these alterations are mostly confined to nervous systems, while plants self-maintainingly adapt their bodily forms to support photosynthesis; and both, like vehicles, rely on a larger ecology for the resources to do so, and so on. Pursuing these analogies raises issues concerning how widely distributed, "socially" interlocking and interactively open an agent's body may be, and conversely how deeply capacity-modifying protheses may be integrated with "natural" agents, and how much does reliance on surrounding ecology for repair differ n-organizationally from reliance on societal economies (cf. notes 9, 10). A second primary difference lies in the external loop: organisms are much more active in responsively regulating their interactions with their environment, and within themselves. This difference is deeply rooted in organism autonomy which provides them a self-orientation to the world that works on integrating many streams of informa-

---

ordered) ones, and often will not be materially closed when doing so (e.g., nucleic acid diffusion across common boundaries among slime molds; the several vitamins that humans cannot manufacture but must import. This requires identifying some other features that characterize closure. Moreno and Mossio (2015) choose dynamical constraints as that characterizing closures and argue that while such entities do no work in a system and have none done on them, they "guide" the reconstitution of autonomous systems (Moreno & Mossio, pp. xxvi–xxx). This turns out to be a challenging set of requirements to sustain (Hooker, Foreword pp. x–xi, Hooker, 2013b). It also leaves the origins of cognition and semantics to be explained (cf. Moreno & Mossio, 2015, Chap. 7). Others attempt to have constructive interaction in the context of autonomous organization bearing the weight of understanding how cognition and semantics emerge within autonomous systems (e.g., Christensen & Hooker, 2000b, 2002). The roots of these approaches lie in C19 biological theorists like Simmel, 1895 (see Coleman, 2002; Hooker, 2013a; von Uexkull, 1926). The approach opens up an integration of cognition and semantics via intentionality as a Merleau-Ponty (1942/1963) close interactive "grip" (see above and e.g. Bickhard, 1993; Christensen & Hooker, 2004; Hooker, 2009), cf. Di Paolo, 2003.

tion (perceptual, proprioceptive, affective, etc.), using them to enrich and modify their anticipative interaction models and the directed responses to which it gives rise. Finally, organisms show a wide variety of boundary forms and defenses, from an identifiable exclusionary membrane offering regulated intake of specific nutrients (e.g., gastrointestinal membrane) to socially constructed maintenance of internal community regulation (e.g., through mating roles) and to a highly inter-penetrating film through which DNA may be directly interchanged. These differences are rooted in differences of n-organization.

This autonomy-based characterization of agency meets all the criteria for being deeply n-organizational: (I) Agency is best understood from within a bio-cognitive n-organizational framework, namely, the inter-locking cycles underlying agent autonomy. (II) There is a key high-level n-organizational characterization of agency, namely, as expressing autonomy. (III) And since every capacity of agency is based on autonomy, whose core is n-organizational, the core of agency is n-organizational. The distinctive n-organizational character of life penetrates deeply into its nature, into universal roots constituting agency. And it will be on this basis that any account of mind as n-organizational will be built and find its place in biology.

Meanwhile, there remain 3 + 2 robotics issues. (A) How might the constitution of an integrated internal perspective be achieved, if at all? What role has autonomous n-organization in the construction of robotic focused and responsive bodies? What might their perspective on the world be? (B) How does the manufacture of tools by tools and commodities by commodities proceed and how, if at all, does it include all elements (And how is it shown that the manifest tool improvement that does in historical fact occur within it, can actually occur within it?) (C) What is the biology and sociology of boundaries, how can these be constructed robotically and with what consequences for internal n-organization? Each of these presents a deep and subtle problem. They are left for the reader to consider, as is their impact on the n-organizational character of these aspects of natural and artificial existence. Only after these issues have been addressed will there be a proper platform for addressing the issue. (D) What are the limits to autonomy? And then (E) can there be a truly autonomous robotics?[10]

---

[10] There are differences among researchers as to how relationships among autonomy, agency, and cognition are properly drawn and this impacts development of a dynamically based account. Compare, e.g., Moreno and Mossio (2015), Chaps. 1, 4, and 7, where each new concept represents an elaborated aspect of the preceding one, with one where autonomy, agency, and cognition are each aspects of the same core n-organizational development (Christensen & Hooker, e.g., 2002, 2004). This latter account needs a problem-solving, as opposed to logic rule applying, conception of rational process and a similar "n-organized focus" account of intentionality that unifies its development with that of cognition and both with agency (note 9), obtaining a unified, dynamically based, n-organizationally characterized, core framework. The specific dynamical ontology that potentiates this framework, and may support a transition from cognition to a broader mentality, is left open here (cf. notes 8, 9, 16, 17). On the matter of the constitution of an internal perspective and artificial robotics, see Christensen and Hooker (2004), cf. Moreno and Mossio (2015), Chap. 7, and Nolfi (2011). On the biology and sociology of boundaries, see, e.g., Rayner (1997), Nolfi (2011), Bickhard (2011, Sect. 3.1), Bishop (2011, Sect. 3.5), and Hooker (2011b, Sect. 3).

Returning to the main argument, its overall structure is as follows. Self-directedness and anticipativeness are two fundamental cognitive capacities harbored by autonomy. Mutually supporting one another, these capacities form the central cognitive process of self-directed anticipative learning (SDAL). SDAL in turn provides the foundation of the deepest, most powerful forms of problem-solving, that is, of cognition, and of tracking, that is, of intentionality. Thus, intention and cognition are provided their common n-organizational root.

Auto(self)-Directedness   Auto- or self-directedness [the latter, more common, version will be used] is the capacity to self-modify interaction in the light of its evaluation by the directing organism. Changing behavior to acquire newly available food (e.g., spring flower nectar) is one example, and changing behavior to manage pain is another. Such sensitive, conditionalized attention forms the intertwined root capacities of intention and cognition (cf. Christensen & Hooker, 2004). A mosquito has one known such process (whether or not to initiate search for a blood host – Klowden, 1995), and a mammal has a vast number of such conditionalizing processes, especially within its motor regulatory system. Cycles (the n-organized aspect) of signaling and initiating specific actions within the external interaction cycle, and evaluating their outcomes against autonomy support through the internal interaction cycle, provide the basic autonomy n-organization with strong outcome-led self-directedness. In appropriate context, something about the direction of value increase (i.e., autonomy support) is also provided (e.g., by testing small departures in various directions from the present setting to see which is more rewarding). In more sophisticated form, self-direction allows the interplay of multiple evaluative signals, combining, compromising and conditionalizing them when arriving at which values are appropriate for guiding action in the context, recognizing corresponding multiple streams of information as relevant to those decisions, and in that light follows their integration for regulation of decision-making. The more mutually convergent guiding values and streams of information the learner has about performance, the more effective its actions can be. Initially, guidance will be limited because of learner ignorance, while at the concluding stage, information will have been sufficiently enriched, focused and integrated into the interaction cycles as to allow the learner to converge on a solution.

Anticipation   There is another, closely related, feature that the mosquito and the mammal share (very unequally): anticipation. Reflex and random actions aside,

---

Christensen and Hooker (2004), followed by Barandiaran and Moreno (2006), provides a critical perspective on formal robotics – dynamical systems theory (DST, van Gelder variants) and autonomous agent robotics (AAR, Brooks/Braitenberg variants) – and analysis of cognitive theory that could integrate with them, while DiPaolo (2003) provides a complementary examination of AAR as anchored in the Maturana/Varela tradition. More widely, Ruiz-Mirazo, together with Moreno and others, pursued the related issues of how minimally artificial chemical cells could be constructed and how they would need to be additionally internally constituted if they are to form evolving communities (e.g., Moreno, 2007; Ruiz-Mirazo et al., 2008; Ruiz-Mirazo & Moreno, 2012; Arnellos et al., 2014).

every action anticipates its outcome. At its most primitive, anticipation is the form-ing (learning) of a simple association between current features and an outcome of an action. The bee dance anticipates re-locating ephemeral nectar supplies as out-come; it would not be attended to unless that outcome and its attendant resource availability were frequently enough the consequence of the dance.[11]

Anticipative learning is where the organism learns to anticipate a goal achieve-ment by employing an action sequence, thus associating receiving goal satisfaction with doing an action sequence. Elementary associative learning such as neural con-ditioning provides the simplest anticipative associations. More sophisticated ver-sions of this process can be elaborated as learning capacities widen. For instance, though much more sophisticated than the mosquito, the cheetah swinging right-wards chasing a dodging gazelle with a right-swing bias is still doing so anticipating a desirable outcome (a kill). But with the cheetah, all the mammalian apparatus of planning ahead, tracking trajectories for oneself and others and so on is put to use managing these interactions fluidly and at high speed. The cheetah's many associa-tions – approach downwind, remain camouflaged where possible, maintain prey separation from herd and so on – have come to be integrated in richly associated models (here of the hunt). Bringing all these capacities together, in n-organizationally mutually supportive ways, provides the close attentiveness to problem-solving that is the core of intentional cognition.

Self-directed Anticipative Learning [SDAL][12] The combination of self-directedness and anticipative action provides the basis of fluid self-steering. An action is undertaken in anticipation of achieving a goal; if it does so, the anticipation is entrenched, and if it does not, the action may be repeated, extended or modified, at the actor's self-direction. In this way, the actor steers itself through a process of learning its environment. The capacity this invests in its agents is adaptability. The ultimate goal of external adaptation is internal regulation, i.e., to be able regulate the operation of the twin autonomy cycles so as to continue to satisfy autonomy, in the environmental circumstances obtaining. However, in a dynamic environment, where creatures are constantly changing (e.g., their current location and posture), often across many fronts and on many time scales, detailed adaptation is momentary and only approximated. Instead, it is necessary to be adaptable, able to adapt once useful adaptations as new conditions emerge (run from a predator, switch diet, migrate, etc.) There are limited physiological adaptabilities, most subconscious and of fixed operation (e.g., callous formation, switching to burning visceral fat to extend flight in extremity). But the largest, most variable and most rapidly adaptable are the behavioral adaptabilities, from singing to flying to technology construction, regu-lated by the central nervous system and largely expressed through the motor system.

---

[11] For a sensitive and powerful exposition of steering, goal-directed regulatory n-organization in mind, see Sommerhof (1974).

[12] See, e.g., Christensen and Hooker (2000a, b, 2002, 2004) and Christensen (2004).

These latter features (largest range, most variable, most rapidly adaptable) do not in themselves constitute more than small augmentations of cognitive power. Fluid adaptation ranges from the superficial to the deep, and these add finesse to the superficial capacity. That the flatworm withdraws into the shadows in a larger range of ways and circumstances, more variably, and faster, when a light is shone on it, does not modify its simplicity, or its fixity, of response. Superficial adaptation offers fixed information channels and evaluation routines that provide only first-order fixed responses to changing situations, the whole working off an n-organized algorithm without the need of higher-order regulation, something that fairly cheap route planners and guided missiles, along with rafts of insects, worms and others, can provide. Moving toward greater capacity involves increasing numbers of conditionalizations, supporting increasing spread and discrimination of judgment. Though always useful, this level of fluid but fixed regulation cannot surmount significant shocks such as failure to recognize anticipated response sequences, or interaction dynamics altering mid-action.

Beyond this impasse lies the introduction of increased layers of higher-order conditionalizations, offering increasing orders of responsiveness and increasing integration of responses. Sufficiently developed, such higher-order, integrated judgment formation underlies powerful new dimensions to problem-solving, for instance, the capacity to respond to a "mis-match" signal as indicating, not merely a new trial in response, but a change in investigative methods used. Consider discovering through a mis-match signal (e.g., unexpected viral outbreaks) that the present testing method has an unexpected high false-negative rate (say in pharyngeal swab testing for a viral infection), requiring a change in testing method to achieve greater reliability in estimates of infection rates and hence in demand for healthcare resources, and so on. As well as method change, consider also bringing about reformulations of the problem to hand ("It's not the measuring process, but it's the modeling of sub-population interactions we are using"), changed criteria for successful outcomes ("predictions of infection breakout locations and frequencies accurate to within 10%"), changed external constraints framing the problem ("rural sub-populations are much more constrained by travel times") and changed criteria for "cleaned" data supporting these judgments ("estimates of false positives as well as of false negatives are required"). As will appear, supporting the integration of these features will provide deep fluid adaptation, or deep adaptability, the mark of truly human intelligence. (In this respect, we are a long way yet from deeply intelligent robotics.)

To see how these features work together, consider a detective conducting a murder investigation. She uses clues from the murder scene to build an initial proposed profile of the suspect and then uses this profile to focus the direction and methods of the investigation. Lipstick on a glass suggests a crime of passion, with the suspect female, in a personal or sex worker relationship to the victim. The profile tells the detective what the murderer may be like and what characteristic types of clues to pursue. For a crime of passion, look for further personal effects – special clothes, whips or other "technologies" in producing sexual effects, etc. Look too for places nearby, possibly frequented for romantic assignations, a romantic bar, a brothel, etc.

The chosen profile in turn sets new intermediate goals, for example, narrow down the nearby places frequented, eliminate or reduce the likelihood of the suspect being a male cross-dresser, but conversely try to obtain an estimate of how many women might be involved. If the chosen profile is at least partially accurate, and with a little luck, the modified investigation will uncover further evidence that in turn further refines the search process, ultimately culminating in the capture of the murderer, and resolving the nature of the investigation.

But such searches are not fixed; a good detective will have in mind other possible profiles awaiting supporting evidence. Further search of the murder site, for example, may uncover a gambling note for a substantial sum. This turns attention to enforcing debt default as the kind of crime involved. The lipstick cue does not fit comfortably into this version; the culprit is more likely a male, with a history of criminal activity and likely enforcer violence. This profile redirects the search from sexual partners to gambling associates and perhaps money laundering and the like. From this point of view, the lipstick is mis-directing; perhaps it belonged to an attempt to persuade the victim to settle his debts, or was indeed worn by a cross-dresser, but just as a personal quirk, irrelevant to the financial issues at stake, or planted to "throw the investigation off the scent."

It is the interplay between the discovery of clues, the construction of a suspect profile and subsequent modification of the investigation that makes the process self-directing. It is powerful self-direction because it encompasses re-thinking the nature of the investigation (here from sex to gambling), contextual assumptions (here from assignations to debt collections), data (lipstick from evidence of lover's presence to irrelevance), and solution types (from identification of passionate conflict, murder process and culprit motive to identification of debt association and assassin presence and actions). As an organism interacts in an SDAL process, its improving anticipative models and model-based interaction processes allow it to (a) improve its recognition of relevant information, (b) perform more focused activity, (c) evaluate its performance more relevantly and precisely, and (d) learn about its problem domain more effectively. Indeed, in this setting, error itself can be a rich source of context-sensitive information that can be used to further refine these four features.[13] The richer the system's anticipative and norm structures are, the more directed its learning can be, and the more potential there is that learning will improve the system's capacity to form successful anticipative models of interaction. To this the detective adds an additional kind of learning, higher-order learning about the entire domain of murders. It is the capacity to learn across many such investigations what sorts of profiles there are; what are their general features and rare exceptions; what their associated kinds of investigatory methods, timetables and costs; and so on and to recognize when there is more to learn and how to be alert to doing it that makes the process such a powerful problem-solving tool.

---

[13] Popper, who emphasized the importance of falsification ("signal mis-match") in scientific method, missed this power to scientific cognition by confining himself to just immediate logical structure, where indeed a falsification conveys no more than "something is false somewhere." See Popper (1980), cf. Hooker (1995, 2010), Hoffmaster and Hooker (2018).

When that kind of "double-loop" learning occurs, the detective is both learning what works, or not, in the immediate investigation to hand and at the same time using that experience to improve general knowledge of detecting murders, and crimes more generally – knowledge that will in turn be used to improve the next investigation. In short, by learning a higher-order characterization of the problem class (murder investigations), she will have been *learning how to learn* about investigations in that domain while learning how to solve specific problems to hand. Just this is the fundamental bootstrap required for all learning to be improvable. It forms the key to understanding the n-organization, and thence the general power, of the learning process. Indeed, this process allows the rational resolution of initially ill-defined problems, problems whose formulation and structure are vague, gappy or ambiguous, or tacitly internally conflicted, or whose valid methods are unsettled, like how to detect or marry well, or validly test a scientific theory in a new domain. Such problems of necessity lie at the root of every new problem-solving domain.

It is possible to synthesize a model process for such learning-how-to-learn-while-learning processes. Each of the five foci or nodes of learning identified above (method, problem formulation, solution formulation, constraints, data) are represented. As each specific learning process is under gone, attention shifts from one node to another, or to several nodes in parallel, as the potential consequences of experimenting with alternatives are explored. (Cf. the different investigations formed by the detective's various crime profiles.) Eventually (and with some luck), the investigations are reduced to one, the one that resolves the core detecting problem. Although all investigations share the same n-organizational form, the non-organizational features play their decision structuring and making roles alongside them and varying from incidental to central. A measure of their importance is the degree to which they must be appealed to at each choice point. For this reason, there is no specifiable model, let alone algorithm, for the order in which nodes are visited, nor for what changes are consequently made, nor for how these changes in turn spread across the nodes, nor for what kinds of compromises are made in reaching for an enriched solution, and so on.[14]

Yet the model does capture the fundamental n-organization of kinds of actions that deliberative problem-solving centrally involves. In its lowest form, this n-organization is expressed in the cyclic processes of specific trial-and-error interrogations. Moving to higher-order organization, these n-organized trial-and-error processes are nested within sharings of information about how to coordinate the findings from several such trials covering these kinds of crimes. Every cheetah hunt, and every detecting, is unique in its qualitative details but they are also all the same

---

[14] The detective, for example, has been developing the crime-of-passion profile, impressed by the initial lipstick clue and visits to nearby gay bars, but it has proven increasingly hard to find further useful clues. Several lines of investigation have been proposed and their consequences pursued, for example, that the lipstick belongs to a relative of the victim, leading to tracking down family members and examining any tensions in these relationships, and so on. As these lines dried up, the pressure mounted to look elsewhere, for example, to business dealings, with trial options ranging from gambling debts to defaulting debtor relatives.

as n-organized hunting processes. In particular, they all share the higher-order prospects of reformulating the problem and/or the solution, changing and nested again inside more general formulations of investigating crimes of these general kinds. This tri-layer of nested cyclings is of the same general form as the Krebs cycle (above), but here its n-organizational depth is much greater because, for example, at each of its nodes, it stores structured content about the domain related to that node, and stores cross-nodal interrelations pertinent to the domain involved, both contents increasing their richness as problem-solving multiply, none of which the Krebs cycle has available. In this enriched form, the problem-solving model has deeply illuminated the 30 years of research into the linguistic capacities of apes, even how (pace Kuhn) rational deliberative problem-solving can proceed through scientific revolutions.[15]

5.4   In Conclusion

It remains to reiterate that this model of problem-solving is primarily n-organizational. No matter the domain concerned, this moderately n-organized, moderately ordered process successfully models the general problem-solving process. The underlying sense of agency on which the problem-solving SDAL process is built is fundamentally n-organizational, satisfying the three criteria for being essentially n-organizational: (I) best understood from within a bio-cognitive n-organizational framework, (II) has a key high-level n-organizational characterization, and (III) its core is n-organizational. The distinctive n-organizational character of life penetrates deeply into its nature, into its universal roots constituting agency. And now it is on this basis that the roots of cognition have also been revealed to be in essence n-organizational. (I) Problem-solving is best understood from within a bio-cognitive n-organizational framework, here that of autonomy-based bio-agency, with its distinctive accounts of identity and normativity. (II) There is a key high-level n-organizational characterization of problem-solving, namely, that of the general SDAL problem-solving process model. (III) The core of problem-solving is n-organizational because it lies within the improvable, enrichable tri-nested cyclicities of the general problem-solving process model. Goal-pursuit is an inherent, if moderate, n-organized process; it marshals the steering sub-processes – anticipation and self-directedness – to orient to the goal and to explore self-improving ways to move toward it. SDAL, equipped with higher-order regulation, is inherently this n-organization. Global-level n-organization is emphasized by the steering processes in SDAL, which are typically higher order. This completes the n-organizational "golden thread" running throughout biology, ultimately integrating mind into living being.

---

[15] For the general model of problem-solving, see Hooker (2017, 2018). For investigation of research in ape language capacities, see Farrell and Hooker (2007a, b, 2009). For the relationship to design problem-solving, see Farrell and Hooker (2013, 2014, 2015).

Such n-organizational principles evidently have but small extension beyond life to the cosmos at large. While the inanimate world has n-organization – wherever "mechanical" cyclicities operate, for instance in rolling boil formation (Introduction: Benard cell) – n-organization evidently does not lie deep throughout the cosmos as it does throughout biology. The inanimate makes more use of order than n-organization. No doubt this reflects the simplicity of orderedness and the priority in time of the inanimate world with the emergence of life within it. This makes living n-organization, autonomy, the more remarkable.[16]

The details of the general problem-solving process will vary across subject matters. The golden thread of n-organization abstracts from these differences to locate a fundamental n-organizational category: life. It is thinkable that it might not have been so. Understanding how that n-organizational category is possible will involve tracing it back to the fundamental qualities as we know them, the quantum and relativistic qualities: mass, spin, charge, and so on, along with those that structure irreversible thermodynamics. There is at present no neat accepted story here, and the problems are so deep and unresolved as to make it thinkable that there is none for finite mortals to have. The complications rise further if those qualities associated with mind, the perceptual and emotional qualities, are included. It is always possible to try for a purely process account of these, or for a more n-organizational one, though how complete they can be also currently remains open.[17]

Acknowledgments Special thanks to Alvaro Moreno and Matteo Mossio for careful, detailed, and critical appraisal of draft versions and to Hal Brown for challenging comments from a wider philosophical perspective – all of which resulted in a substantially improved essay. The defects remaining surely derive in substantial part from my willful refusal to respond to all comments as their authors intended. With so tricky a topic, I have tried to keep the notion at issue – organization – always at the central focus. In addition, there undoubtedly remain defects not yet appreciated because we all – author and commentators – remain blind to them.

References

Arnellos, A., Moreno, A., & Ruiz-Mirazo, K. (2014). Organizational requirements for multicellular autonomy: Insights from a comparative case study. *Biology & Philosophy, 29*, 851–884.

Barandiaran, X., & Moreno, A. (2006). On what makes certain dynamical systems cognitive: A minimally cognitive organization program. *Adaptive Behavior, 14*(2), 171–185.

Barandiaran, X., Di Paolo, E., & Rohde, M. (2009). Defining agency. Individuality, normativity, asymmetry and spatio-temporality in action. *Adaptive Behavior, 17*(5), 367–386.

Bechtel, W. (2006). *Discovering cell mechanisms: The creation of modern cell biology.* Cambridge University Press.

---

[16] The status of biochemically, dynamically characterized autonomy and the scope for inanimate autonomy, in relation to the positions mentioned in notes 2, 8, and 9 are left as issues for the reader.

[17] That there is no neat account of the fundamental metaphysics of mind on offer here emphasizes that naturalist fallibilism remains. Abstraction simply stops where principle is bracketed along with detail; there is no commitment to a formalist n-organizational idealism here.

Bechtel, W. (2007). Biological mechanisms, organised to maintain autonomy. In F. Boogard, F. Bruggeman, J.-H. Hofmeyr, & H. Wesyerhoff (Eds.), *Systems biology: Philosophical foundations*. Elsevier.

Bechtel, W., & Abrahamson, A. (2011). Complex biological mechanisms: Cyclic, oscillatory and autonomous. In Hooker (2011a).

Bennett, C. (1985). *Dissipation, information, computational complexity and the definition of organization*. In D. Pines (Ed.), Emerging syntheses in science. Proceedings of the founding workshops of the Santa Fe Institute. Addison West Publishing.

Bickhard, M. (1993). Representational content in humans and machines. *Journal of Experimental and Theoretical Artificial Intelligence, 5*, 285–333.

Bickhard, M. (2011). Systems and process metaphysics. In Hooker (2011a).

Bishop, R. (2011). Metaphysical and epistemological issues in complex systems. In Hooker (2011a).

Christensen, W. (2004). Self-directedness, integration and higher cognition. *Language Sciences, 26*(6), 661–692. Cognition and Integrational Linguistics, special edition.

Christensen, W., & Hooker, C. (2000a). An interactivist-constructivist approach to intelligence: self-directed anticipative learning. *Philosophical Psychology, 13*(1), 5–45.

Christensen, W., & Hooker, C. (2000b). Organised interactive construction: The nature of autonomy and the emergence of intelligence. In A. Etxeberria, A. Moreno, & J. Umerez (Eds.), *Communication & Cognition 17*(3 & 4), 133–157. Special Edition, The contribution of artificial life and the sciences of complexity to the understanding of autonomous systems.

Christensen, W., & Hooker, C. (2002). Self-directed agents. In J. MacIntosh (Ed.), *Contemporary naturalist theories of evolution and intentionality, Canadian Journal of Philosophy*, Special Supplementary Volume 19–52.

Christensen, W., & Hooker, C. (2004). Representation and the meaning of life. In H. Clapin, P. Staines, & P. Slezak (Eds.), *Representation in mind: New approaches to mental representation* (pp. 41–69). Elsevier.

Coleman, M. (2002). Taking Simmel seriously in evolutionary epistemology. *Studies in History and Philosophy of Science, 33*, 59–78.

Collier, J., & Hooker, C. (1999). Complexly organised dynamical systems. *Open Systems and Information Dynamics, 6*, 241–302.

Di Paolo, E. (2003). Organismically-inspired robotics: Homeostatic adaptation and teleology beyond the closed sensorimotor loop. In K. Murase & T. Asakura (Eds.), *Dynamical systems approach to embodiment and sociality* (pp. 19–42). Advanced Knowledge International.

Eronen, M. I., & Brooks, D. S. (2018). Levels of organization in biology. In E. N. Zalta (Ed.), *The Stanford encyclopedia of philosophy* (Spring 2018 Ed.), https://plato.stanford.edu/archives/spr2018/entries/levels-org-biology/. Accessed 30 Sept 2020.

Farrell, R., & Hooker, C. (2007a). Applying self-directed anticipative learning to science I: Agency and the interactive exploration of possibility space in Ape language research. *Perspectives on Science, 15*(1), 86–123.

Farrell, R., & Hooker, C. (2007b). Applying self-directed anticipative learning to science II: Learning how to learn across 'revolutions'. *Perspectives on Science, 15*(2), 220–253.

Farrell, R., & Hooker, C. (2009). Error, error-statistics and self-directed anticipative learning. *Foundations of Science, 14*(4), 249–271.

Farrell, R., & Hooker, C. (2013). Design, science and wicked problems. *Design Studies, 34*(6), 681–705.

Farrell, R., & Hooker, C. (2014). Values and norms between design and science. *Design Issues, 30*(3), 29–38.

Farrell, R., & Hooker, C. (2015). Designing and sciencing: Reply to Galle and Kroes. *Design Studies, 37*(1), 1–11.

Hoffmaster, B., & Hooker, C. (2018). *Re-reasoning ethics*. MIT Press.

Hooker, C. (1995). *Reason, regulation and realism*. State University of New York Press.

Hooker, C. (2004). Asymptotics, reduction and emergence. *British Journal for the Philosophy of Science, 55*, 435–479.

Hooker, C. (2009). Interaction and bio-cognitive order. *Synthese, 166*(3), 513–546. Special edition on interactivism, M. Bickhard (Ed.).

Hooker, C. (2010). Rationality as effective organisation of interaction and its naturalist framework. *Axiomathes, 21*, 99–172. Special edition on advances in interactivism, M. Bickhard (Ed.).

Hooker, C. (Ed.). (2011a). *Philosophy of complex systems* (Vol. 10: Handbook of the philosophy of science). Elsevier.

Hooker, C. (2011b). Introduction to philosophy of complex systems. Part A: Towards a framework for complex systems. In C. Hooke (Ed.). (2011a), pp. 3–92.

Hooker, C. (2011c). Conceptualising reduction, emergence and self-organisation in complex dynamical systems. In C. Hooker (Ed.). (2011a), pp. 197–224.

Hooker, C. (2013a). Georg Simmel and naturalist interactivist epistemology of science. *Studies in History and Philosophy of Science, Part A, 44*(3), 311–317.

Hooker, C. (2013b). On the import of constraints in complex dynamical systems. *Foundations of Science, 18*(4), 757–780. https://doi.org/10.1007/s10699-012-9304-9

Hooker, C. (2017). A proposed universal model of problem solving for design, science and cognate fields. *New Ideas in Psychology, 47*(December), 41–48.

Hooker, C. (2018). Re-modelling scientific change: Complex systems frames innovative problem solving. *Lato Sensu: revue de la société de philosophie des sciences, 5*(1), 4–12.

Hooker, C., & Hooker, G. (2018). Machine learning and the future of realism. In C. Forbes (Ed.), *The future of the scientific realism debate: Contemporary issues concerning scientific realism. Spontaneous Generations: A Journal for the History and Philosophy of Science, 9*(1), 174–182.

Kauffman, S. (2000). *Investigations*. Oxford University Press.

Klowden, M. (1995). Blood, sex, and the mosquito: Control mechanisms of mosquito blood-feeding behavior. *BioScience, 45*, 326–331.

Li, M., & Vitànyi, P. (1990). Kolmogorov complexity and its applications. In J. van Leeuwen (Ed.), *Handbook of theoretical computer science*. Elsevier.

Maturana, H., & Varela, F. J. (1980). *Autopoiesis and cognition: The realization of the living*. D. Reidel Publishing.

Merleau-Ponty, M. (1942/1963). *The structure of behaviour* (Trans. A. L. Fisher). Methuen.

Moreno, A. (2007). A systematic approach to the origin of biological organisation. In F. Boogerd, F. Bruggeman, J.-H. Hofmyer, & H. Westerhoff (Eds.), *Systems biology: Philosophical foundations*. Elsevier.

Moreno, A., & Mossio, M. (2015). *Biological autonomy: A philosophical and theoretical enquiry*. Springer.

Moreno, A., Ruiz-Mirazo, K. & Barandiaran, X. (2011). The impact of the paradigm of complexity on the foundational frameworks of biology and cognitive science. In Hooker (2011a).

Nolfi, S. (2011). Behavior and cognition as a complex adaptive system: Insights from robotic experiments. In Hooker (2011a).

Pattee, H. (1993). The limitations of formal models of management, control and cognition. *Applied Mathematics and Computation, 56*, 111–130.

Pattee, H. (1995). Evolving self-reference: Matter, symbols, and semantic cloaure. *Communication and Cognition - Artificial Intelligence, 12*(1–2), 9–28.

Pattee, H. (2007). Laws, constraints and the modelling relation - History and interpretations. *Chemistry and Bio-diversity, 4*, 2272–2278.

Penrose, R. (1989). *The emperor's new mind: Concerning computers, minds and the laws of physics*. Oxford University Press.

Popper, K. (1980). *The logic of scientific discovery*. Hutchinson. (First published as *Logik der Forschung*, Wien, 1934).

Rayner, A. (1997). *Degrees of freedom: Living in dynamic boundaries*. World Scientific.

Rosen, R. (1991). *Life itself: A comprehensive inquiry into the nature, origin, and fabrication of life*. Columbia University Press.

Ruiz-Mirazo, K., & Moreno, A. (2012). Autonomy in evolution: From minimal to complex life. *Synthese, 185*(1), 21–52.

Ruiz-Mirazo, K., Umerez, J., & Moreno, A. (2008). Enabling conditions for 'open-ended evolution'. *Biology and Philosophy, 23*(1), 67–85.

Russell, B. (1927). *The analysis of Matter*. Kegan Paul.

Simmel, G. (1895). Ueber eine Beziehung der Selektionslehre zur Erkenntnistheorie. *Archive fur systematische Philosophie, 1*, 34–45. (English translation: part 2 of Coleman 2002).

Smithers, T. (1995). Are autonomous agents information processing systems? In L. Steels & R. Brooks (Eds.), *The artificial life route to artificial intelligence: Building situated embodied agents*. Lawrence Erlbaum.

Sommerhof, G. (1974). *Logic of the living brain*. Wiley.

Stewart, J. (1996). Cognition = life: Implications for higher-level cognition. *Behavioural Processes, 35*, 311–326.

Varela, F. (1979). *Principles of biological autonomy*. Elsevier/North Holland.

Varela, F. J., Maturana, H. R., & Uribe, R. (1974). Autopoiesis: The organization of living systems, its characterization and a model. *BioSystems, 5*, 187–196.

von Uexküll, J. (1926). *Theoretical biology*. Harcourt, Brace.

Chapter 6
Does Organicism Really Need Organization?

Olivier Sartenaer

Abstract The main purpose of the present chapter is to argue in favor of the claim that, contrary to what is usually and tacitly assumed, organization is not necessary for organicism. To this purpose, I first set up the stage by providing a working characterization of organicism that involves two free parameters, whose variations allow for covering the rich and diverse conceptual landscape of organicism, past and present. In particular, I contend that organization is usually construed as a "mean to an end" notion, or as a tool put at the service of vindicating organicism's twofold defining assumption, namely, that organisms are determinative entities in their own right, to the effect that (organismic) biology is epistemologically autonomous from physico-chemistry. After a short detour devoted to show that organicism generally collapses on a spectrum of variants of emergentism, I take inspiration from a recent account of emergence called "transformational emergence" to put forward a transformational version of organicism. For such a version meets organicism's defining standards in a way that is free of any commitment to organization, arguing for its very conceptual soundness finally allows for legitimizing the claim that organicism doesn't really need organization.

Keywords Organicism · Organization · Downward determination · Emergence · Diachronic emergence · Transformational emergence · Transformational organicism

6.1 Introduction

That biological entities are organized, and in a rather intricate way to be sure, is a somewhat bland and commonplace observation. Contrary to certain physical or chemical objects, like the solar system or methane molecules, it is uncontroversial that even the most elementary living entity – whatever it is – consists of an exquisitely complex web of spatially and temporally integrated interactions.

O. Sartenaer (✉)
Département Sciences, Philosophies et Sociétés, Université de Namur, Namur, Belgium
e-mail: olivier.sartenaer@unamur.be

© The Author(s) 2024
M. Mossio (ed.), *Organization in Biology*, History, Philosophy and Theory of the Life Sciences 33, https://doi.org/10.1007/978-3-031-38968-9_6

As the editor of this volume indicated in his introduction, the history of biology is marked by a profound and recurring antagonism as to the exact significance and reach of such an observation. On the first side of this antagonism, we find those reductionist biologists or philosophers who tend to consider biological organization as a mere quantitative prolongation of the kind of structuring of matter and energy that can be found in the physicochemical world. According to them, living organisms just are particularly complex bundles of molecules, for which a proper scientific understanding should not in principle require a substantially different treatment from the one(s) already at stake in physics or chemistry – perhaps issues of computational power apart. On the opposing side of the antagonism, there is a (probably minority) community of thinkers who are rather willing to take biological organization *seriously*, as what essentially marks a dividing line between biological and physicochemical entities, as well as, incidentally, between the biological and the physicochemical sciences. These thinkers, regardless of their specific allegiances and in sharp contrast with their opponents of a reductionist temperament, share as a common rallying sign some specified measure of antireductionism, according to which "there is more in biological objects than physico-chemistry alone could ever tell."

It goes without saying that organicism, a perspective on biology and biological objects that essentially grew as an articulated scientifico-philosophical doctrine in the early days of the twentieth century, is to be counted as a particular instance of such an antireductionist attitude toward biological organization. According to organicists indeed – and in a way that will of course be further explicated in this chapter – it is the very fact that biological entities are organized that constitutes the ultimate, empirical ground as to why one should conceive of the objects and/or the science of biology as unique and idiosyncratic.

In the present chapter, and perhaps a little bit provocatively I'm afraid, I would like to question what is often taken as self-evident, that is to say – and more particularly – I intend to scrutinize the apparently inexorable association of organicism with organization, by raising the following question: does organicism *really* need organization? Of course, for answering this question positively would not be that exciting, I plan to give it a (certainly more remarkable) negative response. So to put it bluntly – and before feeling the need to add a few rhetorical provisos – I'll claim that organicism could actually stay true to its promises by *not* taking organization seriously, by playing the game, so to speak, of their lifelong, reductionist opponents.

As with any philosophical endeavor that proclaims itself provocative, it actually isn't as much as it would like. From the outset, I must indeed temper my enthusiasm of setting the cat among the pigeons by explicitly disclosing what I will *not* claim in this chapter. So here it is. In this chapter, I will *not* claim that:[1]

- Organisms are not organized*
- Organization* doesn't play a crucial role, explanatory or otherwise, in (organismic) biology
- Organization* does not crucially participate in making organisms what they are

---

[1] I use the categories of organism and organization* here for reasons that will be clear from Sect. 2 onward. The reader can then come back here in due time to benefit from a quick reminder.

What this essentially amounts to is this: I will neither endorse nor try to argue for the idea that organicists (or any other biologist/philosopher with an antireductionist penchant) should downplay or neglect biological organization. Rather and more modestly, I'll simply claim that they *can*. The point of such a contention is that, should the reductionist side of biology's recurring antagonism finally have the upper hand, this would not necessarily mean the end of organicism and its uncompromising plea for an autonomous or irreducible biology.

Here is how I intend to structure the upcoming discussion. In Sect. 2, I first propose a general definition of organicism. In particular, after emphasizing the role that organization usually plays in organicism (Sect. 2.1), I'll articulate a characterization of the view in which organization appears as a free parameter (Sect. 2.2). I then turn to a discussion of the close relationship that organicism has with emergentism in Sect. 3. I'll first claim that organicism generally collapses on emergentism (Sect. 3.1) and then proceed by showing how a recent, nontraditional theory of emergence allows for defining a nontraditional, "transformational" version of organicism (Sect. 3.2). In Sect. 4, I'll then be in a position to provide an articulated answer to the core question of this chapter, by showing that such a transformational organicism, which eschews any serious commitment to organization, is conceptually sound (Sect. 4.1). As a bonus, I will provide some preliminary thoughts as to why one should consider transformational organicism as deserving further philosophical scrutiny (Sect. 4.2). I will finally address a possible objection to the whole endeavor of this chapter and use it as a stepping stone to point in the direction where future elaboration on transformational organicism should be made (Sect. 4.3).

6.2   When Is Organicism?

As a preliminary, I first set up the stage by formulating a general, working characterization of organicism that contains two free parameters. Varying both these parameters will allow for covering the very diverse conceptual landscape of organicist variants, past and present.

6.2.1   Organization as a Mean to an End

To this aim, I take inspiration from Nicholson and Gawne (2015)'s recent identification of three "general ideas" that are taken to "unite" the different trends of organicism. These ideas – organism, autonomy, and organization – together with the conceptual connections between them can be explicated as follows:[2]

---

[2] Most of the quotations in this section are directly extracted from Nicholson and Gawne's paper, even if, for the sake of proper reference, I only mentioned their actual origin. The way in which the three tenets are presented and organized is my own.

- *Organism* – It is a recurring theme among organicists that living organisms are idiosyncratic entities. As such, they are supposed to possess some unique (set of) trait(s) that allows for unambiguously distinguishing them from other, non-organismic entities like molecules, stones, or stars. Typically, organisms are considered peculiar in that they are – contrary to molecules, stones, or stars – "unified wholes" that are "more than the sum of their parts." Although this traditional, holistic idea can be given very different interpretations, there seems to be a widespread tendency in organicism to read it in *determinative* terms. In a nutshell: organisms are idiosyncratic wholes to the extent that they are determinative *in their own right*. Put differently, organisms are those united wholes that happen to be determinatively effective *qua* wholes, that is, not only insofar as they are made of underlying entities – cells, molecules, atoms, or ultimately, non-composite physical units – that are themselves determinative. A typical way of making sense of such an overarching determinative effectiveness of organisms is to consider that organisms are the kind of entities that are able to "make a difference" as to how their own constitutive parts behave. In the words of notorious organicists: "The whole enters always into the determination of the activities of the parts" (Woodger, 1929, 247); or "the behaviour of an isolated part is [...] different from its behaviour within the context of the whole" (von Bertalanffy, 1952, 12).
- *Autonomy* – A second core tenet of any particular variant of organicism is the relentless contention that the biological sciences – broadly understood as, among other things, the scientific study of organisms – are autonomous from, or irreducible to, the physical sciences. There are many different ways in which such an autonomy or irreducibility can be precisely construed, depending on one's preferred view about the general nature, goals, and methods of science. These can range from, for instance, the inability to adequately *represent* biological phenomena through the formal machinery of physics to the inability to *deduce* biological laws – if any – from physical laws plus some other assumptions. Perhaps the most widespread way of capturing the putative autonomy of biology is the one that directly pertains to explanation, i.e., biology would be an autonomous discipline insofar as they are biological phenomena out there that cannot be properly *explained* from the exclusive vantage point of physics.[3] As Haldane evocatively put it: "[t]hose who aim at physico-chemical explanations of life are simply running their heads at a stone wall, and can only expect sore heads as a consequence" (1908, 696).

It is noteworthy that both these first ideas – organism and autonomy – are not conceptually independent. If one is ready to take seriously the claim according to which "[a]ll of the organicists shared the conviction that the distinctiveness of organisms demanded a unique set of theoretical tools for their elucidation" (Nicholson & Gawne, 2015, 366), it then appears that the idea of organism is what

---

[3] Although this is a generic way in which the tenet of autonomy will be broadly construed here, it is by no means the only one possible. For different takes on the issue, see, e.g., Moreno and Mossio (2015) or Varela (1979).

enforces, entails, or grounds the idea of autonomy. It is indeed essentially *because organisms are the way they are* – unified whole that are determinative in their own right – that the science that studies them, biology, happens to be autonomous from the science that study their constitutive parts, physics (or chemistry). As it were, it seems that "[b]iology must retain the courage of its own insights into living nature" (Weiss, 1969, 400).

* *Organization* – The third and last core ingredient that should be definable of organicism is the rather commonsensical idea that organisms are organized entities. Obviously, for such an idea not to be overly trivial, or for it to be of any philosophical significance, it has to be considerably sharpened and refined (as it is uncontroversial that molecules, stones, and stars *also* are, to some extent, organized entities, and not mere unstructured clusters of elementary physical units). Because it is not the place to speculate about how exactly one should achieve this sharpening or refinement, that is, how one should precisely conceive of the idiosyncrasy of organismic organization, let's imagine that such a sharpening is possible, and let's refer to its result as "organization*." Under this hypothesis, organization* just is the kind of organization that is typical of organisms. As such, it allows for unambiguously distinguishing organisms from other organized – though not organized* – entities like molecules, stones, and stars.

As with the two first ideas of organism and autonomy, the third tenet of organization is not conceptually freestanding. Rather, organization must essentially be taken as a *mean to an end*, that is, as a tool that supports the organicists' main contention that organisms are unique entities that require an autonomous science to be dealt with. As things stand, "organisms are what they are by virtue of their organization[*]" (Nicholson & Gawne, 2015, 364). That is to say, organisms are (supposed to be) determinative entities in their own right, precisely in virtue of the fact that they happen to be the kind of entities that are organized in a very idiosyncratic way, i.e., they are organized*. In the words of contemporary organicists: "The principle of organization states that biological systems realize a closure of constraints. The organization of constraints realizing closure achieves a form of 'self-determination'" (Mossio et al., 2016, 7). As it appears, it is assumed here that the unique determinative dynamics of organisms – referred to as "self-determination" – turns out to be brought about by the realization of a unique mode of organization, here a "closure of constraints."[4]

This being said, I am now in a position to fully articulate Nicholson and Gawne's three unifying ideas in order to provide a general, working characterization of organicism.

---

[4] It should be noted that organicists – on the model of those quoted here – don't necessarily restrict themselves to considering *organisms* as organized*. More generally, they are often open to extending the scope of organization* to the broader category of "biological systems."

6.2.2   Defining Organicism

Here is a first attempt, formulated as a claim [O^φ], that any organicist, and organicists only, should take as true:[5]

> [O^φ] – Organization* makes organisms what they are – determinative entities in their own right. That organisms are such makes biology an autonomous science.

Of course, presented like this, organicism is nothing but a speculative philosophical view (hence the superscript "φ"). Should one want it to have some actual bearing on science – something that I suspect most, if not all, organicists certainly want – then the following companion claim is also to be endorsed (with the superscript "em" standing for "empirical"):

> [O^{em}] – There is some restricted class of entities in nature, namely, organisms, that are organized*.

In order to generalize this twofold characterization of organicism, three preliminary remarks are in order.

First, it should be noted that the three defining ingredients of organicism are associated with claims that are not on equal footing. In particular, the tenet of organism is associated with a claim – "organisms are determinative entities in their own right" – that is *ontological* (*modulo* a proviso to be found below); the idea of autonomy comes with a contention – "biology is autonomous from physics" – that is essentially *epistemological*; and the claim that corresponds to the tenet of organization – "organisms are organized* – turns out to be *empirical*. By being committed to [O^φ] and [O^{em}], the argumentative structure of organicism is then rather sound, as well as, incidentally, quite widespread in the philosophy of science literature. It consists in identifying a class of entities in nature that appear to share a very special feature, for then extracting some putative metaphysical consequences therefrom, which are believed to have some impact on our way of doing science.

Second, thesis [O^φ] may appear at first glance as overly restrictive, to the effect that, from the outset, some strands of organicism would be excluded from its scope. More particularly, I wouldn't be surprised if some organicists considered the requirement that organisms are determinatively effective in their own right as being too ontological, hence inconsistent with the supposedly *exclusively* epistemological version of organicism they want to promote. I think this hypothetical concern is misguided for two interrelated reasons. The first is that, in a nutshell, it is really far-fetched to consider that there could be versions of organicism that are free of *any* ontological commitment, no matter how thin. As I take it, organicism at least endorses the idea that organisms do exist as "wholes" of a somehow *unique* kind, to the effect that there is a principled way in which they can be classified in a separate category from non-organismic things. "[A]fter all, [and contrary to stones and stars,] organisms are not just heaps of molecules" (Weiss, 1969, 400). Organicism then

---

[5]The (certainly) ambiguous and polysemic notion of "determination" employed here will be unpacked below.

generally comes with some appetite toward an (at least very shallow) "ontologization" of organisms, or some minor degree of "biochauvinism" (Wolfe, this volume).[6] Without it, biology's irreducibility would lack its main rationale and appear accordingly as exquisitely gratuitous. Should it indeed turn out that organisms just are "the sum of their parts," on the model of – as the story goes – molecules, stones, or stars, then one could justifiably wonder why biology's relation to physics should be *that* different from the other special sciences, like chemistry, geology, or astrophysics. This brings us to the second, related reason: as such, the very notion of "determination" is in itself highly noncommittal. This is of course why I opted for this term to begin with: determination denotes a neutral relation that can come with various intensities of ontological oomph, ranging from "thin" to "meaty" ones (Beebee, 2000). The "only epistemological organicists" can then rejoice, for, perhaps contrary to appearances, they have not been left ignored. Among the possible interpretations of the idea of determination that appears in the proposed definition of the view, the first of the following should actually satisfy them:

- Ontologically thin organicism:
  Logical determination – Organisms are determinative in the sense that facts about them entail facts about their parts (to the effect that the deduction/explanation of some facts about parts requires knowledge of some facts about organisms).
- Ontologically modest organicism:
  Noncausal determination – Organisms are determinative in the sense that they noncausally make a difference as to how their parts behave (e.g., organisms constrain the way in which their parts behave).
- Ontologically meaty organicism:
  Causal determination – Organisms are determinative in the sense that they contribute in bringing about their parts' behavior (e.g., organisms possess irreducible causal powers and exercise them for making their parts behave in certain ways).[7]

Third and finally, the core question to be addressed in this chapter – "does organicism really need organization?" – requires us to seriously ponder the hypothesis, *pace* Nicholson and Gawne, that organization* be *not* an integral part of organicism's very *definiens* (otherwise this would simply begs the question at hand). The (organicist) reader is then kindly asked to at least leave open the possibility that the concepts of organicism and organization* are not analytically connected (so organicism is not to be defined as the claim that organisms are organized*).

With these preliminaries, I can now propose a revised version of the aforementioned characterization of organicism, on the following model:

---

[6] The term itself comes from Di Paolo, E. (2009). Extended Life. *Topoi*, 28, 9–21.

[7] For the sake of simplicity, I consider causation as a monolithic concept here, which reduces to "efficient causation" as construed under the compulsion of a productive account (e.g., Dowe's (2000) transfer theory). Accordingly, possible alternative forms of causation – typically "formal causation" – are here considered noncausal. This choice is purely terminological and should therefore not afflict too much causal realists with an Aristotelian penchant.

[O$^\wp$] – X makes organisms what they are – determinative^y entities in their own right. That organisms are such makes biology an autonomous science.

And:

[O^{em}] – There is some restricted class of entities in nature, namely, organisms, that have X.

As it appears, this definition involves two free parameters, X and Y. While the former is meant to cover scientifically kosher empirical means – among which organization* certainly occupies a prominent place – that would be conducing to the uniqueness of organisms, the latter fixes the strength of the ontological oomph one wants to give 1to the view – respectively thin, modest, or meaty.

In the light of such a characterization of organicism, the main question that will keep us busy here is the following: for all possible values of Y, is it possible to vindicate the truth of [O$^\wp$] without considering X as some variant of organization*? In other words, is there a viable, empirical mean different from organization* that organicists could exploit in order to ground the uniqueness of organisms and, with it, the autonomy of biology? In order to address this question and, more particularly, to answer it positively, it is necessary beforehand to make a short detour.

6.3   Organicism and Emergence

The main goal of this section is to show that the varieties of organicism as defined through [O$^\wp$] collapse to a spectrum of emergentist positions. Put differently, I'll argue that, in order to live up to its promises, organicism necessarily has to be committed to some nontrivial form of emergence. This claim will not be defended here for mere informational purposes (though it may have some interest for that sake). Rather, it will open the door for exploiting some recent resources of the emergence debate, which will turn out to be helpful for addressing the central question of this chapter.

6.3.1   Emergence and Organization*

In and of itself, emergence is a very general and uninformative concept. In a nutshell, it captures any kind of relation between two *relata*, usually referred to as an "emergent" and its "emergence basis," such that the emergent *depends on*, is *grounded in*, or *arises from* its basis, and yet, in spite of such a dependence, the former is also to be considered *autonomous from, novel with regard to*, or *irreducible to* the latter (see, e.g., Sartenaer (2016)). Of course, these ideas of dependence and autonomy are (i) very vague and (ii) mutually conflicting, to the effect that emergence is a notoriously ambiguous and unstable idea. Accordingly, putting it to philosophical work previously requires (i) clarifying these ideas in a way that (ii) they are rendered compatible.

That organisms are emergent entities is not a new idea. Actually, it has been explicitly endorsed by the founding fathers of emergentism themselves.[8] For instance, in the words of Lloyd Morgan:

> What emerges at any given level affords an instance of what I speak of as a new kind of relatedness of which there are no instances at lower levels (1923, 15–16); and:

> I accept with natural piety the evidence that there is more in the events that occur in the living organism than can adequately be interpreted in terms of physics and chemistry, though physico-chemical events are always involved. Changes occur in the organism when vital relatedness is present the like of which do not occur when life is absent. This relatedness is therefore effective (1923, 20–21).

There is much to unpack in these quotes, though it is not the place to do it extensively here. Suffice it to emphasize that, according to the characterization put forward in Sect. 2, Morgan's emergentism could be seen as an organicism of some sort, as for him the empirical realization of some kind of organization* – "vital relatedness" – is what renders organisms determinatively effective, to the effect that they cannot be "adequately interpreted" in physicochemical terms only.[9]

Apart from any particular historical episode, three features make organicism generally collapse on emergentism. The first – which is actually sufficient in itself – is an obvious definitional convergence. That organisms are determinatively effective in their own right make them somehow autonomous from their parts – logically, noncausally or causally, according to one's preferred version of the view – parts on which they are also supposed to depend. Second, organisms are ultimately to be considered determinatively effective and emergent because of organization*. More particularly, it actually is some *reification* of organization that provides the necessary ontological oomph for both views to get off the ground. As organicists and emergentists alike would put it, respectively:

> In essence, organization has become a *thing* (Rosen, 1991, 117; emphasis in the original); or:

---

[8] Although emergentism has some deeper historical roots, one usually considers that the first, fully-articulated defense of the doctrine appeared around the 1920s, mainly in the combined works of philosophers and biologists (see, e.g., Blitz (1992)). Apart from George Henry Lewes who coined the term "emergence" in 1875, the first systematic, philosophical use of the concept is to be found in Lloyd Morgan's works. Other notorious early emergentists were Samuel Alexander, Roy Wood Sellars, and Charlie Broad.

[9] True, Morgan's position could also be considered as a (monistic) form of vitalism (Sartenaer, 2013), for (i) it is arguable that a "vital relatedness" that needs to be accepted with "natural piety" is not a scientifically kosher mean to vindicate the uniqueness of organisms, or (ii) such a view requires to be committed to the existence of nonphysical, "configurational forces" (McLaughlin, 1992). As it will appear below, I do not intend to fight over this point, which ultimately hangs upon the boundary between meaty organicism and materialism-friendly vitalism being somewhat blurry. In emergentist terms to be explicated below, one generally considers that Morgan was endorsing a "strong" form of emergence.

All through the argument of this book, we have proclaimed the *reality* of form (Sellars, 1922, 329; emphasis is mine).

Third, it is this very ontologization of organization that provides both views with the opportunity to occupy the conceptual space between reductive physicalism – "no ontological oomph" – and hard-nosed vitalism, "too much ontological oomph" – allowing for a reconciliation between some degree of antireductionism and the naturalistic demands of modern science. As it appears, both organicism and emergentism then conceive of organization* as the very key to their commonly targeted "third way." Besides the intrinsic determinativeness of physical entities themselves, with organization* becomes available indeed an alternative, scientifically legitimate source of determinative effectiveness in the world – *contra* reductive physicalism – source which has nothing to do with the putative determinative potency that a separate realm of nonphysical entities would have intrinsically, *contra* (substantial) vitalism.

Table 6.1 summarizes this collapsing of organicism on emergentism along the possible variations of the parameter Y, X being fixed on organization.* It should be noted that, whereas every possible variety of organicism amounts to a particular declination of emergentism, the converse is not true. This is unsurprising, given the very high generality of emergence, together with the fact that most traditional emergentists, including Morgan, were considering natural entities other than organisms as putative candidates for emergence (for instance, the products of chemical reactions or, typically, mental and conscious states).

Table 6.1 calls for some comments. First, a relatively peripheral one: given that both organicism and emergentism are usually formulated within the framework of a layered ontology of "levels," where an organism is supposed to occupy a higher level than its underlying, lower-level parts, the determination that is at stake is generally to be considered "downward," that is, oriented *from* the higher level of the whole *to* the lower level of the parts. This is why, though varieties of downward determination are typically coextensive with emergence – the most widespread being so-called downward causation (see, e.g., Kim (2006)) – they are also pervasive in organicists' debates.[10]

Second, the reader should not be startled by the diversity of concepts of emergence referred to in the table. Weak and strong (ontological) emergences are actually commonplace in the literature (see, for instance, Wilson (2015)). They are usually distinguished in that the second entail, while the first doesn't, the coming into being of new higher-level causal powers, something which makes the second, though not the first, inconsistent with physicalism (in the minimalist sense according to which "all worldly causal powers are physical"). "Modest emergence" is certainly more unusual and is here only meant as a label to paste on any account of emergence that would allow for

---

[10] See, e.g., Arnellos and El-Hani (2018), where the authors construe modest determination under the category of "medium downward causation" (to be contrasted with "strong downward causation," following Emmeche et al. (2000)). This prevalence notwithstanding, it should be noted that some recent works in the organicist tradition *do* eschew any commitment to the idea of downward determination as construed here (see, e.g., Mossio et al. (2013)).

Table 6.1 Varieties of organicism along dimension Y, together with the type of emergence they are committed to

Organicism	Determination (Y)	Emergence	Physicalism
Thin	Logical	Weak	Yes
Modest	Non-causal	Modest	Yes?
Meaty	Causal	Strong	No

reconciling physicalism with a decent measure of ontological antireductionism – something that, notwithstanding claims to the contrary, has still to this day not been achieved uncontroversially (I take this as an open endeavor whose fruitfulness has, in any case, no bearing on my current objective).

Third and finally, though I'm certainly sympathetic to the idea that nothing should in principle prevent us from considering meaty organicism as a genuine variant of the view, I also don't have any good reason to contest the widespread idea that this looks "dangerously too much like vitalism" to deserve being properly named organicism – Lloyd Morgan's strong emergentism being a vivid illustration of such an uncomfortable borderline situation. Accordingly, and in order not to raise unnecessary matters of controversy, I would be ready to leave aside meaty organicism/ strong emergentism/monistic vitalism out of the scope of "respectable" organicism.[11]

It is now time to close this section by putting forward a third formulation of the characterization of organicism, in the light of what has been just said. Here it goes (with $[O^{em}]$ remaining unchanged):

[O^{φ}] – X makes organisms emerge^Y from a physical basis. That organisms emerge^Y makes biology an autonomous science.

Parameter Y now corresponds to possible variations as to the kind of emergence involved. My central question then becomes: is there an empirical mean different from organization* that would be conducing to emergence^Y? As it turns out, recent works on emergence can be called to support the claim that there is.

6.3.2   Emergence and Transformation

Since its very inception in the 1920s, emergence has been almost exclusively construed and discussed with, in the background, two interrelated assumptions. The first is that the dependence relation that connects an emergent to its basis is to be considered *synchronic*, that is, it is assumed to obtain between the putative emergent and its basis as they are instantiated at the very same time. The second is that emergence is an intrinsically *hierarchical* relation that connects lower-level to higher-level entities. Both these assumptions are implicit in the traditional organicist/

---

[11] For a finer-grained analysis of the relationship between organicism, emergentism, and vitalism, the reader can refer to Sartenaer (2018a). See also Wolfe (2011).

emergentist slogan, according to which "the (higher-level) whole is more than the sum of its (simultaneous, lower-level) parts."

Yet, recent developments have shown that there actually exists a bona fide (family of) concept(s) of emergence that is free of both these assumptions and which is referred to as "transformational emergence" (Humphreys, 2016; Guay & Sartenaer, 2016; Guay & Sartenaer, 2018).[12] In contrast with traditional emergence, transformational emergence is *diachronic*, the putative emergent being typically instantiated later than its emergence basis, and *not hierarchical*, both the emergent and the basis belonging to the same level. What matters for us here is that, as its name suggests, transformational emergence is not driven by organization*, but rather by transformation. In a nutshell, the uniqueness or distinctiveness of emergents doesn't come about because unchanging entities are organized* in a very idiosyncratic way. Rather, it comes about because these entities themselves are *transformed* in a very idiosyncratic way.

Let us illustrate the contrast that is at stake here by considering the case of a putatively emergent organism.[13] In the traditional perspective, an organism at time t emerges from a basis made of cells at t, for the organism is considered both dependent on and autonomous from these cells. For instance, in the ontologically modest declination of organicism, one could argue that the organism at t is constituted by its cells at t and that the former is able to downwardly constraint the behavior of these cells at t. In this first perspective, that the organism is able to do so proceeds from the fact that the organism is an organization* of its constitutive cells.

Things are different in the transformational perspective. It is indeed rather considered there that an emergent organism at t both depends on, and is autonomous from, a basis made of cells at a previous time t'. For example, it could be contented that the organism at t is causally or nomologically dependent on the cells at t' and that the former exercises at t causal powers that are different from any combination of the causal powers that the cells had at t'. In this second speculative scenario, that the organism has new causal powers at t proceeds from the fact that the cells that make it up at t are ontologically different from the cells at t', for the latter have been properly transformed. As it appears, the organism at t is nothing "over and above" a sum of cells at t – it actually is a mere organization, and not an organization*, of these cells at t – though it is ontologically distinct from any sum of cells at t'. Under the form of a slogan: with transformational emergence, "the whole just is the sum of the parts that have been transformed." It is noteworthy that, in such a view, the

---

[12] Of course, these developments have some historical precedents, among which Humphreys (1997)'s own "fusion emergence." Epistemological variants of transformational emergence have also been proposed, for instance, by Bedau (1997) or Rueger (2000). Despite the fact that "transformational emergence" is a label that is sometimes used interchangeably with the one of "diachronic emergence," I will stick here to the convention that consists in considering transformational emergence as a subspecies of diachronic emergence, which has the peculiarity of being flat and ontological.

[13] Just to be clear: I do not offer here the slightest argument to support the claim that organisms are in fact emergent (synchronically or transformationally).

Table 6.2 Varieties of organicism along both dimensions X and Y, together with the type of emergence they are committed to. "S" and "D" subscripts stand for "synchronic" and "diachronic," to the effect that the corresponding causal determination is to be considered downward and flat, respectively

X	Organicism	Determination (Y)	Emergence	Physicalism
Org.*	Thin	Logical	Weak	Yes
	Modest	Non-causal	Modest	Yes?
	Meaty	Causal	Strong$_S$	No
Transf.	Transf.	Causal	Strong$_D$	Yes

organism and the sum of the transformed cells are one and the same thing, to the effect that the emergence at stake is "flat" or nonhierarchical.

Now, should organisms be transformationally emergent entities, their determinative effectiveness would be of a causal nature. As a result, the emergence at play could be considered as "strong," according to the terminological convention adopted above. Yet, it is important to emphasize that such a strong emergence would not be inconsistent with physicalism, to the extent that, in a diachronic and flat scenario, the newly acquired causal powers are unambiguously physical (for the physical level is the only level there is)[14]. Accordingly, and in contrast with the corresponding synchronic scenario, the very idea of transformationally emergent organisms – though strong – doesn't encounter the risk of any detrimental acquaintance with putatively disreputable forms of vitalism.

Table 6.2 summarizes the upshot of this discussion. "Transformational organicism" just is the view according to which [O$^\varphi$] and [O^{em}] come out as true when X is fixed on transformation rather than on organization*.

At this stage of the discussion, answering the main question of this chapter requires a last step, which is to be taken in the next, last section. It only remains to be shown that transformation is a legitimate scientific process, which indeed leads to transformational emergents being (causally) determinative in their own right.

6.4 Transformational Organicism and the Autonomy of Biology

This last section, at the term of which I'll finally be able to answer the question I started with, is structured in two moments. First, I'll show that transformational emergentism is a conceptually viable view – both in general and, in particular, in its organicist declination – to the extent that there is at least one proper construal of transformation that does the job of securing the irreducible determinative

---

[14] For more detail on that point, see Sartenaer (2018b). In a nutshell, transformational emergence is immune to Kim-style exclusion arguments and, as such, allows for consistently combining in a same package causal irreducibility and the causal closure of the physical world.

effectiveness of transformational emergents and, in so doing, grounding the autonomy of the science that study them. In and of itself, this first endeavor is sufficient to answer my main question, whose nature is essentially conceptual. Second and as a bonus, I will briefly explore the idea that there could well be transformational emergents at stake in organismic biology, consistently with claims made by organicists themselves.

6.4.1 Transformational Organicism Is Conceptually Sound

In order to support the claim that transformational organicism is a conceptually viable view, I offer here what essentially amounts to an argument by analogy.

Let us suppose that "condensalism" is a view that is conceptually analogous to organicism, the chauvinist claim of uniqueness being merely shifted from organisms to condensed materials. Coherently with the previous discussion, a possible twofold definition of condensalism would thus be as follows:

[C^φ] – X makes (some) condensed matter emerge^Y from a physical basis, that is, it makes (some) condensed matter what it is – a determinative^Y entity in its own right. That (some) condensed matter is such makes condensed matter physics an autonomous science.

And:

[C^{em}] – There is some restricted class of entities in nature, namely, (some) condensed matter, that has X.

In what follows, I argue that there is a suitable, scientifically kosher transformation X that makes both [C^φ] and [C^{em}] true, with Y being then fixed on transformational or strong$_D$ emergence, or, equivalently, "flat" causal determination. On this basis, I then simply exploit the hypothesized conceptual analogy to support the truth of the following claim:

[O^φ] – Transformation makes organisms transformationally emergent from a physical basis, that is, it makes organisms what they are – (causally) determinative entities in their own right. That organisms are such makes biology an autonomous science.

At this stage, my main objective will be met: organization*, and organization a fortiori, will have been shown not to be necessary for organicism. As an extra, I'll also propose some further considerations in the next section that will provide some preliminary reasons to also take as true the further, empirical claim:

[O^{em}] – There is some restricted class of entities in nature, namely, organisms, that are the product of a transformation.

This being said, I now turn to providing support for the truth of [C^φ] and [C^{em}], with X being fixed on transformation. For this purpose, I here exploit the results of some previous works, in which the emergentist position of a prominent figure in contemporary physics, namely, the 1999 Nobel Prize winner Robert Laughlin, has been philosophically reconstructed in a transformational perspective (Guay & Sartenaer, 2016; Guay & Sartenaer, 2018). In a nutshell, what has been shown there is that the organicist's general methodology, as described in Sect. 2.2, can also be found at play in the debate that pertains to the putative reducibility of condensed matter physics to particle physics. More particularly, it is possible to argue that the physics of some worldly phenomena supports the following argumentative schema: there is transformational emergence in condensed matter physics (empirical claim). Therefore, condensed matter is determinative (*qua* condensed matter; ontological claim). Therefore, condensed matter physics is autonomous from particle physics (epistemological claim).

It is not the place here to extensively develop the way in which such an argumentative structure can be uphold (the interested reader is kindly asked to look into the relevant papers for more detail). I content myself with highlighting its most relevant steps:

- There is a well-documented phenomenon in physics, called the quantum Hall effect, that occurs when some piece of conductor, in which an electric current flows, is placed in a strong, orthogonal magnetic field at very low temperature. The effect in question manifests itself through the existence of plateaus of constant Hall resistance (associated with the transverse current induced), which occur for certain values of the applied magnetic field. These plateaus can be ordered according to a certain filling factor, which can take either integer values only – we then speak of the "integer quantum Hall effect" [IQHE] – or fractional values, the effect is then referred to as the "fractional quantum Hall effect" [FQHE].
- The FQHE is generally associated with the coming into being of a new type of (quasi)particle called "anyon" (Laughlin, 1999, 863; Laughlin doesn't use that term, which comes from Wilczek). Anyons have a striking peculiarity: they obey fractional statistics, making them neither bosons nor fermions. As it appears, anyons are to be counted among the elementary particles of nature, on the model of photons (which are bosons) and electrons (which are fermions).
- The FQHE can be seen as the result of a transformation of a state of a physical system that involves electrons, to a state of the same system with anyons. Such a transformation leads to anyons being transformationally emergent from electrons, in the sense that the former both (diachronically and nonhierarchically) depend on and are autonomous from the latter. In particular, although anyons are a product of a transformation of electrons, they have new determinative powers and obey new laws. These powers are new in a strong sense: they are forbidden to exist in the pre-transformation phase according to natural laws. This striking observation has a theoretical counterpart: the quantum electrodynamical model that best captures the pre-transformation state lacks the resources for "talking about" anyons (it can actually only describe bosons and fermions). In the words of Laughlin himself: "[The discoverers of the effect – his Nobel co-laureates

Tsui and Stormer – found something] which should have been impossible" (Laughlin, 2005, 77).
- Accordingly, the science that study materials in which anyons arise is autonomous[15]. Rather than a "Theory of Everything" that will serve as the final and unique basis for explaining all there is – including the behavior of anyons – science is rather made of many irreducible "theories of things" (Laughlin & Pines, 2000, 30).

The upshot of this is condensalism is a viable view (though it can of course be mistaken as a true description of our world), so is therefore transformational organicism.

6.4.2   Is Transformational Organicism More than Just Conceptually Sound?

I see two ways in which a transformational version of organicism can be claimed to be more than just a conceptually consistent view. Without going as far as supporting the idea that there actually are proper transformations at stake in the biological world, together they at least provide hints that transformational organicism would deserve philosophical scrutiny.

A first way to go in this respect is to make use of an a fortiori argument. In the previous section, it has been claimed that there could be a proper transformation leading to transformational emergence in physics, to the effect that [C^{em}] was given some plausibility. Should such a claim go through, it would indirectly support the truth of [O^{em}], for, a fortiori, if some relevant transformation does take place in the physical world, one could expect it to also occur in the biological world. After all, biological entities just are (or are also) physical entities. Without claiming of course that something like the FQHE occurs within organisms, it would be unsurprising that transformations of a similar nature occur among the very elementary constituents of organisms, for then "percolating up," so to speak, to the organisms themselves.

This apart, another (certainly stronger) case can be made that trans-formational organicism deserves further exploration. It essentially rests on the fact that organicists themselves may have advocated – implicitly to be sure – something along the lines of transformational emergence. As an example, I here consider Soto et al. (2008)'s approach to emergence as it would be substantiated in developmental biology.

As I see it, the core of their approach can be captured through the three following ingredients: diachrony, downward causation, and the breaking of the causal closure

---

[15] Obviously, as the considerations developed here make it clear, the idea of autonomy understood under the transformational perspective is slightly different from the one associated with traditional, synchronic emergence. In the transformational approach, the usual epistemic cutoffs have to be understood diachronically. For instance, non-derivability, unpredictability, or non-explainability has to hold between antecedent and posterior states, and not between higher- and lower-level states. For more details about the impact of such a way of construing autonomy onto the structure of science, see Sartenaer (2019).

of the physical world. For any philosopher that is well-versed in the arcane of the emergence literature, this for sure appears as a quite odd combination. Indeed, if one takes these ingredients at face value, they together delineate a position that happens not to take the best of two worlds, as it were. In the pursuit of biology's autonomy, buying into diachrony alone could actually be enough – this is the main message of transformational emergence – to the extent that there is no need to endorse an extra, possibly controversial commitment to downward causation and the correlative breaking of the causal closure of the physical world. Similarly, biology's autonomy could very well be advocated on the basis of a commitment to the existence of downward causation and the correlative breaking of causal closure, without adding a nonconventional, diachronic twist to the picture.

In the face of what thus appears as an unnecessary metaphysical inflation, two interpretative options are available. First, what *really* matters to Soto et al.'s emergence is downward causation (and the correlative breaking of closure), diachrony being a rather peripheral, extra ingredient. If this is the case, then their account happens to collapse on O'Connor and Wong (2005)'s theory of emergence, which itself is to be taken as conceptually isomorphic to *synchronically* strong emergentism, its self-proclaimed diachronic nature notwithstanding (Wilson, 2015). The issue with this first option is blatant: as it was emphasized in Sect. 3.1, it "dangerously" looks like full-fledged vitalism.[16]

Hopefully, a second option is available. It consists in taking diachrony seriously while not sticking to the letter of the traditional way of framing downward causation (and its purported implications on closure). There are reasons to believe that this actually is Soto et al.'s implicit strategy. For one thing, they are adamant about the importance of taking time seriously – "Time is acting [...]. This action is real and has an ontological meaning" (Soto et al., 2008, 271). But furthermore, they also seem to adopt a conception of downward causation that is very different from what emergentists usually have in mind when appealing to the notion (most of the time critically). As they put it: "[B]asic properties are changing [...]. This is the meaning of downward causation" (*Ibid.*, 272). That the very determinative effectiveness of emergents is to be understood through the changing of basic properties should ring a bell. This is indeed nothing else than the defining claim of transformational emergentism. Keeping in mind that the main motto of transformational emergence is indeed that novelty comes from the "parts" changing through time (rather than being organized*), the following kind of claim renders the association rather legitimate:

> By the time the tissue is formed, the 'parts' that we identify in them are no longer the parts that interacted in their formation. The cellular components now present did not pre-exist the tissue itself – they are interacting in a particular way that is reciprocal. When we artificially separate the components of the tissue, for instance the cells forming epithelium and its

---

[16] It is noteworthy that such an association is not a source of great trouble for O'Connor and Wong, as they are explicitly willing to defend a version or property dualism at the service of a libertarian agenda. And what doesn't seem (apparently) *that* unreasonable when it comes to the obscurities of the mind turns out to be less legitimate when it comes to biological phenomena.

subjacent stroma, cells cease to perform the functions they executed when together in their proper three-dimensional arrangement (*Ibid.*, 268).

Though it is not the time and place to initiate a new philosophical exploration, it is noteworthy that the viability of the transformational emergence that is at stake here rests on the assumption that (at least some) biological kinds should be functionally individuated, to the effect that a given biological entity, say a cell, is to become a new individual as soon as it begins or ceases to exercise its proper function in the organism. Such an idea is certainly not heretical, especially with respect to organicism, as the words of one of the founding fathers of the view indicate:

If a large bomb is dropped upon a populous town we might apply the term 'town-plasm' to the debris which remained, but it would be a little absurd to say that towns were composed of such town-plasm, and that from a sufficient knowledge of such debris it would be possible to gain an adequate knowledge of the organization of towns (Woodger, 1929, 294).

In such a (unnecessarily morbid) scenario, it is indeed contended that the very process of a town's explosion is not to be construed as the disruption of an organization*, the unchanging parts – town-plasm – being once organized* and then not anymore. Rather, the town's explosion is a proper *transformation*: what was at some point an individual (let's say, the roof of a bank and the gates of a school) simply stops being such once the explosion occurred.

6.4.3   A Possible Objection and the Way Forward

Before wrapping up, it is important to defuse a possible objection that might be raised against the conceptual possibility of a genuine form of transformational organicism[17]. In a nutshell, the objection goes as follows: as it has been shown in Sects. 4.1 and 4.2, it might be the case that transformational emergence does occur in nature, be it within some (possibly restricted classes of) physical or biological systems. Accordingly, transformational emergence might serve as a possible tool for arguing in favor of the autonomy of certain scientific fields within the physical and the biological sciences, as well as, incidentally, within other areas of science (e.g., the chemical, psychological, or sociological sciences). But, should that indeed be the case – and here actually lies the objection – what appears as transformational emergentism's very high degree of *generality* is largely outbalanced by its concomitant low degree of *specificity*. As such and with regard to what really matters here, the previous discussion does not provide legitimate reasons why organisms (or other biological systems for that matter) should be transformationally emergent in a specific way and, accordingly, why the specific science that study them, namely, biology, should be autonomous.

---

[17] I would like to thank a reviewer of this chapter as well as the editor of the volume for having drawn my attention to this possible issue.

I think this objection actually picks up on an important point. It reveals a possible blind spot in the reflection carried out so far, whose origin lies, I think, in the very conceptual, generic, and decontextualized methodology that has been adopted to drive home the chapter's main point. Indeed, when organicism is conventionally characterized through [O^φ], "X makes organisms emerge^Y from a physical basis. That organisms emerge^Y makes biology an autonomous science" – it reduces the inherent subtlety of the view to its core autonomist *end*, irrespective of the specificity of the *mean* to reach it. And with such a chosen focus, it is the very "chauvinist" flavor of some variants of organicism – intimately associated with the claim that organisms are somehow of a quite unique and remarkable nature – that is downplayed.

In the face of such an objection, I see two countermoves. First, in the spirit of *avoiding* – rather than defusing – the issue, one might be willing to bite the bullet. For all we know, not all organicists, past, present, and even future, need to be chauvinists about organisms and might actually find themselves happy with the idea that the autonomy of biology is not intrinsically tied to some empirical fact – be it organization* or transformational emergence – that should *only* obtain within organisms. But, in the spirit of reaching to as many strands of organicism as I could, I'd rather adopt a second, more interesting strategy.

Though this has not been frontally addressed so far, I actually believe that there are ways in which organicists' chauvinism might be safeguarded along the lines of the empirical thesis [O^{em}] established in Sect. 2.2 ("there is some *restricted* class of entities in nature, namely, organisms, that are the product of a transformation"). Put differently, there actually is a story to tell about the possible specificity of transformational emergentism as applied to biology. Although exploring this line of thought in detail is certainly the topic of a completely different paper, it is worthwhile mentioning here two possible strategies in that regard, which both take inspiration from Dobzhansky's (1973) famous *dictum*: "Nothing in biology makes sense except in the light of evolution."

A first way to go would be through the following steps:

- (i) Natural selection, which is the driving force of evolution, has the ontological status of a law of nature (Reed, 1981).
- (ii) Natural selection is a fundamental, nonderivative law of nature (Rosenberg, 2006)[18]
- (iii) There is room in all the available metaphysical frameworks about laws of nature for the possibility that the set of fundamental laws changes through time (Sartenaer et al., 2021). In particular, there is room for thinking that (the law of) natural selection wasn't "preformed" before some instant in time, at which it actually "appeared" together with whatever constituted the first units of selection.

---

[18] A claim that is often cast in terms of natural selection being a "force" that cannot be reduced to other forces (be they evolutionary or not). That Rosenberg actually embraces some form of reductionism, that he considers the law of natural selection as being (oddly) chemical in character, or that he doesn't necessarily embrace some form of nomic realism isn't what matters here. The point is that considering the law of natural selection as fundamental isn't completely heretical.

- (iv) That the set of fundamental laws actually at play at some instant in time can change is taken to be a coextensive with transformational emergence (Humphreys, 2016; Guay & Sartenaer, 2016).
- (v) The first units of selection – *and only them* – are transformationally emergent.

It is noteworthy that this argumentative schema is only claimed to have *programmatic* validity. Rather than properly legitimizing a chauvinist version of transformational organicism – which it clearly doesn't – the proposed schema reduces to what merely constitutes a possible research agenda for the view.

Given that this first strategy is intimately associated with the ontological category of laws of nature, and given that such a category might plausibly be totally out of place in biology (Smart, 1959), one might be willing to envision an alternative approach that rather relies upon an ontological category that is better suited for biology, and to which the category of laws of nature may be taken to reduce, namely, *individuals* or *objects* (and their properties, in the spirit of, e.g., Machamer et al. (2000)). Such an alternative approach, which also has the validity of a research plan, goes as follows:

- (i) Biological evolution has been the occasion of several "major evolutionary transitions" (Szathmary & Maynard Smith, 1995).
- (ii) While some of these transitions were "organizational" or holistic, some were rather "transformational" (Jablonka & Lamb, 2006) – like the one from RNA to DNA – in the sense that they were diachronic and flat (or "rank-free"; Okasha, 2011).
- (iii) Such transitions were the occasion of the coming into being of new biological individuals (Godfrey-Smith, 2011; Clarke, 2013).
- (iv) The advents of such new individuals are to be construed as instances of transformational emergence.

Obviously, a lot should be said in order to provide (iv) with some plausibility, though it strikes me as perfectly congruent with Soto, Sonnenschein, and Miquel's line of though as described in Sect. 4.2, where the ontological nature of emergence at stake is to be grounded in a change in objects rather than laws.

As it appears, in the face of the objection according to which transformational emergentism might fail to be specific to biology, and hence fall short of appropriately grounding some (chauvinist) variants of organicism, there exist some prospects – to be fully fleshed out to be sure – of rendering the view sufficient for vindicating the autonomy of biology.

6.5   Conclusion: The Good Fortune of Organicism

Contrary to the received wisdom, I showed in this chapter that organicism could actually fare well even if, for some reason, it finally turned out that organization* is an illegitimate notion. That organicism doesn't really need organization in order to remain chauvinist about organisms and autonomist about biology doesn't entail, to be sure, that organization is not a good way, or even the best way, to meet such standards. The claim made in this chapter is of a purely conceptual nature: there certainly is a possible world in which, although organisms are not organized*, organicism is a flourishing doctrine that happens to be true.

Though certainly not conventional, organicism without organization*, or transformational organicism, has some prima facie interesting features, which together concur to make it an option in the reductionism/antireductionism debate that deserves further exploration. To begin with, transformational organicism has no need to reify "levels of nature," so it happens to be consistent with the currently rising deflationism about levels in the philosophy of biology (see Potochnik and McGill (2012) and Eronen (2015)). Incidentally, as it eschews any commitment to putative forms of downward determination, there is no risk for transformational organicism to be assimilated to dualistic forms of vitalism, nor to be undermined by Kim-style exclusion arguments, which notoriously cast doubts on the very viability of (even thinly) ontological forms of (synchronic) emergence (Kim, 2005).

As it appears, with such a backup plan, the prospects of organicism look very good indeed.

Acknowledgments I would like to thank Argyris Arnellos, Charbel El-hani, Giuseppe Longo, Carlos Sonnenschein, Ana Soto, Charles Wolfe, and more broadly, the audience of the Paris workshop: *Organisation as a Theoretical Principle for the Life Sciences*, for helpful comments and discussions on an earlier version of this chapter. I also thank the editor of this volume, Matteo Mossio, for his numerous advices, comments and suggestions, which considerably helped improve my thoughts on this topic. Finally, I gratefully acknowledge the financial support of the Alexander von Humboldt Foundation.

References

Arnellos, A., & El-Hani, C. N. (2018). Emergence, downward determination and brute facts in biological systems. In E. Vintiadis (Ed.), *Brute facts* (pp. 248–270). Oxford University Press.

Bedau, M. A. (1997). Weak emergence. *Philosophical Perspectives, 11*, 375–399.

Beebee, H. (2000). The non-governing conception of laws of nature. *Philosophy and Phenomenological Research, 61*(3), 571–594.

Blitz, D. (1992). *Emergent evolution: Qualitative novelty and the levels of reality*. Kluwer Academic Publishers.

Clarke, E. (2013). The multiple realizability of biological individuals. *The Journal of Philosophy, 110*(8), 413–435.

Dobzhansky, T. (1973). Nothing in biology makes sense except in the light of evolution. *The American Biology Teacher, 35*, 125–129.

Dowe, P. (2000). *Physical causation*. Cambridge University Press.

Emmeche, C., Koppe, S., & Stjernfelt, F. (2000). Levels, emergence and three versions of downward causation. In P. B. Andersen, C. Emmeche, O. Finnemann, & P. V. Christiansen (Eds.), *Downward causation: Minds, bodies, and matter* (pp. 13–34). Aarhus University Press.

Eronen, M. (2015). Levels of organization: A deflationary account. *Biology and Philosophy, 30*, 39–58.

Godfrey-Smith, P. (2011). Darwinian populations and transitions in individuality. In B. Calcott & K. Sterelny (Eds.), *The major transitions in evolution revisited* (pp. 65–81). MIT Press.

Guay, A., & Sartenaer, O. (2016). A new look at emergence. Or when after is different. *European Journal for Philosophy of Science, 6*(2), 297–322.

Guay, A., & Sartenaer, O. (2018). Emergent quasiparticles. Or how to get a rich physics from a sober metaphysics. In O. Bueno, Fagan, & R.-L. Chen (Eds.), *Individuation, process and scientific practices* (pp. 214–235). Oxford University Press.

Haldane, J. S. (1908). The relation of physiology to physics and chemistry. *British Medical Journal, 2*, 693–696.

Humphreys, P. W. (1997). How properties emerge. *Philosophy of Science, 64*(1), 1–17.

Humphreys, P. W. (2016). *Emergence. A philosophical account.* Oxford University Press.

Jablonka, E., & Lamb, M. J. (2006). The evolution of information in the major transitions. *Journal of Theoretical Biology, 239*, 236–246.

Kim, J. (2005). *Physicalism, or something near enough.* Princeton University Press.

Kim, J. (2006). Emergence: Core ideas and issues. *Synthese, 151*(3), 547–559.

Laughlin, R. B. (1999). Nobel lecture: Fractional quantization. *Reviews of Modern Physics, 71*(4), 863–874.

Laughlin, R. B. (2005). *A different universe: Reinventing physics from the bottom down.* Basic Books.

Laughlin, R. B., & Pines, D. (2000). The theory of everything. *Proceedings of the National Academy of Sciences, 97*(1), 28–31.

Machamer, P. K., Darden, L., & Craver, C. F. (2000). Thinking about mechanisms. *Philosophy of Science, 67*(1), 1–25.

McLaughlin, B. P. (1992). The rise and fall of British emergentism. In A. Beckermann, H. Flohr, & J. Kim (Eds.), *Emergence or reduction? Essays on the prospects of nonreductive physicalism* (pp. 49–93). de Gruyter.

Moreno, A., & Mossio, M. (2015). *Biological autonomy. A philosophical and theoretical enquiry.* Springer.

Morgan, C. L. (1923). *Emergent evolution.* Williams & Norgate.

Mossio, M., Bich, L., & Moreno, A. (2013). Emergence, closure and inter-level causation in biological systems. *Erkenntnis, 78*, 153–178.

Mossio, M., Montévil, M., & Longo, G. (2016). Theoretical principles for biology: Organization. *Progress in Biophysics and Molecular Biology, 122*(1), 24–35.

Nicholson, D. J., & Gawne, R. (2015). Neither logical empiricism nor vitalism, but organicism: What the philosophy of biology was. *History and Philosophy of the Life Sciences, 37*(4), 345–381.

O'Connor, T., & Wong, H. Y. (2005). The metaphysics of emergence. *Noûs, 39*, 658–678.

Okasha, S. (2011). Biological ontology and hierarchical organization: A defense of rank freedom. In B. Calcott & K. Sterelny (Eds.), *The major transitions in evolution revisited* (pp. 53–63). MIT Press.

Potochnik, A., & McGill, B. (2012). The limitations of hierarchical organization. *Philosophy of Science, 79*(1), 120–140.

Reed, E. S. (1981). The lawfulness of natural selection. *The American Naturalist, 118*(1), 61–71.

Rosen, R. (1991). *Life itself: A comprehensive inquiry into the nature, origin, and fabrication of Life.* Columbia University Press.

Rosenberg, A. (2006). *Darwinian reductionism. or, how to stop worrying and love molecular biology.* The University of Chicago Press.

Rueger, A. (2000). Physical emergence, diachronic and synchronic. *Synthese, 124*(3), 297–322.

Sartenaer, O. (2013). Neither metaphysical dichotomy nor pure identity. Clarifying the emergentist creed. *Studies in History and Philosophy of Biological and Biomedical Sciences, 44*(3), 365–373.

Sartenaer, O. (2016). Sixteen years later. Making sense of emergence (again). *Journal for General Philosophy of Science, 47*(1), 79–103.

Sartenaer, O. (2018a). Disentangling the vitalism-emergentism knot. *Journal for General Philosophy of Science, 49*(1), 73–88.

Sartenaer, O. (2018b). Flat emergence. *Pacific Philosophical Quarterly, 99*(S1), 225–250.

Sartenaer, O. (2019). Humeanism, best system laws, and emergence. *Philosophy of Science, 86*(4), 719–738.

Sartenaer, O., Guay, A., & Humphreys, P. (2021). What price changing laws of nature? *European Journal for Philosophy of Science, 11*, 12. https://doi.org/10.1007/s13194-020-00327-4

Sellars, R. W. (1922). *Evolutionary naturalism*. Russell & Russell.

Smart, J. J. C. (1959). Can biology be an exact science? *Synthese, 11*(4), 359–368.

Soto, A. M., Sonnenschein, C., & Miquel, P. A. (2008). On physicalism and downward causation in developmental and cancer biology. *Acta Biotheoretica, 56*(4), 257–274.

Szathmary, E., & Maynard Smith, J. (1995). The major evolutionary transitions. *Nature, 374*, 227–232.

Varela, F. J. (1979). *Principles of biological autonomy*. Elsevier.

von Bertalanffy, L. (1952). *Problems of life: An evaluation of modern biological and scientific thought*. Harper & Brothers.

Weiss, P. A. (1969). The living system: Determinism stratified. In A. Koestler & J. R. Smythies (Eds.), *Beyond reductionism: New perspectives in the life sciences* (pp. 3–55). Hutchinson.

Wilson, J. (2015). Metaphysical emergence: Weak and strong. In T. Bigaj & C. Wurthrich (Eds.), *Metaphysics in contemporary physics* (pp. 251–306). Brill.

Wolfe, C. T. (2011). From substantival to functional vitalism and beyond. *Eidos, 14*, 212–235.

Woodger, J. H. (1929). *Biological principles: A Critical study*. K. Paul, Trench, Trubner.

Wolfe, C. T. (this volume). Varieties of organicism: A critical analysis. In M. Mossio (Ed.), Organization in biology. Springer.

Chapter 7
Organisms: Between a Kantian Approach and a Liberal Approach

Philippe Huneman

Abstract The concept of "organism" has been central to modern biology, with its definition and philosophical implications evolving since the nineteenth century. In contemporary biology, the divide between developmental and physiological approaches and evolutionary approaches has influenced the definition of organism. The convergence between molecular biology and evolutionary biology has led to the term "suborganismal biology," while the return to the organism has been characterized by animal behavior studies and Evo-devo. The philosophical approach to the concept of individual is divided between a Kantian understanding of organism, which defines necessary and sufficient conditions for any X to be a "natural purpose," and an evolutionary approach, which considers what a biological individual is and confers natural selection a key role in this definition. While the former aims to find necessary and sufficient conditions for an organism, the latter thinks in terms of conceptual spaces, being much more liberal in pointing out organisms in the world. The paper examines possible connections between these two approaches and assesses the prospects of a reconciliation between them.

The notion of organism stands between self-evidence and inscrutability: self-evidence, because someone outside of theoretical biology would easily agree that most of the living things are organisms or, in other words, that whatever life is, it comes under the form of "organisms;" and inscrutability, because when one wants to make sense of organisms, difficulties are innumerable: What do make them different from other complex systems? Should they be principally understood as products of evolution, as, according to Huxley's phrase, "bundles of adaptation?" Are they just an instance of *an organization* or something specific that requires more than "organization" to be understood?

In current days, this difficulty appears even more pressing, for at least two reasons. Borrowing the usual distinction between functional biology and evolutionary biology that Ernst Mayr has drawn based on a difference between proximate and

P. Huneman (✉)
Institut d'Histoire et de Philosophie des Sciences et des Techniques (IHPST, CNRS/Paris1 Sorbonne), Paris, France

© The Author(s) 2024
M. Mossio (ed.), *Organization in Biology*, History, Philosophy and Theory of the Life Sciences 33, https://doi.org/10.1007/978-3-031-38968-9_7

ultimate causation (Mayr, 1961), let's survey these reasons. In molecular biology, the pervasiveness of network thinking (e.g., Barabasi, 2018, Newman, 2010) challenges the idea that biological systems should be investigated mechanically or in a reductionist way (i.e., starting from the parts – cells or molecules – and their dispositions). Thus, it contributed to the rise of systems biology (Kitano, 2002, Green, 2013), an approach designed to address organisms as wholes irreducible to the effects of their parts, that is, as a set of instructions given by the genes. While the hope of elaborating the basic explanatory repertoire of biology at the suborganismal level of macromolecules vanishes, the notion of the organism itself, engaged by systems biology, requires a novel theoretical framework.

On the other hand, it has been repeatedly said that the Modern Synthesis in evolutionary biology tended to confer organisms an ancillary status because the basic evolutionary processes stand at the levels of genes and populations. Organisms were something to be explained, as Dawkins (1976) suggests by wondering why genes tend to coalesce into organisms instead of living by themselves; or they were an instance supposedly left aside by evolutionary biology, whereas developmental biology or Evo-devo rightly take organisms as a structuring concept: this omission of organisms was the target of the famous "spandrel paper" by Gould and Lewontin, whose major claim is the inability of the current evolutionary biology to soundly handle organisms. Walsh (2017) sees evolutionary theory as a "suborganismal" account (opposed to a potential organismal one). But for more than a decade, evolutionary theory has been undergoing major controversies about the necessity to revise or expand (Gould, 2002) the modern synthesis framework, and one important issue arising here concerns the status of organisms (Bateson, 2005, Huneman, 2010). Several dimensions of the claim of a return of the organism coexist:

- The idea that organisms contribute to causing their environment (named *niche construction*, Odling-Smee et al., 2003).
- The idea that some variation can be heritable and directed toward adaptation, for instance, based on *phenotypic plasticity* (West-Eberhard, 2003; Sultan, 2015, Walsh, 2015).
- The relevance of organismal development to evolution, while it has been separated from evolution on the ground of various concurring conceptual distinctions, such as development vs inheritance, somatic vs germinal lineages, or even lately theso-called "central dogma of molecular biology".

Evo-devo has been built around this call to reintegrate developing organisms in evolution (e.g., Raff, 1996; Gilbert et al., 1996), and the developmental systems theory (Griffiths & Gray, 1994; Oyama et al., 2001) is a general account intending to replace genes by developmental cycles as units of selection or evolution.

All these critiques were noticeably led by philosophers (e.g., Walsh, Stotz, Griffiths, Oyama) and biologists alike.

Granted, claims that organisms have been neglected from evolutionary biology seem unfair, to the extent that behavioral ecology is the science of the traits of organisms as adaptations and in general conceives of evolution at the level of organismal phenotypes – named "strategies" (see Grodwohl, 2019). However, what's left out from behavioral ecology is the sense of the integration of all the strategies within one organism – hence, the specific sense of the organization of the organism, which was precisely the target of Gould and Lewontin (1978) under the name of *Bauplan*, a term borrowed from the German tradition of transcendental morphology in the late nineteenth century.

From the viewpoint of either functional or evolutionary biology, this organismal organization should therefore be the object of a theorizing effort. This is not to say that such an effort does not exist. On the contrary, most of what labels itself "theoretical biology," from Rashevsky on, thought intensively about what the organization of an organism is, using ideas forged by some inaugural figures of this tradition (Rashevsky, D'Arcy Thompson, Rosen, Ganti, to name a few). In this paper, I will consider that two threads of thought about organisms coexist in biology and will leave out this tradition of theoretical biology; the question of explicitly articulating these two trends to such tradition should be the object of another paper. One of these threads is mostly found in the circles of developmental biology or Evo-devo and philosophically owes a lot to Kant; the other is mostly elaborated by evolutionary biologists and I will argue that it is much more liberal than the former one. They give room to two general ways of thinking about how the two notions of biological *individual* and *organism* are connected. After having presented these two accounts of what organisms are – and their highly different methodologies – I will say a word about their respective conditions of validity.

7.1 Making Sense of Organisms: The Kantian View

7.1.1 Purposiveness

Among philosophers who addressed biology before the Darwinian turn, Immanuel Kant cannot be overlooked. His *Critique of Judgment* provides an "analytic of teleological judgment" that has often been interpreted as an inquiry into the conditions of possibility of biology.[1] And his key claim that "organisms" (or organized being, *Organisierte Wesens*) are the "natural purposes" (*Naturzwecke*) directly connects with the idea that 'organism' became a crucial concept for biology at the times of Kant's philosophy. Developmental biology or embryology developed after Caspar Wolff's seminal *Theorie der Generation* (1764) into a science of the developing organisms whose key figures in the nineteenth century have been exposed to Kantian thinking, as had been made clear by several historians of biology (e.g., Lenoir, 1982; Richards, 2001; Sloan, 2002). Among these biologists, Blumenbach was in epistolary contact with Kant, and major names such as Pander or Von Baer belonged to the same tradition (Von Baer authored the *Entwicklungsgeschichte den Thieren* (1828), arguably the most important nineteenth-century biology book, as Darwin himself acknowledged).

---

[1] For this claim, Lenoir (1982), Zumbach (1984), McLaughlin (2000), Huneman (2008), and Ginsborg (2004) against Zammito (2018) and Richards (2001).

Comparative anatomy built itself on two key principles, the "principle of the conditions of existence" and the "principle of the connections," leading to the "principle of unity of type." The former was advocated by George Cuvier, whose *Leçons sur l'anatomie comparée* (1805) were a milestone in this science; the latter is developed by Geoffroy Saint-Hilaire, who was a young colleague of the former. Even though they are often contrasted as two divergent ways of making biology, the former focusing on *function* and the latter on *form* – and this is the influential reading by Russell (1911) that Amundson (2005) endorsed later on – both their principles target the structured organism.

The fact that Kant's analytics of biology centers on the notion of organisms – which I will explicate quickly – therefore matches with the new role of organisms in the nineteenth-century biology.[2] Thus, it sounds natural that those who vindicate a return of the organisms in evolutionary biology through the evolutionary theory of development (or Evo-devo) trace back their key concept to Kant's view of organisms (e.g., Gilbert & Sarkar, 2000)

When Gould and Lewontin (1978) use the German term *"Bauplan"* to label what – in an organism – resists the adaptationism proper to the Modern Synthesis, in a paper that has been heavily quoted by Evo-devo people when arguing against the treatment of organisms by this Modern Synthesis, the connection between developmental thinking in evolution and a Kantian tradition in biology becomes obvious. It is thus natural that I sketchily expose this Kantian view now. Since this is not a piece of Kantian scholarship, I' will be fast in reconstructing Kant's reasoning, citing materials likely to back up my claims (including my work).

Let's start with this key notion of purposiveness since Kant's main object is the judgment that ascribes finality to natural systems. Such a judgment explains something by invoking a preexisting concept of this thing – a concept standing "at the root of the production of the object," as Kant says. This is the most general concept of purposiveness. When Kant adds *"natural* purposiveness," he means such things that are judged purposive but that are at the same time naturally produced, in contrast with artificial and technical items.

It is often reminded that Kant contrasts mechanisms and teleology – those are his two terms, and a section in the *Critique of Judgement*, the "Antinomy of the teleological judgment," intends to articulate them. *Mechanism* is about explaining a whole from the parts, and *teleology* is therefore the opposite, explaining the parts from the whole. This latter characterization of purposiveness instantiates the most general definition I gave above (in terms of a kind of causality of a presupposed concept). Saying that I explain the part based on a knowledge of the whole is to say that the part needs the idea of the whole to be understood, and this constitutes the presupposition of a concept at the root of the item to be explained.

Regarding biology, this becomes clear when we turn to a classical example also considered by Kant: the eye. In the eye coexist lots of distinct parts – retina, cones, rods, crystalline, cornea, etc. Each of them follows its proper, distinct "laws", as he

---

[2] A claim defended at length in Huneman (2008, forth).

argued in the First Introduction to the *Critique of Judgement*. If they were arranged differently, for instance, the retina slightly more to the left, then we would see nothing. To explain why all the parts are where and as they are, one has to posit something: the notion of a seeing device, at the root of the production of this organ. Otherwise, it seems that pure chance distributed all these elements in such a way; whereas if the concept of vision is the ground for the construction of the eye, this harmony between the proper laws of each organ becomes necessary. That is a teleological judgment.

This allows us to understand a decisive phrase Kant uses to explicate what is purposiveness: it is the "lawlikeness of the contingent as such" (*First Introduction to the CJ*). What is *necessary* cannnot be otherwise, it follows the laws of nature. Whatever X is *contingent* on could be otherwise – X is contingent upon the antecedent state of the universe, in the sense that, had this state been different, X would be different. But some systems are such that to understand them one needs to consider that the contingency of their states (as contingent upon other antecedent states) can be bracketed in favor of a sort of necessity of their features. The "concept" alluded to by a teleological judgment is an instantiation of this necessity, a sort of rule for behavior proper to such systems. This pattern holds for biology: even if contingent on the laws of physics, as contingent, living entities have some lawlikeness of their own when taken as living entities. For instance, it is merely physically contingent that the development of a chick embryo ends up in a chicken or a monster, yet from the viewpoint of a biologist, these two states are not at all on a par. We call this difference viability vs. teratology. *Viability is a norm.* While physics knows no norms, biology does; for instance, besides the norms of development, any *function* in biology states a norm – "functioning vs malfunction" is normative: a kidney that does not eliminate toxins is abnormal even though (or rather: because) its function is to eliminate toxins.[3] These norms constitute a "lawlikeness for the contingent as such"; and the "concept" assumed in any teleological judgment instantiates such a norm.

7.1.2   Regulative Principle?

Purposiveness, that is, normativity as a lawlikeness of the contingent as such, assumption of a concept at the root of production, and epistemic precedence of the whole over the parts: those are the main elements of Kant's idea of teleological judgment. And for this reason, an organism, an organized being, is a natural purpose: a purpose, because it has parts that are only understandable based on the whole, hence on the concept of the whole. Thus, from the viewpoint of science, a concept stands at the roots of their production, hence the contingency of the agreement of the parts shows, as such, a lawlikeness – and this lawlikeness is not in the things themselves; it is in the eye of the beholder. This latter point is the other major

---

[3] For the normative interpretation of Kant, see Ginsborg (2004, 2014) and Huneman (2014a, b, c).

aspect of this view of organisms as natural purposes – namely, such lawlikeness stems from our project of understanding life as such: it is a "regulative" principle for our cognition. Kant writes: "This principle does not pertain to how such things are possible themselves through this kind of production (things considered themselves as phenomena) but pertains only to the way our understanding can judge them" (CJ § 77, 408). This regulative character makes perfect sense with the requirement that the "concept" at the root of their production is posited by the teleological judgment to the aim of guaranteeing this lawlikeness at the biological level, which will then constitute the object of inquiry for the biologist.

Three things must be written now:

(a) The "concept" of vision, used to allow an investigation of the eye, and more generally any of these concepts that take the role of norms in biological inquiry behave like *attractors*. What does it mean?

Suppose a complex system in phase space. If the system, when faced with a small range of initial conditions, lightly changes its final state, we have a classic case of predictable determinism. But if in the same situation the system hugely changes its dynamics and final state, it is unpredictable, since the error margin on measuring the initial conditions is mapped onto a very large margin of error regarding the final result. The range of final states now is too large to predict anything from the knowledge of an initial state.

But in addition to these two situations, we can consider a third pattern where, whatever the initial conditions, the system will always end up in the same final state. Such a state is called an *attractor*; an example turned into a metaphor is a valley at the bottom of a mountain: whatever place one lets a stone roll down from the top, it will end up at the same location – down in the valley.

The concept of vision somehow turns a pattern of the type "sensitivity to initial condition" into the pattern of the type "attractor." Physical conditions of the embryogenesis of the chick can vary a lot, but then one can (most of the time) safely assume that the chick will develop an eye, since the whole development for the biologist is supposed to produce an eye. There are many different obstacles in embryogenesis but in the end the animal mostly sees. Embryologists have a concept to name the way the developmental process almost always reaches the adult type as a target, even when the initial genes are mutated: "canalization," a term famously coined by Conrad Waddington. Canalization is a form of attractor thinking.

(b) The "concept" (supposed at the root of the purposive system) being a concept of the whole, one has to emphasize Kant's shift between the concept pair "means-ends" (which involves references to utility and intentionality) and "whole-parts" in the very meaning of purposiveness (see Huneman, 2007, 2017). Arguably, Kant detached the notion of purposiveness from the notion of utility and intentionality – while it intends to remain scientific.[4]

---

[4] Notice that this idea of design is quite different from the English tradition: hence, Kant tends to detach organism design from natural theology.

(c) Philosophically, Kant's analysis puts the difference between things that are likely to be explained by pure physics and things that can't – namely, physics and organisms, to say it bluntly – in epistemology rather than in ontology. Physical systems not organized should be explained starting from the parts; in organisms the direction of explanation is inverted, or at least, an "ideal" causation, from the whole to the part, is articulated to a real mechanical causation (i.e., from the parts to the whole). This difference is obviously epistemic, not metaphysical. It's about what is required for an explanation to be possible.

The concept at the root of the production of the purposive thing is part of the teleological judgment rather than "within" the thing. Many philosophers and scientists in the early nineteenth century, while interested in Kant's account of the whole-parts relationship, will give up on this notion of regulative principle: Blumenbach to begin with, and then Kielmayer or Meckel. Historical epistemology is cleaved about that question: Lenoir, who initially considered what he called the "vital-materialists," namely Kant, Blumenbach, and other German biologists, lumped all of them in a sort of Kantian tradition in which teleology constitutes a regulative framework to search for mechanisms (what he called "teleomechanists"). Larson (1979), Richards (2001), and later Zammito (2018) on the contrary argued that Kant was alone in his view of regulativeness and that biologists will consider that purposiveness as described by Kant (including this focus on whole-parts relationship) is objective; with Zammito, they often also see this disconnection with Kant as a source of the fruitfulness of the attitude. This stance is also a feature of current views of organisms, such as the ones defended by researchers in the wake of Varela, Rosen, or Maturana, who start with the notion of self-organization understood as *autopoiesis* – e.g. Moreno and Mossio (2015), Montévil and Mossio (2015), or Saborido et al. (2011): there is something objective in organims' being purposive. Yet for Kant, the concept of purposiveness, because of its constitution – namely, positing, within the judgment, a concept at the source of the production of the object – is necessarily regulative, in the sense that it concerns the modalities of the judgment rather than the thing about which one judges.

7.1.3 Natural Purposes and Self-Organization

Up to now, I unpacked the notion of purposiveness; but organisms are "natural purposes." What does natural stand for here? Purposes can be artificial – in this case, the "concept" at the root of the production of the thing is simply the idea that the maker, the craftsman, or the artist has when she makes the product. The antecedence of wholes over parts is clearly here taking place. But natural items are such that they have no makers; they seem to be produced by themselves. In this case, when a natural purpose is found, the parts are what exist, so they create themselves in accordance with an idea of the whole – which, says Kant, is merely a "principle of cognition" and not a "principle of production" (CJ §65). This idea of the whole is not the principle of their *making* but the condition of our understanding of organs and traits as parts of an organism, i.e., as involved in the development and functioning of a living entity.

Kant precisely says:

> In such a product of nature each part, at the same time as it exists throughout all the others, is thought as existing with respect to [*um...willen*] the other parts and the whole, namely as instrument (organ). *[1]* That is nevertheless not enough (because it could be merely an instrument of art, and represented as possible only as a purpose in general); the part is thought of as an organ *producing* the other parts (and consequently each part as producing the others reciprocally). *[2]* (CJ §65).

Condition [1] for being a natural purpose characterizes a purpose in general, as I explicated it until now. It is not proper to organisms, and it is where arise functions and functionality (as playing a role in a whole). I call it (Huneman, 2014c, 2017) the *design* criterion, since it fits any system that is designed and/or has a design. And criterion [2] specifies what makes a *natural* purpose. Kant then develops his view of what a "part producing another part" means: "Thus, concerning a body that has to be judged as a natural purpose in itself and according to its internal possibility, it is required that the parts of it produce themselves [*hervorbringen*] together, one from the other, in their form as much as in their binding, reciprocally, and from this causation on, produce a whole." I call this criterion [2] the *epigeneticism* criterion (see also Huneman (2017)). It distinguishes organisms from artifacts because their design, in the sense of an arrangement of the parts according to an idea of the whole, is not achieved by some external agent considering precisely such idea of the whole as a building plan – the process of building organisms is rather done by the parts themselves; hence, they produce themselves: as a consequence says Kant, "organized beings are self-organized beings" (ib.). This essential character of such systems accounts for the sort of triadic phenomenology of organized beings proposed by Kant just before (CJ §64), namely, its self-production as individuals, when the tree grows; as a set of parts, when it grows leaves; and as a species, when it disperses seeds that grow.

This original occurrence of the word self-organizing will be quoted later by Kauffman (1993), who sees Kant as a father of the theories of self-organization, even though we now have a galaxy of "self-X" terms, such as self-assembly, self-maintenance, self-building, etc., within which "self-organization" stands rather on the side of physics. Even though Kant would not acknowledge the formal apparatus of Kauffmann or Santa Fe style theories, he indeed held this strong thesis that living organization is self-organization. But his claim was rather tied to nascent embryology theory, namely, Wolff's epigeneticism, than to the mathematics of nonlinear differential equations, fractals, and Boolean networks, as it is now (see Ruelle, 1989).

More generally, I don't refer (with my labels) to the current notion of "epigenetics" (namely, whatever touches on the regulation of gene expression) but only to the notion opposed to preformism, namely, the capacity of living systems to build themselves through all the interactions with their external environment, without a preexisting template, and based on the activity of their parts producing other parts (for instance, we would now talk about cells). This criterion (2) provides us with a grasp between a general account of organization and the specificity of biological organization, which can't rely on an extant template – a difference being addressed in the introduction of this volume.

7.2   Making Sense of Organisms? From Kant to the Modern Synthesis

For Kant, the design criterion (1) and the epigeneticism criterion (2) are two criteria for ultimately capturing the instantiations of one concept: they are unified through the unity of the concept of purposiveness as a transcendental presupposition, a concept of which they are the two facets. Both refer to an "idea of the whole" as "principle of cognition"; and it is the *same whole* in each case, and Kant analyzes at length the justifications for this claim in his "transcendental deduction" of the concept of purposiveness, undertaken in the Dialectics of the third *Critique*. But does this have any relevance for anyone now interested in the concept of organism? I will argue for the affirmative since, as indicated above, Kant's views of organization in general (criterion 1 above) and of *biological* organization (criterion 2 above) are mentioned as a philosophical foundation for thinking of organisms by Evo-devo people (e.g., Raff, 1996; Caroll, 2005), or theoretical biologists interested in self-organization, or, indirectly, by critiques such as Gould and Lewontin in their spandrels paper.

Thus, I will quickly consider what are these two criteria in the context of current evolutionary biology and developmental biology.

7.2.1   Design Criterion

The design criterion consists in presupposing that organisms are wholes in which parts fit the needs and demands of the persistence of these wholes. It is easy to see that such a criterion suits well the practice of behavioral ecology, namely, the subdiscipline of evolutionary biology that studies traits of organisms as adaptations or equilibrium strategies. Clearly, "traits" are more general than parts, but we can consider behavior as something of the organism, and then behavioral traits as parts of this dimension of the organism.

Methodologically, behavioral ecology is generally adaptationist as made clear by Reeve and Sherman (1993) or Krebs and Davies (1995), namely, it starts by assuming that a trait results from natural selection, either by maximizing fitness, or inclusive fitness, or (in the case where fitness payoffs depend upon the frequencies of traits) by realizing an "evolutionary stable strategy," which is a kind of equilibrium in evolutionary contexts (Maynard-Smith, 1982).

Behavioral ecology asks questions such as the following: "Why do passerines or great tits lay four or five eggs by nest?"; "Why do gorillas of this region change mate every three years? "; and "Why are the leaves of the cypress of this size?". Behavioral ecologists center on traits, hence on phenotypes; they demand that traits are somehow *heritable*, which means that some alleles make a difference in the value of the trait. Such heritability is assumed; it may be very low but, in most cases, the precise genetic makeup involved in traits under focus is unknown. More than a hundred genes are involved in a phenotype as seemingly simple as the size of mammals, so one should not expect that traits such as foraging behavior, studied in behavioral ecology, rely on a knowable genetic circuitry.

In this approach, parts – traits – are assumed to fulfill environmental demands. One often uses here the method labeled "reverse engineering" – namely, assuming that a part is an adaptation, and trying to reconstitute the environmental demands it was designed to fulfill. For instance, the horn of the *Parasaurolophus* has been intensively studied, and many hypotheses about the environmental demands it addressed have been emitted until one reached a consensus on the idea that it was used as a communication tool in sea or river shores (Turner, 2000) (Fig. 7.1).

To this aim, the structure of the horn, and especially its hollowness, has been taken into account, to deduce what the effect of such a horn could be – emitting recognizable sounds has been declared a much more probable selected effect than fighting competitors with the head.

Assuming that in a given system the conditions for natural selection that population genetics can unravel are met, then the traits we see are adaptations, which means that they fulfill environmental demands. By examining them, we

Fig. 7.1 Parasaurolophus and its hollow horn

may reconstitute these environmental demands. Such reverse engineering is thereby a legitimate method. This justifies that the design criterion is still nowadays legitimate since we've seen that reverse engineering is a clear instance of this criterion.

But the process of adaptive evolution involves a process of allele frequency change. Population genetics models such a change. In this approach, evolution is due to forces that act on populations modeled as gene pools[5]; those forces are migration, mutation, natural selection, and random genetic drift (which is, to say it quickly, a sort of error sampling, whose intensity – by definition – decreases with population size[6]). Among them, natural selection is the only one that creates adaptation, hence its epistemic primacy for evolutionary biologists. But nothing guarantees that in a given population, natural selection will overcome the other forces: if it is a very small population, or if migration is too strong, natural selection will be superseded by other forces and evolution will not yield adaptation.[7]

More precisely, natural selection understood as the "survival of the fittest" means that it tends to increase fitness, understood as the expected number of offspring. This maximization generally produces an adjustment between organisms and the environment, since meeting environmental demands allows one to survive and reproduce optimally.[8] The general idea is that being more adapted than others involves surviving more and reproducing more so that the organisms or the traits that maximize their fitness tend to be optimized regarding environmental demands. In a given environment, for instance, the leaves of the cypress will have a size that allows them to maximally photosynthesize, and produce more trees, and more seeds, than if it were having smaller leaves; otherwise, genetic variants with other leaf sizes would thrive against the resident trees and would invade the population. It is such a process, at the level of gene dynamics, which justifies the reverse engineering, hence grounds the design criterion.

However, things are more complicated. Is it really the case that natural selection in principle optimizes and then creates the environmental fit with organisms? Or at least that it tends to optimize[9]? Birch (2015) and Okasha (2018) have indeed shown that there is no satisfying a priori proof that selection by itself and alone always optimizes fitness or inclusive fitness and that equating selection with optimization and adaptation can only be locally legitimate and often waits for empirical corroboration.

---

[5] See Sober (1984) for a canonical formulation of this account.

[6] On drift see Plutynski (2007), Abrams (2007).

[7] Of course, this is the simplest case and I bypass here issues regarding social evolution and then kin selection and inclusive fitness, (Hamilton, 1964) as well as population structure or maternal effects. Suffices to say that natural selection tends to produce adaptation because maximizing fitness entails optimizing traits with regard to environmental demands.

[8] Fitness is as we know a much-discussed and controversial concept; but this is not my point here.

[9] As to the prospects of this optimization, see also Huneman (2014a, b, 2019b).

To this extent, the design criterion cannot be seen as what should any organism satisfy on the grounds of the fact that a population fulfills the classical conditions of heritability, variation, and fitness that Lewontin (1970) famously formulated as conditions for potential evolution by natural selection (or any other version of the characterization of the conditions of evolution by natural selection).

7.2.2 Epigeneticism Criterion

Cell theory, which is, along with molecular biology and evolutionary theory, the third major global theory of life underpinning modern biology (Gayon & Petit, 2019), provides us with a clear instance of this criterion (as explicated in Sect. 7.1.3): cells are producing cells, and this production leads to the organism. But this is also happening in accordance with an "idea of the whole," as Kant required. Why? Early molecular biology, in the enthusiasm of the discovery of DNA and the genetic code, would easily consider that this idea of the whole is the genotype and therefore exists rather "in" the cell than within the epistemic activity of the researcher. This is a kind of preformationism (as made clear by Müller and Hallgrimson (2003)), but recent developmental biology has increasingly shown that development is more complex than the unfolding of a program.

Granted, cells differentiate according to what Kant calls an "idea of the whole"; yet unlike what I just said, it's not the genetic program that differentiates each cell since all carry the same genotype – in most metazoan and plants; on the contrary, differentiation is an epigenetic process involving the environment of each cell, the activation states of the genome in neighboring cells, and for each gene, the gene regulatory network that regulates its expression according to the states of all elements (other genes, transcripts, etc.; see Davidson, 1986; Oliveri et al., 2008).[10] Thus, we seem to move away from the gene-based preformationism toward a more epigeneticist account of development in which cells produce cells in accordance with a general "idea of the whole" that is less located "in" the genotype than instantiated in a distributed way across genotypes, cell environments, and multiplicity of gene regulatory networks (GRN).

To make this kind of production clearer, remember, in developmental biology, the classical French flag model (Fig. 7.2) (due to Wolpert, 1969). In this account, cell differentiation as the response to a gradient of morphogenetic substance turns a continuous proportion of morphogen into a discrete series of expression states (the "flag"). It realizes an instance of this self-organization conceived of by Kant, to the extent that the organism is created on the basis of cells that respond individually to an overall state of the whole that they locally encounter, as represented by the state of the gradient in a flag.

---

[10] Such process during embryogenesis also involves "programmed cell death" or apoptosis (Kerr et al., 1972), a major dimension of development that I studied in Huneman (2023).

Fig. 7.2 A representation of the French flag model of cell differentiation (Wolpert)

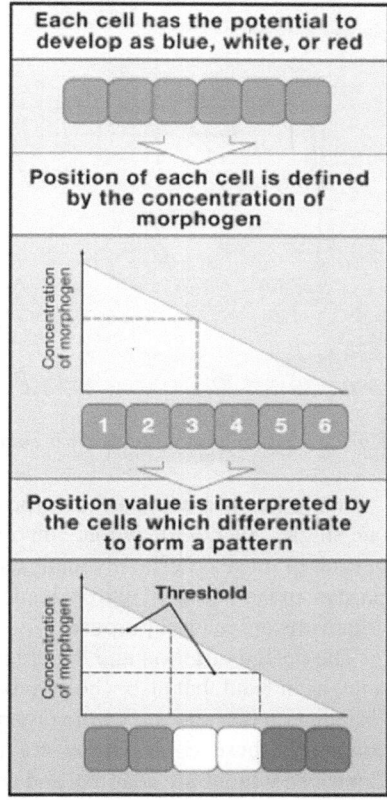

In turn, the GRNs are also, and maybe even more, an instance of this epigeneticism criterion. Developmentalists nowadays think of them as explaining the French flag model among other things (Davidson et al., 2003). They determine the expression of a gene based on the states of hundreds or thousands of other genes or genetic elements in the cell. GRNs react to the state of the organism – which is, for the cell, its environment – and determine in response to the contribution of the focal cell (Fig. 7.3). This corresponds to the self-organizing logic of the epigeneticism criterion.

Moreover, GRNs are implied in both the development of the organism and its functioning, since their dynamics in each case determines what a gene – and then all genes – do, and therefore, what does a cell of such and such genotype within the organism, at each stage of the life cycle. Given that GRNs instantiate the epigeneticist criterion, the fact of their involvement in cell physiology can be interpreted as acknowledging the epigenetic character of organism functioning, which would clearly correspond to the Kantian view of organisms. In any case, we see here a neat intertwining between development and functioning: genetic regulatory networks are involved both in cell specification and pattern formation – and within the regular activity of the cell.

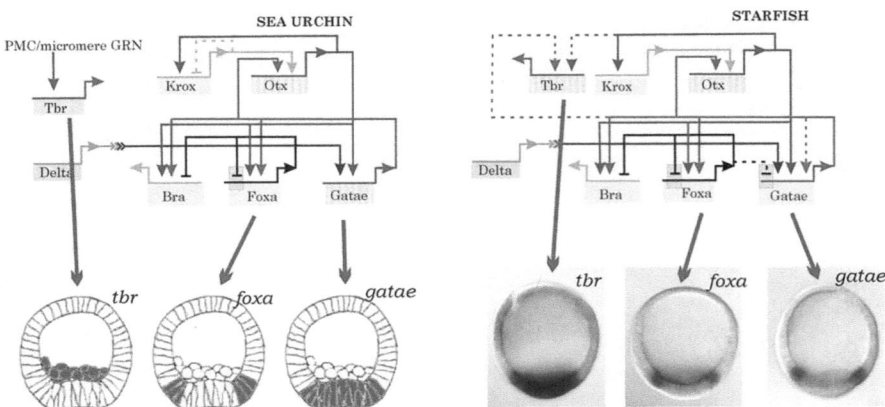

Fig. 7.3 Gene regulatory networks in starfish and sea urchin

While the two criteria for purposiveness were unified by Kant since there was one single "idea of the whole" involved in the teleological judgment and expressed in each of these criteria, my analysis emphasizes differences in the way these criteria can make sense of purposiveness today and possibly reassert the claim that organisms are natural purposes.

The design criterion and the epigeneticism criterion appear indeed not only distinct, as in Kant, but also wholly separated. Emphasizing the design criterion means focusing on the way the whole organism is designed, namely, the way that parts are contrived; these contrivances are the effect and the sign of natural selection. Contrived wholes are adapted, and adaptation results from natural selection, according to the Modern Synthesis. On the other hand, emphasizing the epigeneticism criterion means focusing on development as the proper epigenetic process, and then, more generally, on self-organization of the whole. When authors in the tradition of self-organization as an account of organisms refer to Kant's idea of self-organization, they commit to Kant's idea of parts that create other parts and themselves according to the an "idea of the whole", even though of course they rely on a much richer empirical work, and sometimes they use types of mathematical tools unknown by Kant – Kauffman's Boolean networks or, on another side, Rosen's algebra.

But for Kant, the two criteria were unified as two aspects of the same "idea of the whole" which is involved in the notion of "natural purpose" as a transcendental principle for reflective judgment. This transcendental or criticist dimension of Kant's thinking - visible in the notion of 'regulative principle' - is not adopted by the thinkers interested in self-organization, nor is it related to the modern avatar of the design criterion, namely, the various brands of adaptationism (reverse engineering, etc.; see Lewens (2004)). Hence, the two criteria are divided. They can't be understood as the two criteria of "organisms" (themselves being instantiations of "natural purposiveness").

Through these reflections, I put in a Kantian light the conflict between adaptationism and developmentalism as it is considered by supporters of Evo-devo, and sometimes deemed unsurpassable (e.g., by Amundson, 2005). It is not the only possible reading of such analyses of the two criteria of purposiveness, but it helps put them in the context of current theoretical cleavages.

And as it is, it shows that there is a major issue with any attempt to now use the Kantian analysis of organisms – which is one of the major philosophical analyses of this concept – as a way to make sense of organisms in current biological thought. Maybe that leads to a final verdict of unsurpassable cleavage: focusing on the epigeneticism criterion leads to viewing self-organization while focusing on the design criterion sketches of the whole organism as a set of responses to environmental demands, and the two views of the "whole" that emerges on its side remain separated.

With these words, I turn to a wholly different approach, which is the way an ontological theorizing based on Darwinian principles intends to make sense of organisms understood as a major kind of "biological individuality."

7.3   Evolutionary Individuals: A Liberal Approach Based on Conceptual Spaces

7.3.1   Transitions in Individuality

Biological individuality has been the object of many conceptualizing attempts from philosophers and biologists relying on the Modern Synthesis. I'm not trying to review these accounts or systematize them here since this would require a full paper or a book. But I think that the underlying idea grounding these sometimes conflicting accounts is the connection made by David Hull in his seminal paper "A matter of individuality" (1980) – namely, a connection between individuality and natural selection. In a nutshell, to be an individual is to be a target of selection. Since this latter notion is controversial and not well defined, approaches to the individuality/selection connection are numerous. Yet all share the idea that to see what are individuals in the world, one has to identify what is the object of some selection. Some accounts intend to specify what exact concept of a unit of selection is required to single out individuals (e.g., Folse III & Roughgarden, 2010; Clarke, 2014; Bouchard, 2008), while others are more pluralist, allowing for several types of individuality according to the aspect of selection considered (e.g., Goodnight, 2013). Yet in Sect. 7.2, I will argue that these accounts yield views more liberal than the Kantian-based view of organisms because they don't commit to the idea that something could be either an individual or not an individual, but most of the cases of individuality are graded stages of individuality.

Granted, organisms are individuals; but of course other things can be biological individuals, and this intuition is backed by selection-based accounts of individuality: bacteria (which are unicellular and may not be organisms strictly speaking); genes, given that there exists a selection at the level of genes, for instance, in the case of segregation distorters (Burt & Trivers, 2006); possibly colonies of hymenopteran insects; and perhaps species, if one follows Ghiselin and Hull who famously argued that species are not classes but individuals whose conspecific organisms are genuine parts (Hull, 1980, Ghiselin, 1974). The Darwinian approach intends to make sense of all individuals based on natural selection. But among them, multicellular organisms enjoy a paradigmatic status: first, they constitute most usually our favorite example of individuals, since they fit our intuition more than species or genes; second, they display spatial contiguity and often genetic homogeneity – in the case of most metazoans – which makes easy to talk of the self-containment and indivisibility implicit in the word "individual." If, following Aristotle, "individuality" means the logical inseparability (a horse and a horseman can be separated into two concepts of particulars, but a horse cannot), the genetic homogeneity of something that was born a zygote and then developed based on clonal cell division makes it into something apparently logically indivisible.

Hence, "organisms" as understood by the Kantian approach that I exposed, namely, multicellular organisms, are a paradigmatic but not exclusive kind of biological individual. The question of "organisms" may therefore be summarized by the question raised by Dawkins (1976), namely, why does life on Earth comes mostly under the form of organisms rather than by a total mess of genes as the only individuals? The start of an answer is given by the program called "evolutionary transitions," initiated by Buss (1988) and then Maynard-Smith and Szathmáry (1995) and Michod (1999). The main idea is that throughout evolution, distinct forms of individuality understood as entities that reproduce by themselves and thereby can be targets of selection came into existence. For instance, cells appeared on the basis of macromolecules possibly replicating because of them being autocatalytic and templates; cells that made up life on Earth from 3,5 By ago to 1 By ago evolved into multicellular organisms. And then some of them evolved into forms of individuality that can be composed of individuals and show the division of labor between reproduction and survival/development that is characteristic of multicellular organisms: namely, hymenopteran insects form colonies where a cast reproduces and a cast does defense, territoriality, and foraging without reproducing.

This research program, most generally understood, intends to capture the generic processes leading from groups to individuals made up of a collection of entities. The process of going from prokaryotes or unicellular eukaryotes to multicellularity is one crucial transition. But the same general rules should govern all processes, even though local differences are investigated. As to themselves, multicellular organisms *develop*; the development possibly (and most often) starts with a genetic bottleneck; those individuals contain differentiated cells with identical genomes, hence the need for epigenetic gene expression mechanisms; many recent clades feature sequestration of germ-line paralleling the division of labor in hymenopteran insects (Buss, 1988).

The main process involved in the evolution of forms of individuality is "multilevel selection" (MLS) (Michod, 1999, 2005). It means that selection operates in opposite ways at two levels: the one constituted of entities and the one constituted of groups of these entities – for instance, cells and groups of cells. Among cells, those that reproduce faster or more than the average have better evolutionary success. But at the level of groups, having too many cells that work "for them-selves" may distort the group, and then such a group eventually fares less well than groups where cells are more coordinated. Hence, selfishness – in the sense of repro-ducing more than others – wins *among cells*, but altruism, in the sense of reproduc-ing less than others, or, more formally, having a lesser fitness, wins *among groups*. This is multilevel selection, a concept considering selection as the result of combin-ing intragroup competition and intergroup competition (Sober & Wilson, 1998).

The researchers interested in evolutionary transitions emphasize not only the fact that multilevel selection may foster altruism (while at the intragroup level, altruism always loses); but also that this process may ultimately lead to groups that are likely to reproduce as a single entity. This is exactly what plausibly happened with multi-cellular organisms. Briefly said, altruism among cells is maintained because of group benefits, and in some cases, the group starts reproducing as one, and then emerging policing devices ensure the persistence of this reproduction.

This process can be understood in several ways. Appealing to the useful distinc-tion made by Damuth and Heisler (1988) between two kinds of MLS defined by two kinds of group fitness, labeled MLS1 and MLS2, Okasha (2006) and Michod (1999) argued that a transition is a transition between these two kinds of fitness. In MLS1, fitness is defined by counting the total number of offspring of all the individuals of a given group; in MLS2, it's defined by counting daughter groups of a group. This intuitively fits the transition toward multicellularity: a fitness of a group of cells is the amount of cells after one generation; but the fitness of a multicellular organism is the number of daughter organisms, not the total number of cells at the next gen-eration. The transition toward multicellularity is therefore a transition from one to the other type of MLS, from MLS 1 to MLS2. And formally, what makes this pos-sible is the decoupling between these two kinds of fitness, and it often happens because the trade-off between fecundity and viability in cells becomes a convex function when the group size increases (Michod, 2005).

This explanatory scheme is supposed to account for all kinds of transitions. However, multicellular organisms constitute a paradigmatic transition. For this rea-son, researchers such as Michod and his team extensively investigated a clade in which unicellular and multicellular species coexist – namely, the order *Volvo*cales (*or Chlamydomonadales*), within which *Chlamydomonas* is a unicellular species, Gonium is a colonial species undifferentiated, and Vovox is a colonial differnetiated species (the transition took 35 My to occur). But the key role of multicellular organ-isms for the question of individuality is not only due to the intuitive appeal they have for us, and then our familiarity with metazoan. Within evolutionary theorizing, this is also a salient feature. Take the hierarchy of individuality. What scholars are inter-ested in evolutionary transition research is the generative process that accounts for steps in individuality, as I said. But while individuality is hierarchical, through a hierarchy based on compositionality (chromosomes -> cells-> multicellular organ-isms -> colonies, as Michod (1999) shows), the nature of this hierarchy is complex. Paleobiologist Niles Eldredge argued in the late 1980s that there are at least two hierarchies, one genealogical and one ecological (Eldredge, 1985).

The genealogical hierarchy consists of levels of increasing complexity in repro-duction: each level consists of entities that include entities of the previous level but reproduce by themselves. They have a direct genealogical link. The ecological hier-archy consists of levels of ecological interaction: chromosomes assemble through meiosis; cells interact in microbial ecology; organisms interact within ecological settings; and groups of organisms may compete and cooperate in competitive con-texts. Interestingly, "organism" is the level that belongs to the two hierarchies: it is a main agent in ecological interactions, and it is also a crucial step in genealogy.

For this reason, the multicellular organism is crucial for the notion of individual-ity in Darwinian contexts even though, as Godfrey-Smith (2009) forcefully claimed, not all organisms are Darwinian individuals – since some of them don't reproduce by themselves – and not all Darwinian individuals are organisms.

7.3.2   Conceptual Spaces: Being Liberal

These considerations indicate that such an approach to individuality may not pro-vide a complete account of organisms, even though organisms are individuals. But a closer look at the evolutionary transition programs reveals that this approach is quite different in its spirit from the Kantian approach.

The parallel between bee colonies and organisms, grounded on the division of reproductive labor, comes with a few lessons. Colonies are individuals, in the sense that they can be seen as units of selection under some perspectives, for instance, MLS; but they lack the self-contained character of organisms as well as their capac-ity to reproduce for themselves. Everything happens as if the transition from MLS1 to MLS2, through which the groups have daughter groups that can be counted, did not come to terms. In Huneman (2013), I proposed to distinguish two kinds of tran-sition, depending on whether they come to an achievement (like multicellular organ-isms) or not (like bee colonies). Pandas realize exemplarily complete transitions; bee colonies realize component transitions.

But this is less a binary distinction than two poles of a continuum. There are degrees in "component transition" and inversely some organisms may lack or lose features of complete transition – e.g., cancer as disruption of organisms (see Featherston & Durand, 2012), failure of policing devices in the case of immunity disease, etc.

That gives us a flavor of the liberality proper to the Darwinian approach: systems can be more or less individuals, to the extent that they can come from more or less complete transitions. Organisms are a result of the former, but for the same reason, "being an organism" will come by degrees.

However, this continuum of biological individuality has been even more expanded. In a series of papers, Joan Strassmann and David Queller (Strassmann & Queller, 2010; Queller & Strassmann, 2009) have suggested a view of individuality that is less a gradient than a two-dimensional hyperspace (Fig. 7.4). They argue that individuals require cooperation – in the sense of altruism, as indicated in the context

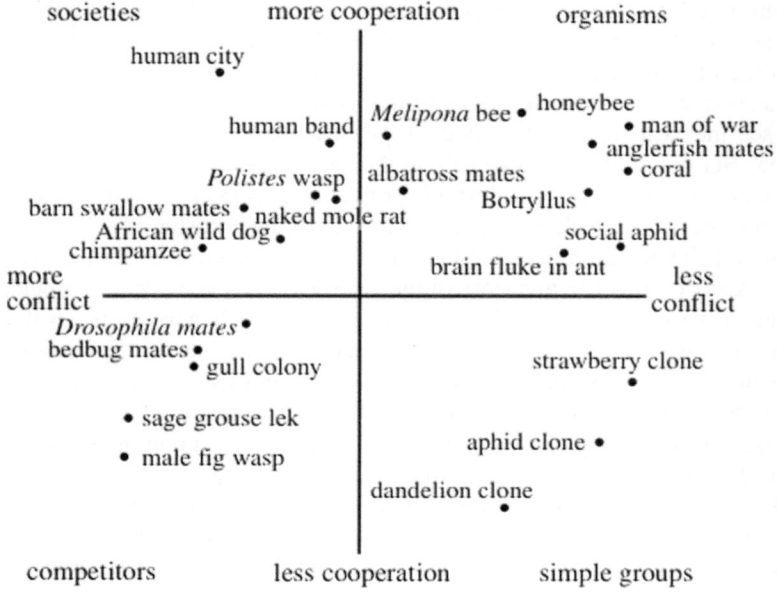

Fig. 7.4 The space of biological individuality, according to Queller and Strassmann (2009)

of multilevel selection – but also a loss of potentiality for conflict. They claim that these are two different things, even though both of them are defined in relation to natural selection, which characterizes this account as a Darwinian account of individuality. Therefore, systems should be situated within a general conceptual hyperspace of organismality – but I would say "individuality" – whose dimensions are the degree of cooperation and the degree of absence of conflict. In honey bees, for instance, there is lots of cooperation, as in corals, but the system of repressing possible alternative queens in bee colonies allows them to have far less conflict than in corals. This means that it is not always possible to say that a system is more an individual than another one – everything depends upon the dimension (decreasing conflict/degree of cooperation) that one favors.

For this reason, I consider the Darwinian approach as the most liberal: not only "being an individual" and then "being an organism" is not a question for which necessary and sufficient conditions (NSC) should be given; but even in the space of individuality, there is some liberality in the dimension supposed to be the most relevant. I call it the conceptual space approach and it philosophically differs from the Kantian approach consisting in building the concept of the organism, thereby setting criteria for being an organism.

Additionally, in this view, there is no requirement for genetic homogeneity or species homogeneity – associations between different species such as aphid and *Buchnera*, or in general host with symbionts, but also ant-plants or kinds of multicellular organisms made up of distinct species can form individual in evolutionary time.

I proposed that the component vs complete transition should be supplemented with another distinction due to Queller (1997) in order to make complete sense of the space of individuality. In effect, most of the transitions in individuality that I talked about are what Queller called "fraternal transitions": the entities that tend to coalesce into a higher-level individual are genetically similar or close or highly related. But many of the individuals in the space of individuality are made up of genetically heterogeneous entities: think of the lichens made up of fungi and algae. And most deeply in evolutionary history, we have the ancestor of eukaryotic cells, supposed to be the result of encapsulation of an archaea into a bacterium (Margulis, 1970). A story similar to this story of the emergence of the nucleus of a eukaryotic cell has also been told (also by Margulis) about the mitochondria, which is the result of the integration within a eukaryote of a smaller prokaryote, through endosymbiosis. In these transitions, the result is an autonomous individual; the components lose their individuality, not only because they don't reproduce by themselves but also because they lose many of their genes since the functions supported by these genes can be done through genes of the host individual, and reciprocally.

There is no definite criterion for being a biological individual, but mostly dimensions in an abstract space, and then the characterization of elements of this space according to Table 7.1. Given that transitions can be egalitarian or fraternal and then can be ranged across a gradient that goes from poorly component transition to complete transition, we have four extreme cases for transitions, summarized in Table 7.1. Importantly, given that most organisms are made of cells but also of many symbionts that constitute their microbiota, the egalitarian transitions are all over the place. As a result, an element in the space of individuality can be understood according to the following:

– How it scores on each of the two dimensions (lack of conflicts, cooperation).
– How much complete it is – and here, the measure of the "completeness" is given by the norm of the vector (**OA**) where point A is the putative individual with measures x and y on each axis of the space, and O is the origin (the norm U is computed in the ordinary scalar way, $U^2 = x^2 + y^2$): the largest is this norm, the more complete is the transition.
– To what extent it is egalitarian or fraternal.

Table 7.1 Four types of transition (a quadripartition that structures the space of biological individuality)

The four kinds of transitions		
	Complete transition	Component transition
Fruternal	Transition toward multicellular organisms	Colony of *Melipona* bees (high level of potential conflict makes them different from organisms; see Queller & Strassmann, 2009); *Bacillus subtilis* bacteria
Egulitarian	Transition toward eukaryotic cells (mitochondria as symbionts); Termite mounds by Macrotermes (Turner, 2000); Lichens	Some fig-pollinator wasp mutualisms

This conceptual space approach is clearly much more liberal than the Kantian approach; it requires one to be pluralist regarding the sense of individuality and therefore to give up the hope to capture what an organism is. Granted, organisms are in the space of individuality; one can require that they have a high degree of completeness in transition, but this leaves lots of room for many possibilities for organismal structures, features, and functions, as I will consider in the next section.

7.3.3 Ecosystems, Individuals, and Organisms

Additionally, taking into account egalitarian transitions raises a complex issue, namely, the individuality of ecosystems. Ecosystems are made up of many individuals of many species and include the overall abiotic element. They are generally not seen as units of selection, given that they don't display (obvious) heritability; thus, they would hardly respond to selection. To this extent, they could not pretend to be individuals in the Darwinian liberal view.

However, ecologists still often think that some ecosystems are more individual than others. The intuition behind this idea is that while some ecosystems are a loose assortment of species, whose unity is in the eye of the beholder, others are quite cohesive sets of entities likely to persist in time.[11] As Evelyn Hutchinson – a key figure in modern ecology – tended to say, these communities or ecosystems show much stronger interactions within themselves than with others, and that's why they are ontologically more robust.[12]

I gave a formal characterization of the individuality that such ecosystems feature, called "weak individuality."[13][d] But my only point here is to show that these ecosystem individuals may enter the space of individuality, even though they are not obviously part of them since concerning them selection cannot be appealed to. I suggest that we have here a local instance of ecological individuals that appears as a limit of evolutionary biological individuals when the degree of egalitarianism of the transition is extremely higher than the coefficient of "fraternity."

But following the indications I gave while discussing "weak individuality" (Huneman, 2020), one can sketch another conceptual space, proper to ecological individuality. The axes then would be the relative strength of the major interactions

---

[11] This view was held by the very influential treatise *Principles of Animal Ecology*, published in 1949 by prominent ecologists Clyde Allee, Thomas Park, Orlando Park, Alfred Emerson, and Karl Schmidt. They thought that a selection at the level of the group of species fosters the unity of an ecological community exactly as natural selection fosters the unity and individuality of organisms. This view faded away in the 1950s with the emergence of behavioral ecology, which mostly relies on natural selection, (for instance Lack, 1954) and then with the devastating critique of group selection by George Williams (1966).

[12] See Hutchinson (1957). On the problematic ontological character of ecological communities, see Sterelny (2006).

[13][d] I developed a conception of weak individuality to make sense of these accounts of individuality (Huneman, 2014b, c, 2020).

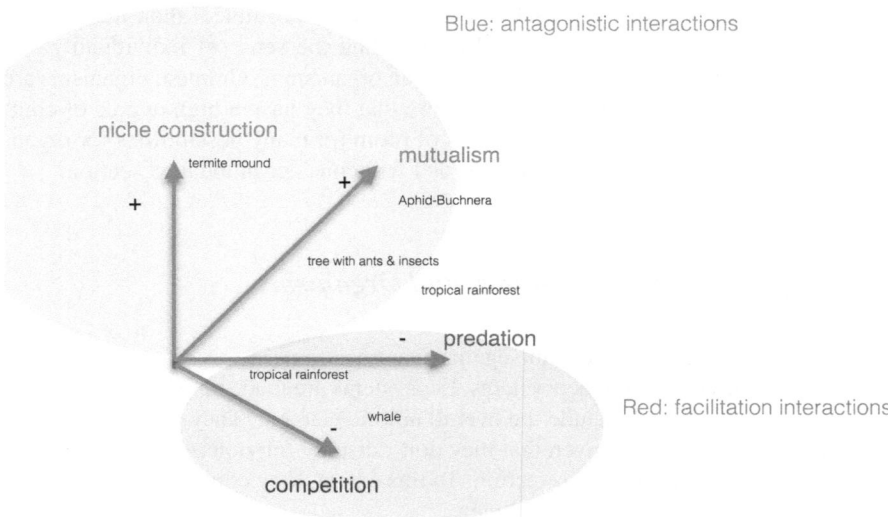

Fig. 7.5 The space of ecosystemic individuality. Each axis represents a structuring interaction. The situation of a given ecosystem depends upon the relative strengths of the interactions. (After Huneman, 2020)

that hold together an ecosystem: competition, mutualism, parasitism, predation, and niche construction.

This "space of ecosystemality," so to say, pertains to the same conceptual space approach as the space of individuality. It is not wholly orthogonal to it, since ecological interactions are the ground of the selective pressures, which in turn make up the selective force, which drives the constitution of evolutionary individuals. In Fig. 7.5, I present a version of this space of ecosystemality. My only concern here is in showing that the conceptual space approach *generaliter* allows one to conceptualize quite extensively issues related to individuality.

7.4 Confronting the Approaches

Is it possible to go further in confronting these two approaches?

To summarize, we have two distinct approaches toward organisms and individuals in general:

(a) In the wake of Kant's *Critique of Judgment*, the necessary and sufficient conditions (NCS) approach provides *the two criteria for organisms*: design and epigeneticism.

In a non-Kantian context, the issue is are they unified? What could be the unification principle? Especially, what happened if only one criterion is fulfilled? As I wrote, the transcendental perspective implied that the idea of "natural purpose" instantiated by organisms is unified. But outside this perspective, things change. Especially, the two criteria can be fulfilled independently; whereas in the Kantian view, once something satisfies the epigeneticism criterion, an "idea of the whole" is presupposed by the biologist as what guides the epigenetic process, and therefore this idea of the whole is also the design of the system. But independently of Kant, one may find things that satisfy the design criterion – design with no self-production (artifacts) – and things that satisfy only the epigeneticism criterion, especially many of the systems investigated by the so-called science of complex systems, starting with the iconic Bénard convection cells, which are not at all alive and don't include functional parts.[14]

(b) On the other hand, we have a liberal attitude: the conceptual space approach (CSA). Here, *organisms inhabit a space of individuality*. The axes are defined by cooperation and by lack of conflict, and the transitions feature several degrees of completeness.

This approach meets its proper issues: first, *how to define "degrees" of individuality in the absence of total order*, assuming that the scalar norm is a too-rough measure? Another issue concerns the axes: are they the only ones? And what is the relation with the space of ecosystematicity addressed in Sect. 7.3.3? And finally, given that some modeling of organisms appeal to ecological concepts, by seeing processes in terms of predation and competition rather than execution of a genetic program (e.g., Costello et al., 2012), or sometimes reintroducing ecological concepts such as niche (Scadden, 2006), would it be possible to think of the organism as ecosystems first, before being something else (I investigated the plausibility of this proposition in Huneman (2020))?

One may be dissatisfied with having two accounts of organisms, distinct but with obvious overlaps. Granted, an option could consist in saying the following: organisms are one thing; they pertain to several biological investigations often lumped under the label "functional biology" (sensu Mayr); and they are integrated, develop, and feature adaptations, but may not necessarily be under natural selection. On the other hand, evolutionary biology handles entities that could be counted, so that the concept of fitness can be instantiated, since fitness is a mathematical construct based on a probability distribution over offspring numbers.

Thus, a reasonable pluralism could say that there are two concepts of organisms: the direct concept in functional biology and the evolutionary concept according to which organisms represent an important kind of individuals, and then individuals are thought of in evolutionary terms.

---

[14] Trying to recover from within the set of self-organizing system such as Bénard cells, candle flames or whirlpool, the subset of things that are alive and therefore that feature functions, and then would be in Kantian terms fulfilling the design criterion: this endeavor has been undertaken by Mossio, Moreno and Saborido in a set of papers and by Mossio and Moreno (2015) Their notion of "closure of constraints "would play the role of what unifies the two criteria, in the present perspective.

This reasonably pluralist option is not dismissing the major claim I want to make in this paper, namely, the difference between an NSC approach and an approach via conceptual space (CSA), as I have exposed them. There is a principled distinction between these two approaches, and I tried to show their respective justifications. And the NSC and the CSA approach could be let as it is: they would coexist as two distinct approaches, each favored by one theoretical school, evolutionists being massively interested in the CSA. This strategy fits the view of Godfrey-Smith (2013), who tends to see organisms and individuals as two conceptual elements proper to two distinct explanatory projects, which can sometimes overlap. But the two approaches do not exactly match with, respectively, a developmentalist concept of the organism used in functional biology and an evolutionary-based concept of individual, according to the reasonable pluralism just sketched. Why? Because the Kantian concept is in itself divided between a developmentalist and an adaptationist understanding.[15]

Thus, how can we articulate the NCS approach of the organism and the CSA liberal evolutionary approach to individuality? I will sketch two strategies successively. I label the first one the threshold strategy and the second one the pragmatic strategy.

7.4.1 Threshold Strategy

According to the threshold strategy, one has to specify a boundary (in terms of a scalar norm) above which X in the space of individuality is a genuine organism. This gives way to articulate individuals and organisms, organisms being a proper subspace of the space of individuality (Fig. 7.5). But how to justify the values of the thresholds? That is the main issue with this otherwise attractive approach. It is hard to do it without some arbitrariness, for instance, by saying that this individual (quaking aspen) is an individual but this other one (ant colony? Dandelion field?) is not an organism. Or, if one wants to avoid being arbitrary, one should provide a concept of organism – which implies an obvious case of circularity, since the whole point here is about determining organisms within the space of individuality (and not extrinsically) in order to make it correspond to the NSC approach to the organism.

But even though we accept arbitrariness or circularity, there is a more pressing issue with this approach. Consider the epigeneticism criterion, which characterizes organisms as natural purpose (*qua* natural) according to the Kantian approach: "Thus, concerning a body that has to be judged as a natural purpose in itself and according to its internal possibility, it is required that the parts of it produce themselves [*hervorbringen*] together" (CJ § 65). We said that this criterion easily fits the development of multicellular organisms. However, in the CSA liberal approach, we

---

[15] See Jaeger, (this volume), which suggests another route toward this problem.

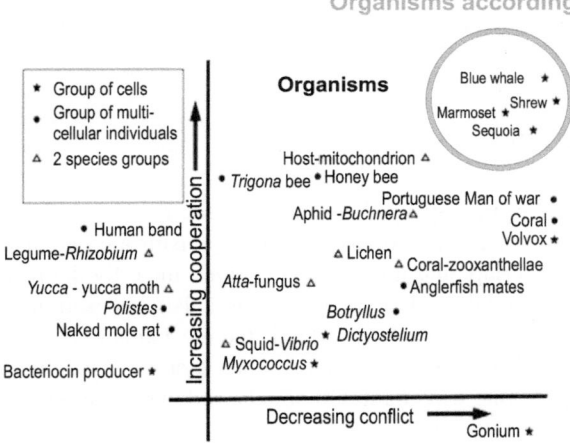

Fig. 7.6 The threshold strategy: organisms according to NCS approach correspond to a subspace in the CSA approach, delimited by a threshold

included products of the "egalitarian" transitions: here, there are heterospecific individuals; along evolution, different species coevolved and concurred in producing such individuals; and along development, a given individual accepts and recruits bacteria (see Nyholm and McFall-Ngai (2004) on the *Vibrio fischeri* bobtail squid with its luminescent bacteria). But this recruitment is not production, at least in the way cells produce new cells. Thus, even within a "zone" of the space of individuality supposed to fit "organisms," the epigeneticism criterion fails, so the two approaches don't match (Fig. 7.6).

An answer to the objection would consist in deflating the sense of "production" in the criterion. Hence, any kind of causation, for instance, counterfactual causation, would suffice – and since bacteria recruited in an organism, for instance, in the bobtail squid, are counterfactually cause of the form of the squid – in the sense that if the bacteria were different (its species being different from the set of species recruited by the squid), they wouldn't end up in the squid –, this case fulfills the epigeneticism criterion.

Ecologists talk of "facilitation" when species A increasing in abundance makes another species B increase in abundance, for instance, by eating more of the predators of B (Bruno et al., 2003). Thus, if causation means something like facilitation, heterospecific individuals could fulfill the epigeneticism criterion. But the drawback is significant: one loses the distinction between the two criteria of organism in Kant's sense since both of them are now about "causing" (in a nonproductive sense) the form of the parts! Therefore, the design criterion and epigeneticism criterion collapse, and the whole approach loses its benefits.

Thus, whether one accepts or rejects this conceptual alternative, namely, weakening the epigeneticism criterion, the threshold strategy raises massive issues.

7.4.2   Pragmatic Strategy

If now we turn to the pragmatic strategy, what are the prospects?

As a strategy, pragmatism means that individuals and organisms are two distinct concepts, whose extensions overlap, and this overlap is determined by the respective uses of these concepts in models and theories. So, "organism" is used in some disciplines and allows to ask questions such as "what is the mechanism of this life-sustaining function?", but not in others: for instance, when there is a specific process of natural selection at play, one will count individuals, and in case of multilevel selection, models will include two kinds of individuals (e.g., cells/organisms or organisms/herds) rather than two kinds of organisms. Some philosophers came to contest the legitimacy of "organisms" in individuality biology, arguing that it is enough to talk about "individuals" (Haber, 2013; Bouchard, 2008); pragmatists will reject this option and keep the duality of concepts while accepting that in many contexts, "individuals" is the only useful and legitimate concept.

This strategy accounts for the differences between individuals and organisms; it makes sense of the fact that there is no point in inferring from "organisms" traits likely to define individuals in general, or that "organisms" cannot be a subclass of "individuals" since what matters is the context of use of these concepts rather than their purported rigid reference.

However, one major issue is the lack of complete separation between these two concepts: as I said, the evolutionary models account for some aspects of the design criterion (see also Huneman, 2017). Thus, the specificity of organisms with respect to biological entities is to some extent acknowledged within the concept if individuality, defined from the liberal viewpoint of evolution.

A pragmatist strategy assumes that the concepts are independent, in terms of their meaning. It happens that they coincide, in the sense that one concept picks out in some contexts the same thing as the other concept employed in other contexts – even though they are different. The strength of a pragmatist strategy is that the two concepts, as I tried to show, are very different, to the extent that one is defined in terms of NSC and the other comes from a liberal CSA approach.

But since there is an overlap in their signification, to the extent that the design criterion belongs to the organism concept but can be accounted for in the CSA in the context of a discourse about biological individuality, the pragmatist's strategy may not be perfectly legitimate. One may discuss whether the conditions of applying a pragmatist strategy are met in general. Yet, if the two frameworks, NSC and NCA, are not independent, then they are competing when it comes to accounting for several classes of biological realities; and therefore, they cannot be handled through a pragmatist strategy.

7.5   Conclusion

In this paper, I have reviewed two approaches to the concept of organisms. In the former one, organisms are understood through a "necessary and sufficient conditions" approach, via two criteria inherited from the Kantian approach as an early investigation in the ontology of organization (itself embedded in the birth of descriptive embryology (Lenoir, 1982, Huneman, 2007; forth)), an investigation that the present volume continues. The latter derives from an ontology grounded on evolutionary models and ideas. I insisted on the difference in the logical structure of the two approaches. A conceptual space allows for much more liberality, while the NCS approach is precise in the sense that it supposedly picks up the "organisms" and nothing else. The conceptual space approach does not provide us with an idea of what are organisms and therefore cannot provide identity conditions or truth-makers of the sentence "X is an organism." It is therefore much more deceptive from an ontological or metaphysical viewpoint. But it may fit some scientific practices, within which sharp boundaries between extensions of concepts and the rest are often hard to find, given the model-relativity of many of the propositions uttered by scientists. Its liberality may be appreciable in other cases; therefore, the distinction I found here between Kantian and evolutionary approaches, namely, a distinction between NCS and CSA, may be relevant for metaphilosophy and helps address other, unrelated, issues.

Clearly, organisms are at the crossroads of several hierarchies of individuals and ways to talk about individuals in biology; Eldredge's two hierarchies here are indicative of the multiplicity of takes on organisms and, therefore, of the sense in which "organism" can be seen as a crossroad concept. "Design" as a concept making sense of this organization proper to organisms is torn between a selectionist understanding in terms of bundles of adaptations and ultimately natural selection and functions in an etiological sense (Millikan, 1984, Neander, 1991) – and a concept of an organization that anchors organization in self-organizing processes rather than external selective pressures.

As a consequence, the plurality of conceptual schemes to make sense of organisms will probably cohabit in biology, and the duality between functional biology and evolutionary biology is not enough to distinguish these schemes – given that design is an idea proper to both approaches.

In the last step, I tried to propose some ways the two approaches – the NCS Kantian one and the evolutionary liberal one – can mutually relate. The result is a rather mild skepticism (to be aptly contrasted with Jaeger's chapter, in this volume). I don't think the concept of the organism, given its uses in biology, can be the place where a pacific coexistence between accounts can take place. The skepticism of this paper tends toward an acceptance of the conflictual and fractured essence of the concept of the organism in biology.

References

Abrams, M. (2007). How do natural selection and random drift interact? *Philosophy of Science, 74*(5), 666–679.

Allee, W. C., Park, O., Emerson, A. E., Park, T., & Schmidt, K. P. (1949). *Principles of animal ecology*. W. B. Saunders Company.

Amundson, R. (2005). *The changing role of the embryo in evolutionary thought*. Cambridge University Press.

Barabasi, A. L. (2018). *Network science*. Cambridge University Press.

Bateson, P. (2005). The return of the whole organism. *Journal of Biosciences, 30*, 31–39.

Birch, J. (2015). Natural selection and the maximization of fitness. *Biological Reviews, 91*(3), 712–727.

Bouchard, F. (2008). Causal processes, fitness, and the differential persistence of lineages. *Philosophy of Science, 75*(5), 560–570.

Bruno, J., Stachowicz, J., & Bertness, M. (2003). Inclusion of facilitation into ecological theory. *Trends in Ecology & Evolution., 18*, 119–125.

Burt, A., & Trivers, R. (2006). *Genes in conflict the biology of selfish genetic elements*. Harvard University Press.

Buss, L. W. (1988). *The evolution of individuality*. Princeton University Press.

Caroll, S. (2005). *Endless forms most beautiful: The new science of evo devo and the making of the animal kingdom*. Norton.

Clarke, E. (2014). The multiple realizability of biological individuals. *Journal of Philosophy, 110*(8), 413–435.

Costello, E., Stagaman, K., Dethlefsen, L., Bohannan, B. J., & Relman, D. A. (2012). The application of ecological theory toward an understanding of the human microbiome. *Science, 336*, 1255–1262.

Damuth, J., & Heisler, L. (1988). Alternative formulations of multi-level selection. *Biology and Philosophy, 3*, 407–430.

Davidson, E. H. (1986). *Gene activity in early development*. Academic.

Davidson, E., McClay, D., & Hood, L. (2003). Regulatory gene networks and the properties of the developmental process. *PNAS, 100*, 1475–1480.

Dawkins, R. (1976). *The selfish gene*. Oxford University Press.

Eldredge, N. (1985). *Unfinished synthesis: Biological hierarchies and modern evolutionary thought*. Oxford University Press.

Featherston, J., & Durand, P. M. (2012). Cooperation and conflict in cancer: An evolutionary perspective. *South African Journal Science, 108*(9/10), 1–5.

Folse, H. J., III, & Roughgarden, J. (2010). What is an individual organism? A multilevel selection perspective. *Quarterly Review of Biology, 85*, 447–472.

Gayon, J., & Petit, V. (2019). *The knowledge of life today*. ISTES.

Ghiselin, M. T. (1974). A radical solution to the species problem. *Systematic Zoology, 23*, 536–544.

Gilbert, S. F., & Sarkar, S. (2000). Embracing complexity: Organicism for the 21st century. *Developmental Dynamics, 219*(1), 1–9.

Gilbert, S., Opitz, G., & Raff, R. (1996). Resynthesizing evolutionary and developmental biology. *Development and Evolution, 173*, 357–372.

Ginsborg, H. (2004). Two kinds of mechanical inexplicability in Kant and Aristotle. *Journal of the History of Philosophy, 42*(1), 33–65.

Ginsborg, H. (2014). Oughts without intentions: A Kantian approach to biological functions. In E. Watkins & I. Goy (Eds.), *Kant's theory of biology* (pp. 259–274). De Gruyter.

Godfrey-Smith, P. (2009). *Darwinian populations and natural selection*. Oxford University Press.

Godfrey-Smith, P. (2013). Organisms and individuals. In F. Bouchard & P. Huneman (Eds.), *From groups to individuals*. MIT Press.

Goodnight, C. J. (2013). Defining the individual. In F. Bouchard & P. Huneman (Eds.), *From groups to individuals* (pp. 37–54). MIT Press.

Gould, S. J. (2002). *The structure of the evolutionary*. Belknap Press.

Gould, S. J., & Lewontin, R. (1978). The spandrels of San Marco and the pangloss- ian paradigm: A critique of the adaptationist programme. *Proceedings of the Royal Society of London, B205*(1161), 581–598.

Green, S. (2013). When one model is not enough: Combining epistemic tools in systems biology. *Studies in History and Philosophy of Biological and Biomedical Sciences, 44*(2), 170–180.

Griffiths, P., & Gray, R. (1994). Developmental systems and evolutionary explanation. *Journal of Philosophy, 91*, 277–304.

Grodwohl, J-B. (2019). Animal behavior, population biology and the modern synthesis. *Journal of the History of Biology 52* (4):597–633.

Haber, M. (2013). Colonies are individuals: Revisiting the superorganism revival. In F. Bouchard & P. Huneman (Eds.), *From groups to individuals* (pp. 196–217). MIT Press.

Hamilton, W. D. (1964). The genetical evolution of social behaviour I and II. *Journal of Theoretical Biology, 7*(1), 1–52.

Hull, D. L. (1980). Individuality and selection. *Annual Review of Ecology and Systematics, 11*, 311–332.

Huneman, P. (2007). Reflexive judgement and wolffian embryology : Kant's shift between the first and the third critique. In P. Huneman (Ed.), *Understanding purpose? Kant and the philosophy of biology* (Vol. 2007, pp. 75–100). University of Rochester Press.

Huneman, P. (2008). *Métaphysique et biologie*. Kimé.

Huneman, P. (2010). Assessing the prospects for a return of organisms in evolutionary biology. *History and Philosophy of the Life Sciences, 32*, 341–372.

Huneman, P. (2013). Adaptation in transitions. In F. Bouchard & P. Huneman (Eds.), *From groups to individuals* (pp. 141–172). MIT Press.

Huneman, P. (2014a). Individuality as a theoretical scheme 1. Formal and material concepts of individuality. *Biological Theory, 9*(4), 361–337.

Huneman, P. (2014b). Individuality as a theoretical scheme 2. About the weak individuality of organisms and ecosystems. *Biological Theory, 9*(4), 374–381.

Huneman, P. (2014c). A pluralist framework to address challenges to the modern synthesis in evolutionary theory. *Biological Theory, 9*(2), 163–177.

Huneman, P. (2017). Kant's concept of organism revisited: A framework for a possible synthesis between developmentalism and adaptationism? *The Monist, 100*(3), 373–390.

Huneman, P. (2019a). How the modern synthesis came to ecology. *Journal of the History of Biology, 52*, 635–686.

Huneman, P. (2019b). Revisiting darwinian teleology: A case for inclusive fitness as design explanation. *Dies in History and Philosophy of Science Part C: Studies in History and Philosophy of Biological and Biomedical Sciences, 76*, 101188.

Huneman, P. (2020). Biological individuality as weak individuality: A tentative study in the metaphysics of science. In A. S. Meincke & J. Dupré (Eds.), *Biological identity* (pp. 40–62). Routledge.

Huneman, P. (2023). *Death: Perspectives from the philosophy of the biology*. Palgrave-McMillan.

Hutchinson, G. E. (1957). Concluding remarks. Cold Spring Harbor symposium. *Quantitative Biology, 22*, 415–427.

Jaeger, J. (this volume). The fourth perspective: Evolution and organismal agency. In M. Mossio (Ed.), *Organization in biology*. Springer.

Kauffman, S. (1993). *The origins of order: Self-organization and selection in evolution*. Oxford University Press.

Kerr, J. F., Wyllie, A. H., & Currie, A. R. (1972). Apoptosis: A basic biological phenomenon with wide-ranging implications in tissue kinetics. *British Journal of Cancer, 26*(4): 239–257.

Kitano, M. (2002). *Foundations of systems biology*. MIT Press.

Krebs, J. R., & Davies, N. (1995). *Behavioral ecology: An evolutionary approach*. Blackwell.

Lack, D. (1954). *The natural regulation of animal numbers*. Oxford University Press.

Larson, J. L. (1979). Vital forces: Regulative principles or constitutive agents? A strategy in German physiology, 1786–1802. *Isis, 70*, 235–249.

Lenoir, T. (1982). *The strategy of life. Teleology and mechanism in nineteenth century German biology*. Reidel.

Lewens, T. (2004). *Organisms and artifacts: Design in Nature and Elsewhere*. MIT Press.

Lewontin, R. C. (1970). The units of selection. *Annual Review of Ecology and Systematics, 1*, 1–18.

Margulis, L. (1970). *Origin of eukaryotic cells*. Yale University Press.

Maynard-Smith, J. (1982). *Evolution and the theory of games*. Oxford University Press.

Maynard-Smith, J., & Szathmáry, E. (1995). *The major transitions in evolution*. Oxford University Press.

Mayr, E. (1961). Cause and effect in biology. *Science, 134*, 1501–1506.

McLaughlin, P. (2000). *What functions explain: Functional explanation and self-reproducing systems*. Cambridge University Press.

Michod, R. (1999). *Darwinian dynamics*. Oxford University Press.

Michod, R. E. (2005). On the transfer of fitness from the cell to the multicellular organism. *Biology and Philosophy, 20*, 967–987.

Millikan, R. G. (1984). *Language, thought, and other biological categories*. MIT Press.

Montévil, M., & Mossio, M. (2015). Biological organisation as closure of constraints. *Journal of Theoretical Biology, 372*, 179–191.

Mossio, M., & Moreno, A. (2015). *Biological Autonomy*. Springer.

Müller, M., & Hallgrimson, E. (2003). Preformationism. In O. Müller (Ed.), *Keywords and concepts in evolutionary developmental biology*. MIT Press.

Neander, K. (1991). The teleological notion of function. *Australasian Journal of Philosophy, 69*, 454–468.

Newman, M. (2010). *Networks: An introduction*. Oxford University Press.

Nyholm, S. V., & McFall-Ngai, M. J. (2004). The winnowing: Establishing the squid-*vibrio* symbiosis. *Nature Reviews. Microbiology, 2*, 632–642.

Odling-Smee, J., Laland, K., & Feldman, M. (2003). *Niche construction: The neglected process in evolution*. Princeton University Press.

Okasha, S. (2006). *Evolution and the levels of selection*. Oxford University Press.

Okasha, S. (2018). *Agents and goals in evolution*. Oxford University Press.

Oliveri, P., Tu, Q., & Davidson, E. (2008). Global regulatory logic for specification of an embryonic cell lineage. *Proceedings of the National Academy of Sciences, 105*(16), 5955–5962.

Oyama, S., Griffiths, P., & Gray, R. (Eds.). (2001). *Cycles of contingency: Developmental systems and evolution*. MIT Press.

Plutynski, A. (2007). Drift: A historical and conceptual overview. *Biological Theory, 2*(2), 156–167.

Queller, D. C. (1997). Cooperators since life began. *The Quarterly Review of Biology, 72*, 184–188.

Queller, D. C., & Strassmann, J. E. (2009). Beyond society: The evolution of organismality. *Philosophical Transactions of the Royal Society London Biological Sciences, 364*, 3143–3155.

Raff, R. (1996). *The shape of life*. University of Chicago Press.

Reeve, H. K., & Sherman, P. W. (1993). Adaptation and the Goals of Evolutionary Research. The Quarterly Review of Biology, 68(1), 1–32.

Richards, R. (2001). *The romantic conception of life*. University of Chicago Press.

Ruelle, D. (1989). *Chance and chaos*.

Russell, E. S. (1911). *Form and function*. Cambridge University Press.

Saborido, C., Mossio, M., & Moreno, A. (2011). Biological organization and cross-generation functions. *British Journal for the Philosophy of Science, 62*(3), 583–606.

Scadden, D. (2006). The stem-cell niche as an entity of action. *Nature, 441*(7097), 1075–1079.

Sloan, P. R. (2002). Performing the categories: Eighteenth-century generation theory and the biological roots of Kant's a priori. *Journal of the History of Philosophy*, 40(2): 229–253.

Sober, E. (1984). *The nature of selection*. MIT Press.

Sober, E., & Wilson, D. S. (1998). *Unto others. The evolution and psychology of unselfish behavior*. Harvard University Press.

Sterelny, K. (2006). Local ecological communities. *Philosophy of Science*, 73(2): 215–231.

Strassmann, J. E., & Queller, D. C. (2010). The social organism: Congresses, parties, and commit-
tees. *Evolution, 64*, 605–616.
Sultan, S. (2015). *Organism and environment: Ecological development, niche construction, and
adaptation*. Oxford University Press.
Turner, D. (2000). The functions of fossils: Inference and explanation in functional morphology.
*Studies in History and Philosophy of Science Part C: Studies in History and Philosophy of
Biological and Biomedical Sciences, 31*(1), 193–212.
Walsh, D. M. (2015). *Organisms, agency, and evolution*. Cambridge University Press.
Walsh, D. (2017). Chance caught on a wing. In P. Huneman & D. Walsh (Eds.), *Challenging the
modern synthesis: Adaptation, development and inheritance*. Oxford University Press.
West-Eberhard, M.-J. (2003). *Developmental plasticity and evolution*. Oxford University Press.
Williams, G. C. (1966). *Adaptation and natural selection*. Princeton University Press.
Wolpert, L. (1969). Positional information and the spatial pattern of cellular differentiation.
Journal of Theoretical Biology, 25, 1–47.
Zammito, J. (2018). *The gestation of German biology*. University of Chicago Press.
Zumbach, C. (1984). *The transcendant science: Kant's conception of biological methodology*.
Martinus Nijhoff.

Chapter 8
The Fourth Perspective: Evolution and Organismal Agency

Johannes Jaeger

Abstract This chapter examines the deep connections between biological organization, agency, and evolution by natural selection. Using Griesemer's account of the reproducer, I argue that the basic unit of evolution is not a genetic replicator, but a complex hierarchical life cycle. Understanding the self-maintaining and self-proliferating properties of evolvable reproducers requires an organizational account of ontogenesis and reproduction. This leads us to an extended and disambiguated set of minimal conditions for evolution by natural selection—including revised or new principles of heredity, variation, and ontogenesis. More importantly, the continuous maintenance of biological organization within and across generations implies that all evolvable systems are agents or contain agents among their parts. This means that we ought to take agency seriously—to better understand the concept and its role in explaining biological phenomena—if we aim to obtain an organismic theory of evolution in the original spirit of Darwin's struggle for existence. This kind of understanding must rely on an agential perspective on evolution, complementing and succeeding existing structural, functional, and processual approaches. I sketch a tentative outline of such an agential perspective and present a survey of methodological and conceptual challenges that will have to be overcome if we are to properly implement it.

8.1 Introduction

There are two fundamentally different ways to interpret Darwinian evolutionary theory. Charles Darwin's original framework grounds the process of evolution on the individual's struggle for existence (Darwin, 1859). It is a theory centered around the organism. The neo-Darwinian interpretation of the modern synthesis, in contrast, sees evolution grounded in the shift of allele frequencies in populations. It

J. Jaeger (✉)
Department of Philosophy, University of Vienna, Vienna, Austria

Complexity Science Hub (CSH), Vienna, Austria

© The Author(s) 2024
M. Mossio (ed.), *Organization in Biology*, History, Philosophy and Theory of the Life Sciences 33, https://doi.org/10.1007/978-3-031-38968-9_8

completely brackets out the organism, focusing on the lower level of the gene and the higher level of the population instead (see, e.g., Walsh 2015; or Amundson, 2005, for a historical perspective). This reductionist approach provided much needed clarity for the study of evolutionary phenomena in the early twentieth century. But it hardly does justice to *the complexity of causes underlying evolutionary change* which—through Darwin's struggle for existence—may involve nontrivial contributions of organismic behavior.

Many researchers in the field are aware of this limitation and are trying to move beyond it. The principal aim of evolutionary developmental biology (evo-devo), for example, can be construed as providing causal-mechanistic explanations for the evolution of the complex regulatory processes involved in development (Wagner et al., 2000; Calcott, 2009; Brigandt, 2015; DiFrisco & Jaeger, 2019; DiFrisco et al.,2020). The limits of reductionism have also come to the attention of philosophers of biology, and there is much interesting work on the subject (some of which will be discussed here). Unfortunately, progress toward an organismic evolutionary biology remains slow, in part because of the daunting intricacy of the matter, in part because of the lamentable and still widespread identification of "mechanism" with explanations at the molecular level (Nicholson, 2012), but also because many criticisms of reductionism in evolutionary biology remain wide of the mark, failing to properly engage the problem of organismic complexity in a philosophically grounded manner.

One particularly prominent example of this problem is a recent talk about understanding the "causal structure of evolution" by addressing the role of "constructive development" and "causal reciprocity" in the context of an "extended evolutionary synthesis" (e.g., Laland et al.,2015). *Constructive development*—defined as the ability of the organism to shape its own ontogenetic trajectory—implies some kind of agency, leading to open-ended exploratory evolution, in ways which are never clearly defined. *Causal reciprocity* emphasizes the mutual influence between ontogeny and phylogeny, or an evolving population and its environment. It is claimed to be widespread and to violate Ernst Mayr's (1961) classical distinction between proximate and ultimate explanations in evolution. One problem is that such claims are hardly original. The constructive role of the organism in evolution goes straight back to Darwin himself (see, e.g., Amundson, 2005, or Walsh, 2015), and reciprocal causation is an integral part of many models in classic evolutionary genetics (Svensson, 2018; Buskell, 2019).

However, there is a more serious problem: such shallow theorizing does not even scratch the surface of the causal complexity underlying evolution. It is not wrong. It even goes in the right direction. But it does not go far enough. If we are serious about investigating the complex causes governing evolutionary change, *we must tackle issues such as organismal agency and the fundamentally dialectic nature of evolutionary causation head on.* We must call these problems by their name without avoiding the uncomfortably radical conclusions that might spring from their examination. This is what I am trying to do here.

Tackling the causal complexity underlying the evolutionary struggle for existence is no task for the faint-hearted. The causal structure of evolution is profoundly impenetrable. In fact, I believe that the whole undertaking is completely hopeless, unless it is informed by an adequate ontology and epistemology, specifically developed for the task. Luckily, such a foundation is available in the form of William Wimsatt's *perspectival realism* (Wimsatt, 2007). It properly reckons with the limited nature of our cognitive abilities and the utterly byzantine character of reality. As an added bonus, it takes a differentiated view on the role of reductionism as an epistemic tool in biology and is antithetical to any quixotic quest for a grand synthesis of evolutionary thought.

At the heart of Wimsatt's ontology lies the recognition that the causal structure of the world resembles a rich and dynamic tropical-rainforest ecosystem rather than the eliminativist desert suggested by traditional ontological reductionism (Wimsatt, 1994, 2007). In this lush ontological forest, there are areas that exhibit cleanly separated levels of organization, defined as "local maxima of regularity and predictability in the phase space of alternative modes of the organization of matter" (Wimsatt, 2007, p. 209). Examples are the subatomic, atomic, and molecular levels studied by physics and chemistry. In other areas of reality, however, this compositional hierarchy breaks down into more localized and less resolved causal structures, captured by *perspectives*—defined as "intriguingly quasi-subjective (or at least observer, technique or technology-relative) cuts on the phenomena characteristic of a system" (Wimsatt, 2007, p. 222). Ultimately, even perspectives break down, resulting in *causal thickets*, which are hard to disentangle since they lack any discernible regularity or layering. The causal structure underlying the process of organismic evolution is a perfect example of such an impassable thicket.

Organizational levels and causal thickets require different epistemic strategies. In particular, reductionist methods can be useful, but remain fundamentally limited in the context of *the evolutionary causal thicket* (Wimsatt, 2007). What is needed to assess, complement, and contextualize them is a perspectival approach that aims to cut through the thicket in alternative ways. This refocuses our attention and our limited resources toward important aspects of evolution that are usually neglected in the standard reductionist account of evolutionary genetics.

At first glance, a multiplicity of limited and biased perspectives seems to constitute an insurmountable obstacle for obtaining robust empirical knowledge. There is no way to "step out of one's own head" to gain a truly objective "view from nowhere" (Giere, 2006; Wimsatt, 2007; Massimi, 2016). Upon closer examination, however, an explicitly perspectival approach enriches scientific inquiry into complex causal thickets in two important ways. First, the higher the diversity of perspectives, the wider the range of questions we can ask, and the larger the variety of approaches we can use to answer those questions. Second, comparative analyses of theoretical perspectives yield valuable insights into their respective applicability and limitations, as well as the robustness and consistency of their claims (Giere, 2006; Griesemer, 2006; Wimsatt, 2007; Massimi, 2016). Put simply, more diverse approaches can lead to broader and more trustworthy insights into complex and entangled processes such as evolution. What we need are more varied and valid perspectives rather than some kind of misguided theoretical synthesis, which is the remnant of an earlier— and by now thoroughly outdated—positivist view of evolutionary biology (Smocovitis, 1996; see also Walsh, 2015).

In this spirit, James Griesemer (2006) suggests a radical change of philosophical focus for evolutionary theory, from selecting the best among competing approaches and generalizing it toward a comparative analysis of the strengths, weaknesses, and complementarities of different local perspectives. These perspectives are not right or wrong, better or worse, per se, but succeed or fail to achieve their specific purpose. Griesemer (2006) distinguishes three kinds of evolutionary perspectives: structural, functional, and processual. To this, I will add *a fourth perspective* here, which emphasizes the agency of evolving organisms. A truly comprehensive science of evolution will have to include all four. Together, they yield more inclusive explanations of relevant evolutionary phenomena than each one of them on their own. In addition, a comparative approach allows us to reveal and assess the abstractions, idealizations, and simplifications that each approach is bound to make. Finally, the robustness of specific claims "can only be assessed if a scientific community pursues phenomena from a variety of perspectives... It is not enough merely to compete." (Griesemer, 2006, p. 363). Is it really that surprising that a field centered on biological diversity would profit from a more diversified epistemic approach?

Furnished with these epistemological tools, we will now embark on a journey that explores the importance of organismic organization and organismic agency for the basic principles underlying evolution by natural selection. This journey starts with an introduction to the central concepts of organizational closure and organizational continuity in Sect. 8.2. I then briefly recall Lewontin's (1970) minimal conditions for natural selection in Sect. 8.3. These conditions by themselves may be simple, but their mapping onto the physical world is incredibly complex. To unravel this complexity, we can take structural, functional, or processual perspectives, as described in Sect. 8.4. In Sect. 8.5, I will focus on Griesemer's (2006) reproducer perspective, a processual view demonstrating that genetic replicators must be deeply embedded in a complex and hierarchical life cycle to be able to multiply themselves. Section 8.6 reviews why an organizational account of reproduction is necessary to understand such life cycles and, at the same time, implies organismal agency and self-determination. For this reason, agency itself becomes a fundamental aspect of systems that are evolvable by natural selection. This is why we need a fourth perspective on evolution. Section 8.7 presents a very preliminary exploration of what such an agential perspective would look like in terms of its mathematical and explanatory structure. I conclude with some general thoughts on what this implies, not only for evolutionary theory but also for scientific explanation in general.

8.2   Organizational Closure and Continuity

In this chapter, I focus on processual and agential perspectives on evolution, which revolve around the distinctive *organization* of living systems and how it is maintained—within and across generations—through *continuous regeneration* (Saborido et al., 2011; Mossio & Pontarotti, 2020; DiFrisco & Mossio, 2020; Pontarotti, this volume). Biological organization, of course, is the unifying topic of this volume, and I refer the reader to its introduction for a general overview (Mossio, this volume; see also Moreno and Mossio (2015) and Toepfer, this volume). In this section, I will only briefly revisit those organizational concepts that are particularly relevant to my argument.

The organizational account is founded on the basic insight that the important difference between life and nonlife is not a difference of composition (what organisms are made of) but a difference in the way that system components relate to each other (how organisms are organized). The central defining feature of biological organization is *organizational closure*, a concept introduced by Jean Piaget (1967), which means that all essential parts of a living system mutually depend on each other, could therefore not exist without each other, and must maintain each other through their collective interactions. Organizational closure is complementary to thermodynamic openness; in fact, it can only occur in far-from-equilibrium systems. It leads to a causal circularity that is already reflected in La Mettrie's metaphor of the living body as "a spring that winds itself." Organisms are closed to efficient causation (Rosen, 1991): their organization is maintained *from within*, even though matter and energy constantly flow through the system. It is in this sense that organisms are self-making and self-producing: they are *autopoietic systems* (Varela et al., 1974; Maturana, 1980).

Causal circularity and closure are necessary but not sufficient to account for the organization of autopoietic systems. Biological organization also requires a dialectic relationship between the physicochemical processes that materially compose an organism and the system-level constraints that act upon them (Montévil & Mossio, 2015; Mossio et al., 2016). *Processes* denote various kinds of transformations (such as chemical reactions or the physical rearrangement of cells and tissues) that involve the generation, constitution, alteration, consumption, and destruction of system components. *Constraints* act on processes but remain unaltered by them (at least at the time scale at which the constrained process occurs). Constraints can be external or internal to the system. They reduce the degrees of freedom of the process on which they act. Their effect is strongly context-dependent. Examples of organismic constraints are enzymes, or the vascular system in vertebrates, which catalyze their metabolic reactions and transport blood without themselves being altered at the time scale of the process they constrain.

Just like any other physicochemical component of a living system, its constraints need to be constantly replaced, repaired, and maintained. Enzymes, for example, decay and must be replenished through the processes involved in protein synthesis. This means that constraints can depend on each other. They are generated by processes on which other constraints are acting. In turn, they can generate other constraints by regulating the processes that produce them. The synthesis of enzymes, for example, depends on ribosomes whose synthesis, in turn, depends on enzymes. If each constitutive constraint in a living system is both dependent on and generative for at least one other constraint, then there is *closure of constraints*, which represents a specific kind of organizational closure (Montévil & Mossio, 2015; Mossio et al., 2016). It means that the constrained overall dynamics of the system determine the conditions for the continued existence of the constraints. In this way, the processes and constraints of a living system logically and materially entail each other. One is required for the existence of the other.

This raises the question of how living processes and constraints co-emerge through their dialectic dynamic interactions. Kauffman (2000) argues that living organization must be powered by *work-constraint cycles*. Incorporating this into the account of Montévil and Mossio (2015), we can say that the constrained release of energy by the organized system provides the physical work required to maintain its existing constraints and to constantly generate new ones. In this way, work-constraint cycles can explain various kinds of self-organization far from equilibrium, but are not yet specific enough or sufficient to account for the emergence, persistence, and propagation of organizational closure in living systems. For this, we need the additional concept of *organizational continuity* (DiFrisco & Mossio, 2020; Mossio & Pontarotti, 2020; Pontarotti, this volume). It means that closure at any particular time dynamically presupposes closure of constraints that have operated earlier (see Bickhard, 2000). Organizational continuity represents a specific type of causal continuity. The key point here is that the particular organization of constraints in an organism not only can but *must* continuously change for it to maintain organizational closure and to continue living (Montévil et al., 2016; see also Nicholson, 2018). It must engage in a process of continuous regeneration (Saborido et al., 2011). Hans Jonas (1966) calls this *needful freedom*—the capacity of living matter to change its form—and the *thermodynamic predicament*, the irremissible necessity for it to do so.

On this view, the organism can be seen as a continuously changing but persistently closed organization of constraints that "lifts itself" out of the thermodynamic background of all possible physicochemical processes (see, e.g., Hofmeyr & Cornish-Bowden, 2000; Kauffman, 2000). It does this through work-constraint cycles that recursively actualize a closure of constraints. What this means is that the organization of constraints at any point in time—the channeling of physicochemical processes in certain directions—arises from *within* the organism itself. Due to the condition of organizational continuity, it is a consequence of previous organismic constraints, of earlier organizational closure. For this reason, we cannot predict its temporal evolution from considerations of far-from-equilibrium thermodynamics (or other physical laws of nature) alone. It is in this sense that the organism has a degree of *autonomy* from its environment (Moreno & Etxeberria, 2005; Moreno & Mossio, 2015). The organism generates its own dynamics of unfolding closed constraints. These constraints determine not only the internal constitution of the system but also its interactions with the environment. Simply put, all organisms possess at least some minimal kind of agency—they "act on their own behalf" (Kauffman, 2000; Moreno & Mossio, 2015, esp. Chap. 4).

In this way, the organizational account grounds the teleological notions of *biological function*, as well as self-determination and *agency* on naturalistic principles that lie perfectly within the scope of scientific explanation (see also Sect. 8.7). Functional constraints in living systems are defined as those that contribute to organizational closure and continuity (Mossio et al., 2009a; Moreno & Mossio, 2015;

Mossio & Pontarotti, 2020). Self-determination arises from the recursive and reflexive diachronic emergence of functional constraints from previous realizations of organizational closure (Mossio & Bich, 2017). Agency is defined as the capacity to internally generate causal effects (actions) that involve interactive functions—those constraints subject to closure which mediate the organism's boundaries and exchanges with its environment (Moreno & Etxeberria, 2005; Barandiaran & Moreno, 2008; Barandiaran et al., 2009; Moreno & Mossio, 2015). Put more simply, the organism selects and initiates the kind of interactions it has with its surroundings. This kind of agency is an observable property of an organism—its ability to cope with a particular situation, to pursue its goals in response to opportunities or obstacles present in its perceived environment (Walsh, 2015).

What is most important to point out here is that *all these teleological properties are a direct and necessary consequence of the fundamental self-maintaining organization of living systems*. Anything that is alive can be legitimately described from the perspective of organismic agency and goal-orientedness (Walsh, 2015).

But if such teleological aspects are fundamental—and unproblematic from the point of naturalistic explanation—why do we constantly attempt to explain them away? Why do we ignore them? Why do we not take them at face value, even though they imply profound and radical challenges for our thinking about biological systems and their evolution? What does this mean for what we consider a scientific explanation? These are the kind of questions that will keep us busy throughout this chapter.

8.3 Minimal Conditions for Darwinian Evolution by Natural Selection

To better understand the close and intricate relationship between organismal organization, agency, and the process of evolution by natural elections, I must briefly review the prerequisites for this type of evolution to occur. Ever since Darwin, biologists and philosophers of biology have sought to stipulate the most concise formulation of necessary and sufficient conditions for evolution by natural selection (reviewed in Godfrey-Smith, 2007, 2009). The shortest one I could find is Sober's "Darwinian general principle": "*if* there is heritable variation in fitness, *then* there will be evolution" (Sober, 1984, p. 28, original emphasis).

The most cited core requirements are those first published in Lewontin (1970) and, slightly revised, in Lewontin (1978), which state that the process of evolution by natural selection is based on three fundamental principles or propositions: (1) *the principle of variation*, there is variation in physiological, ontogenetic, morphological, and behavioral traits between individuals in a population; (2) *the principle of heredity*, this variation is (at least in part) inherited such that offspring resemble their parents; and (3) *the principle of differential fitness*, different phenotypic variants vary in their influence on the rate of survival and reproduction of their bearers in different environments, leading to different numbers of offspring in either immediate or remote generations. While these principles hold, a population will undergo evolution by natural selection.

There are many more elaborate formulations of these conditions, and it has been pointed out that any simple enumeration of core requirements provides more of *a recipe for evolutionary change*, rather than a true summary that encapsulates all cases of evolution by natural selection (Godfrey-Smith, 2007). For the purposes of my argument, a recipe will suffice. It is absolutely not my intention to defend any kind of adaptationism stating that evolution occurs by natural selection only. Quite the contrary, my interest here is to explore a new perspective on what makes such evolution possible in the first place. In this spirit, I will continue my argument by reviewing three existing perspectives on evolution by natural selection before adding a new one to the canon.

8.4 Three Different Perspectives on the Evolutionary Causal Thicket

The minimal conditions for evolution by natural selection appear deceptively simple. However, the path to understanding how they map onto physical reality is complex and full of conceptual pitfalls. One particularly important aspect of this problem concerns the connection between population-level descriptions of evolution and the underlying causal structure of the process, which is ultimately rooted in Darwin's famous struggle of individual organisms for their survival (Walsh, 2015; see Sect. 8.1). It seems unlikely that population-level statistical averages (e.g., mean and relative fitness measures) and emergent properties (arising from interactions of individuals and their environment) will suffice to explain all aspects of these highly complex and heterogeneous underlying causal dynamics at the organismic level. But what kind of understanding *can* be gained at this underlying level? Considering the immensity, diversity, and complexity of individual-level causal interactions, is it possible to gain any foothold at all? This question remains not only unresolved but mostly also unasked in current evolutionary biology.

One way to unravel the evolutionary causal thicket is to distinguish different perspectives that can validly be adopted to tackle the central problems related to evolution by natural selection. As already mentioned, we can distinguish three kinds of perspectives on evolution by natural selection (Griesemer, 2006). Each of these perspectives focuses on a different set of questions and has different advantages and limitations. I will briefly review each of them (with examples) and show how they relate to one another.

1. *Structural perspectives* focus on *what* evolves. The most famous debate in this domain is concerned with the units of selection (Lewontin, 1970). Approaches within this perspective consider evolving lineages as organizational hierarchies of compositional levels (molecules, organelles, cells, tissues, organs, organisms,

populations, and species). They ask at which level (or levels) selection applies and attempts to identify the pertinent structural units on which it acts. Such units must meet the minimal conditions for evolution by natural selection through completing the circle of development, reproduction, and selection (Brandon, 1990). Evolutionary change is treated as change in unit structure. Structural approaches are indispensable for the investigation of multilevel selection. They are powerful tools for the formalization of selection and inheritance. Their main weaknesses are that they leave unexamined the evolutionary origin of the multi-level hierarchy they presuppose and that they have difficulties accommodating inter-level processes such as development (which maps changes at the genetic level to phenotypic ones; see Griesemer (2000a,b, 2006) for details).

2. *Functional perspectives* focus on *why* things evolve. A well-known example is the replicator-interactor perspective developed by Dawkins (1976, 1983) and refined by Hull (1980, 1981, 1988). Replicators are entities able to transmit their structure directly and (relatively) intact through a copying process that produces more entities like themselves. Interactors, in contrast, are entities that engage with their immediate environment in ways that lead to differential replication. The focus here is not on the exact structure of either replicators or interactors (even though the former are generally assumed to be genes, and the latter organisms), but on the functional roles they play in generating the minimal conditions for evolution by natural selection. Dawkins and Hull differ on this matter. While Hull acknowledges selection of interactors at multiple compositional levels, Dawkins only recognizes replicators as true units of selection, since they alone are stably and faithfully copied and transmitted through the germ-line from generation to generation (see Griesemer, 2005). Apart from being the cause of inheritance, replicators also determine the development of the interactor's phenotypic traits, and even the construction of environmental features such as beavers' dams and human megacities, as proposed by Dawkins' (1982) perspective of the "extended phenotype." The environment then acts as a filter on populations of interactors, allowing some to survive and reproduce better or worse than others, depending on what kind of (extended) phenotypes are encoded by their replicator genes. This leads to a clean separation of development and inheritance. These processes do not interact directly, even though they share replicators as their common cause. Its conceptual simplicity is the main advantage of this perspective. At the same time, its failure to accommodate causal interactions between the processes of development, selection, and inheritance is also its biggest shortcoming. It leaves the functional separation of these processes unexamined, presupposing an excessive form of genetic determinism instead, which leads to an extremely oversimplified replicator/gene-centered view of evolution. Another problematic aspect is that both replicators and interactors are defined in circular ways that implicitly depend on goal-oriented processes (i.e., replication and the interaction of the organism with its environment), which are simply taken for granted (see Griesemer, 2006, for details).

3. *Process perspectives* focus on *how* evolutionary change occurs. As their name indicates, these perspectives focus on processes as the basic units of evolution. One example of such a perspective is process structuralism, which aims to understand the lawlike behavior of developmental processes that generate biological form (Goodwin, 1982a,b; Webster & Goodwin, 1982, 1996). It describes these generative processes as morphogenetic fields, whose underlying causal structure determines their dynamic behavior and the kind of phenotypic transitions they can produce. A methodical exploration of these structures through dynamical modeling and simulation would result in a rational system of related forms and the transformations between them. This provides an ahistorical "space of the possible," which the historical process of evolution explores. In this sense, process structuralism provides an understanding of the structured variability that provides the substrate for natural selection to act on. This is something neither structural nor functional perspectives can provide. But there are two main drawbacks. First, the rules behind the processes that generate variability are assumed to be universal and time-invariant, an assumption that is no longer tenable (see Sect. 8.6). Second, process structuralism only deals with regular phenomena in evolution. However, the generic forms that are actually realized in evolution are probably only a tiny fraction of all possible forms, which means that contingency probably still plays a dominant role in evolutionary dynamics (Griffiths, 1996).

Developmental systems theory (DST) is another example of a process perspective, which addresses this problem of contingency (Oyama, 1986; Oyama et al., 2001). Its basic unit is a developmental system, a process which is organized through the interactions of a variety of developmental resources in ways that lead to the completion of the life cycle. The precise structure of these interactions is not the focus here, since it is assumed to be a contingent product of evolution by natural selection. Instead, DST emphasizes the distributed and decentralized nature of control in evolving developmental systems. On this view, the transformation of biological form requires not only genetic resources but also epigenetic and environmental factors that are treated as equally important. In other words, the entire developmental system, rather than the gene, is the replicator (Griffiths, 1994). The main weakness of the approach is, however, that the boundaries of a developmental system are extremely difficult to define. This makes it hard to represent different modes of inheritance or to delimit life cycles between parents and offspring (see Griesemer, 2000a, 2006, for more details).

Traditionally, all of these perspectives are seen as competing with each other. DST, for example, explicitly positions itself as an alternative and a replacement for process structuralism (Griffiths, 1996). However, the existence of generic forms and more or less plausible evolutionary transformations does not deny the importance of historical contingency. Both can be seen as complementary aspects of the evolution of form. Similarly, functional approaches—in their unbridled ambition to provide a complete and unified account of evolutionary change—often treat structural or process perspectives as superfluous. However, we have seen how these perspectives can cooperate in a debate about the nature of the replicator. Griesemer (2006) expands on this topic by introducing his own process perspective, which sheds light on the nature of biological multiplication. This *reproducer perspective*—originally developed as a tool to investigate major transitions in evolution (Griesemer, 2000c)—powerfully illustrates how we can identify and transcend the limitations of specific approaches through an inclusive and comparative perspectival framework.

8.5   Reproducers, Evolvability, and the Completion of the Life Cycle

The reproducer perspective takes a closer look at the process of biological replication (Griesemer, 2006). In the previous section, we have seen that Dawkins' ultra-reductionist functional approach considers replicators as the only valid units of evolution. They alone are transmitted stably and faithfully through the kind of template-based copying process—exemplified by semiconservative DNA replication—that is presumed to form the basis for heritable variability. We have also noticed the circularity of their definition: replicators are essentially defined as structures able to replicate, which takes the seemingly goal-oriented process of replication itself for granted, leaving its underlying principles (and its origins) unexamined.

Can this circularity be avoided? To find an answer to this question, we have to examine the rules underlying the copying process. Specifically, to be a proper unit of evolution, an entity must adhere to the following three principles (Szathmáry & Maynard-Smith, 1993): (1) *the principle of multiplication*, entity A must give rise to more entities of type A; (2) *the principle of heredity*, entity A must produce entities of type A (not B); and (3) *the principle of variability*, the copying process is not perfect such that, every so often, entity A will give rise to an entity A' (which, in fact, may be identical to entity B). If we add different copying rates for different entities, we arrive back at Lewontin's minimal conditions for the process of evolution by natural selection as described in Sect. 8.2. What is new in this approach centered on the unit of evolution is an explicit focus on the notion of biological "multiplication."

What does it mean for an entity A to give rise to more entities of type A in an evolutionary context? And which conditions must be met for this process of biological multiplication to result in principles of heredity and variability that enable evolution by natural selection? There are several reasons to suspect that simple template-based replicators fail to meet these conditions. They are all connected to the problem of *evolvability:* the capacity of a system to generate (at least potentially) adaptive variability (Wagner & Altenberg, 1996).

The first of these reasons is that template-based copying by itself is too fragile and imprecise to support the kind of stability that is needed for the evolution of complex living systems. This argument is rooted in Eigen's paradox (Eigen & Schuster, 1977, 1979). In its original formulation, it states that the production of complex enzymes requires a large and complex genome, while the replication of a large and complex genome requires complex enzymes (necessitating a complex and precisely regulated cellular environment). More specifically, the enzymes required for genome replication must be able to proofread, i.e., to correct errors in the copying process. Without proofreading, complex genomes would be too unstable to evolve: copying mistakes would rapidly accumulate over time, inducing an *error catastrophe* that causes the collapse of the organization of a living system. This sets a very narrow upper limit on the size and complexity of evolvable genomes. But even if most mutants would be viable and able to reproduce (as may be the case in viral evolution), the original genome would quickly be lost in a sea of different variants, leading to the inevitable dilution and disappearance of any evolutionary lineage.

How could this fundamental limitation on the evolvability of complex living systems be overcome? One way is through compartmentalization and the hierarchical organization of living systems. Szathmary and colleagues have formulated a *stochastic corrector* model, which shows how compartmentalized probabilistic replicators can overcome Eigen's error catastrophe by constantly being selected at the higher level of the compartment population (Szathmáry, 1986; Szathmáry & Demeter, 1987; Grey et al., 1995; Zintzaras et al., 2002). This indicates that multi-level composition may be required to render an evolutionary unit evolvable.

Alternatively, it has been proposed that autocatalytic processes could lead to stable self-maintenance without complex genomes or hierarchical organization. Based on this general idea, Eigen and Schuster (1979) developed their own minimal autocatalytic model, the *hypercycle*, as a proof of concept. Unfortunately, hypercycles were shown to be extremely vulnerable to "selfish" replicators within them. In the meantime, Stuart Kauffman (1971, 1986, 1993) was proposing more general and robust models for autocatalytic sets. Kauffman's models consist of networks of chemical reactions that are capable of self-maintenance through catalytic closure: every reaction within the set is catalyzed by at least one product of the network itself. Even though this avoids error catastrophes, it is difficult for autocatalytic sets to generate the kind of variability that evolution requires. In stark contrast to the fragility of template-based replication, these sets are too rigid, since any reaction that does not contribute to the self-maintenance of the network is quickly outcompeted. Because of this, the system strongly converges to one particular optimal and invariant set of autocatalytic reactions (an attractor in the sense of being a strongly self-maintaining organization), which leaves very little heritable variability for selection to act upon (Fontana & Buss, 1996).

All of this suggests that Lewontin's minimal conditions on their own are not quite sufficient. They remain ambiguous. Evolution by natural selection not only needs heritable variability but also needs *the right amount of heritable variability in the right context*. Neither systems that are too stable nor systems that are too unstable can evolve. Stuart Kauffman famously illustrated this by his metaphor that evolvable systems must be poised "at the edge of chaos," a dynamic regime including "islands of chaos" among a "percolating network or order" (Kauffman, 1993).

Whatever we make of this metaphor, it is certainly true that the principle of variability imposes more specific and stringent conditions on evolution than is evident at first sight and that some sort of self-organization within the context of a hierarchical organization is required for natural selection to occur.

Based on the argument so far, let us take a closer look at the self-organizing processes able to generate the kind of heritable variability required for evolution. There is another reason why biological multiplication must be more complex than a simple template-based copying process. Copying does not require any material continuity between generations. Copies must resemble their template in form (e.g., similar genetic sequences), but can be made of different material components (the bases that are incorporated into the newly synthesized strand of DNA come from outside the original double helix). In this case, there is a clear separation between copying cycles: we can precisely determine when one cycle ends and another one begins. In contrast, biological multiplication always involves some *material and temporal overlap* between parents and offspring and between reproducer and reproduced. Organisms arise from material components of other organisms, and they do this in a gradual manner.

This implies some kind of *development,* which for the purpose of my evolutionary argument can be defined in a broad and minimal sense as "acquiring the capacity to reproduce" (Griesemer, 2006). Unlike the common (and narrower) definition of development as "embryogenesis" or "morphogenesis," this more general concept applies to unicellular and multicellular life-forms alike (see also Bich & Skillings, this volume, and Montévil & Soto, this volume). To come back to Eigen's example: a mitotic cell must first replicate its genome before it can divide again. This qualifies as "development" sensu Griesemer. To avoid confusion, I will use the term *ontogenesis* to describe the totality of regulatory processes—metabolic, physiological, developmental, and behavioral—that are involved in acquiring the capacity to reproduce.

This brings us to a central point of the argument: it is the process of ontogenesis which must provide the error-correcting capabilities that are needed to produce the kind of heritable variability required for evolution by natural selection. Both template-based replicators and autocatalytic networks lack ontogenesis, which is why they are not properly evolvable. In Eigen's example, a complex genome cannot be faithfully replicated unless it is embedded in the kind of complex and precisely regulated cellular environment which provides the necessary proofreading enzymes. Genome replication only ever happens if it is embedded within the more complex context of a cell cycle (even in those cases where the resulting cells do not separate completely). This illustrates the fundamentally dialectical relationship between ontogenesis and reproduction in evolution. They logically and materially entail each other. This relationship goes beyond mere causal reciprocity (see Sect. 8.1). Ontogenesis and reproduction do not only influence each other, *but cannot exist independently*—they *must* co-emerge for organisms to be evolvable. They dynamically presuppose each other (Bickhard, 2000). The resulting system is a true unit of evolution called *a reproducer* (Griesemer, 2006).

Reproducers are more complex than replicators, since they include ontogenesis and material overlap between generations. Ontogenesis and reproduction together form the life cycle of the reproducer. *This life cycle must be completed* for biological multiplication to continue from generation to generation. Variability cannot disrupt life cycle completion without disrupting evolution. In other words, freshly multiplied entities *a* must be *organized* in a way that enables them to mature into entities *A*, which have the capacity to reproduce. Otherwise, they are not evolvable. This constitutes a *principle of ontogenesis* (or *development*), which we must add to the principles of multiplication, heredity, and variability to define a proper unit of evolution (Griesemer, 2006).

One additional point requires our attention. In principle, ontogenesis could be based exclusively on some robust but spontaneous process of self-organization. However, this alone does not allow for natural selection to occur: in such systems, there is no true heredity of organization (beyond parents and offspring sharing the physical context that enables self-organization) and thus no selectable heritable variability. Moreover, it is quite probable that, even if they could evolve, purely self-organizing reproducers would easily be outcompeted by those possessing some kind of inherent hereditary processes, which lead to a much more efficient and stable propagation of organization across generations. Thus, heritable variability must be reliably regenerated and re-established through ontogenesis during each generation. For reproducers to meet the minimal conditions for evolution by natural selection, they require not only ontogenesis and material overlap between parents and offspring but also some kind of inter-generational *continuity of organization* that allows for heritable variability to be regenerated. The precise nature of this kind of organizational continuity will be the focus of Sect. 8.6.

In the reductionist framework of Dawkins, the source of organizational continuity is located entirely within the replicators themselves: it is the genes alone that are transmitted across generations, and genes alone determine the phenotypic traits of the interactor. This presupposes, however, that replicators are able to reproduce (complete a life cycle) all by themselves. Unfortunately, I have just shown that this is not the case: the simple template-based copying process on which replication is based fails to provide proper principles of heredity and variability for evolution by natural selection. To put it more simply: *replicators can only evolve if they are embedded in the more complex dynamics of a reproducer process* (Griesemer, 2006). The reproducer perspective therefore absorbs and replaces functional perspectives based on replicators as the fundamental units of evolution. The latter may still be useful to study the evolutionary role of genetic replication—but they can no longer serve as the foundation for a comprehensive theory of evolution. What the reproducer perspective offers is no extended synthesis, but rather evolutionary theory put back on its original Darwinian footing (see Walsh, 2015). A couple of examples will serve to illustrate this fundamental point.

The simplest reproducer systems that we currently know of are infective prions and virus particles. Because of their self-assembling structure, they are the closest we have to a "naked" replicator in nature. The mature, infectious forms of these entities self-aggregate from their macromolecular components according to simple thermodynamic principles. Thus, at first glance, they appear to lack proper ontogenesis or organizational continuity as defined above. However, this appearance is misleading. To generate their macromolecular components, prions and viruses rely on the pathways for biosynthesis and the homeostatic maintenance of the cellular milieu in a living host. These cellular processes are necessary to provide the substrates and the appropriate conditions for self-aggregation to occur. Moreover, both biosynthetic pathways and homeostatic mechanisms are central features of the host's self-maintaining organization (Hofmeyr, 2017). They provide the organizational principles required for prions and viruses to acquire the capacity to reproduce. Therefore, prions and viruses are not evolvable at all, if considered as isolated replicators apart from their hosts. In order to evolve, they *must* be embedded within a reproducer—the complex ontogenetic processes that constitute their host's life cycle (see Moreno & Mossio, 2015, Chap. 4, for a more detailed discussion).

As a second example, let us look at multicellular animals, which are vastly more complex than prions and viruses. They are particularly interesting in our context, because of their exceptionally well-defined separation between germ-line and soma. As we have seen in Sect. 8.4, the strict separation of reproduction (germ-line) and development (soma) is a central postulate of replicator-inheritor theory. But even in this case, replicators must be embedded in the larger context of a reproducer process in order to propagate and evolve. In fact, there are replicator processes at multiple levels of organization. We have already seen that the replication of a genome only ever occurs in the context of a cell cycle. At the tissue level, the maintenance and proliferation of germ cells require a specific niche within the context of the larger multicellular body. Finally, at the organismic level, animal reproduction relies (at least to some extent) on the behavior of the organism and its complex and goal-oriented interactions with the environment and other members of its species. I'll revisit this important point in Sect. 8.6.

Let me emphasize the main point of the argument once again: replicators cannot be the fundamental units of evolution unless they are embedded in a reproducer system, which necessarily includes a process of ontogenesis that generates the kind of heritable variability needed for evolution by natural selection. With this in mind, we can now reexamine the nature of a replicator. The ontogenetic process underlying replication must have the specific character of a *coding system* (Griesemer, 2006). To complete the replication cycle, ontogenesis must produce an interactor. The traits of this interactor must somehow be *encoded* in the genome. For this to work, the number of possible states in such a coding system must vastly outnumber the actual states that occur in an evolving population. This kind of coding process is what allows for a separation of genotype and phenotype. Without the cell to interpret and replicate it, however, there is no sense in which the genome carries a code (see Waddington, 1957). Therefore, replication must be seen as a highly specialized and context-dependent ontogenetic process, embedded in a hierarchy of reproductive organizations. As stated in no uncertain terms by Griesemer (2006, p. 359): "Far from being master molecules, genes are prisoners of development, locked in the deepest recesses of a hierarchy of prisons" (see also DiFrisco & Jaeger, 2020).

8.6 Organization, Reproduction, Agency, and Minimal Evolution

The reproducer account blurs the distinction between structural, functional, and processual perspectives. It shifts our focus from replicators to the more general category of reproducers as the fundamental (processual) units of evolution. Considered from a functional point of view, the central question about biological multiplication shifts from a simple template-based copying process (replication) to the propagation of complex *biological organization* across generations (reproduction). What are the heritable organizational principles that enable the reproduced system to acquire the capacity to reproduce? What are the heritable organizational principles that govern its ontogenesis and the completion of its life cycle through reproduction? Griesemer (2006) argues that these principles have to be based on some kind of organized *material propagules*, not mere informational programs as in a replicator perspective. These propagules must not only account for ontogenesis to explain self-maintenance, self-production, and self-regeneration within a life cycle but must also form the basis for *an organizational account of reproduction.*

There are several such accounts in the literature. The ones I will focus on here extend the notions of organizational closure and organizational continuity (see Sect. 8.2) across generations, beyond the temporal boundaries of the individual organism (Christensen & Bickhard, 2002; Saborido et al., 2011; Mossio & Pontarotti, 2020; DiFrisco & Mossio, 2020; Pontarotti, this volume). One option, from a functional point of view, is to treat entire reproductive lineages as organized systems (Christensen & Bickhard, 2002). However, such higher-order organization remains difficult to delineate precisely. Instead, we can take a more focused approach and consider the reproducer-reproduced dyad as a continuously organized system (Saborido et al., 2011). The important point here is to distinguish the boundaries of self-maintaining organization from the boundaries of the individual (DiFrisco & Mossio, 2020).

On the one hand, reproducer and reproduced can be considered the same organized system since there is organizational continuity between the two: closure of constraints must be maintained throughout the process of reproduction. An egg cell, for example, is both the product of the organization of the reproducer and the source of the organization of the reproduced. It exerts its function within the context of a cross-generational organization (Saborido et al., 2011; Mossio & Pontarotti, 2020; Pontarotti, this volume). On the other hand, the reproduced system is not the same individual as the reproducer. In fact, reproducer and reproduced often continue to coexist, more or less independently of each other. In this context, they must be treated as different organized systems. But there is no contradiction here, just a simple distinction: what sets apart reproduction from ontogenesis is not a break in organizational continuity, but a difference in the number of organized systems that are present at a given moment (DiFrisco & Mossio, 2020). Discontinuities between generations are characterized by fission (e.g., nuclear or cellular division, budding, or birth), or fusion (e.g., gametic, in case of sexual reproduction) of organized systems. A fertilized egg cell is the product of the functional organization of two different reproducer systems. It is dynamically presupposed by both of them (Bickhard, 2000).

On this view, the reproductive process is seen as the means of an organism to maintain its organization beyond the boundaries of its individual life cycle despite the discontinuities that characterize reproduction, and reproductive functions are those that contribute to inter-generational closure. But what exactly is meant by organizational continuity across generations? This concept clearly goes beyond mere material overlap through propagules (Griesemer, 2006), and it is more specific than the general notion of shared developmental resources (Griffiths, 1994). We have already seen in Sect. 8.2 that organizational continuity is a special case of causal continuity. Now that we have extended it to organizational closure across generations, it enables a new principle of heredity as *continuous self-maintenance of cross-generational functional organization* (Mossio & Pontarotti, 2020; Pontarotti, this volume). Remember that this does *not* imply that any specific physical structures or components of the system must persist unchanged through reproduction. Structure and material composition are in constant flux. What persists is the organization required to acquire the capacity to reproduce a specific kind of biological entity, a disposition for recurrent ontogenesis where similar functional constraints reoccur at the time scale of each generation (Mossio & Pontarotti, 2020; Pontarotti, this volume).

This organizational account of the reproducer affects several of the principles that underlie the minimal conditions for evolution by natural selection. Most important of all, it renders the replicator obsolete and replaces it with reproducers as the proper units of evolution. Furthermore, it suggests a new organizational principle of heredity: conservation of cross-generational functional organization, which requires organizational continuity. This kind of heredity enables and at the same time also depends on a new principle of ontogenesis: the capacity to reproduce must be reacquired during each generation in order to complete the life cycle (Griesemer, 2006). The flip-side of both principles of ontogenesis and heredity is a revised principle of variability (Montévil et al., 2016): variation can only occur under the general constraint of maintaining organizational closure within and across generations. Without this kind of cross-generational organizational continuity, life cannot go on and evolution cannot occur. In summary, we end up with the following set of tightly interlocked principles: *a new principle of ontogenesis, radically revised and refined principles of heredity and variation (plus Lewontin's unchanged original principle of differential fitness)*. Together, they comprise an extended and disambiguated set of minimal conditions for *an organizational theory of evolution by natural selection, which has the organism (and its struggle for existence) back at its core*, as it was in Darwin's original theory (Walsh, 2015).

But this is not all. Behind this revision of the minimal conditions for evolution by natural selection lies an even more significant implication for evolutionary theory. It is rooted in the simple fact that organizational closure must be retained throughout ontogenesis and reproduction for a life cycle to be completed. And the life cycle must be completed for evolution to occur. In other words, *without organizational continuity and the functional conservation it enables, there is no reproducer, and thus no proper unit of evolution. Without organizational continuity, there are no evolvable systems.*

If we accept this general conclusion, we must face another profound consequence: any proper unit of evolution, any evolvable system, must involve some kind of agency. It must be an autonomous agent with some degree of self-determination, since self-determination and autonomy are fundamental properties of organized systems (see Sect. 8.2). This is even true for the example of prions and viruses from Sect. 8.5: they require a host with organizational continuity (and thus agency) to reproduce and evolve. It may come as a surprise, but *evolution by natural selection is always the evolution of autonomous self-determining agents*, or occurs in higher-level organized systems—such as ecosystems, cultures, or economies—that involve autonomous self-determining agents. This basic insight has important implications for the theory of evolution. But what are these exactly? What does it mean to look at evolution from an *agential perspective*?

8.7 The Fourth Perspective: An Agential Theory of Evolution

Before I outline the possible shape of an agential theory of evolution, it is important to reiterate what agency is, and what it is not. Most importantly, *agency is the capacity of an organism to originate causal effects from within its own boundaries* (Barandiaran et al., 2009), particularly those that define its interactions with its external environment (see Sect. 8.2). These effects are observable as goal-oriented actions—selected from a more or less ample behavioral repertoire—which enable the organism to attain its ends by taking advantage of opportunities or avoiding obstacles in its experienced environment (Walsh, 2015). Biological organization and continuity provide the self-determination and autonomy necessary for true goal-oriented agency (cf. Sects. 8.2 and 8.6). Shadlen and Gold (2004) call this kind of relative autonomy "freedom from immediacy." While the organism's actions arise from its interactions with its environment, they are not directly imposed or determined by it.

This kind of *agential emergentism* (Walsh, 2015) stands in strong opposition to the more traditional approach to "agency" in evolution, which was first introduced by Ernst Mayr, and is still widely shared by biologists today. Mayr (1961) popularized the notion of *teleonomy* to denote preprogrammed behavioral routines that *appear* goal-seeking because they are adapted to their environment through evolution by natural selection. On this view, there are no causal effects (no actions) that are generated within the organism. Organisms have no intrinsic goals. There is only automated processing of external stimuli into responses adapted to a given environment. In other words, *there is no true organismic agency*. Adapted behaviors are explained entirely by factors in the external environment. The environment poses problems that organisms solve through evolution by natural selection (Lewontin, 1978; Levins & Lewontin, 1985). While denying true agency to organisms, this view can lead to the strange result that information-processing becomes interpreted as cognitive ability in simple organisms without a nervous system (e.g., bacteria; see Fulda, 2017, for an excellent critique). Such paradoxical consequences arise because the conventional view does not take agency quite serious enough—as it does not even include true agency in its ontology.

What happens to evolutionary theory if we *do* take agency at face value? Walsh (2015) provides a very thorough philosophical analysis of this question and concludes that a number of implications follow from agential emergentism. First, evolution must be treated as a fundamentally *ecological* or *relational phenomenon* arising from the purposive engagement of the organism with its experienced environment (Darwin's struggle for existence). Second, it is not possible to causally separate the processes of inheritance, reproduction, and development: "fragmented" evolutionary theory is an idealization. Third, there is no privileged control by replicator genes: genetic causation always has to be interpreted in its organismic context (see also DiFrisco & Jaeger, 2020). These insights do not fundamentally differ from claims made by other movements toward a more organismic evolutionary biology, such as the extended evolutionary synthesis (Laland et al., 2015). However, there is a central question that an agential theory of evolution raises, which remains largely unexplored by other approaches: *how does true organismic agency impact evolutionary change?* Timid first steps toward an exploration of this question have been made in studies of phenotypic plasticity (e.g., West-Eberhard, 2003; Moczek et al., 2011; Levis & Pfennig, 2016; Uller et al., 2019) and niche construction (e.g., Odling-Smee et al., 2003; Scott-Phillips et al., 2014; Uller & Helanterä, 2019). But so far, none of these efforts incorporate the dialectic multilevel dynamics underlying biological organization and the goal-oriented behavior of the organism. They remain anchored in a flattened and shallow cybernetic view of "agency" as mere information-processing and feedback-driven goal-seeking.

Why is that so? The problem is as simple as it is fundamental: because of the widespread mechanistic distrust concerning the notion of purposiveness, we do not possess the conceptual and mathematical tools required to appropriately incorporate true organismic agency into models of evolutionary dynamics. This is why we'd rather pretend the phenomenon does not exist, rather than taking it seriously. In fact, there are three related epistemological and methodological issues that need to be considered here.

The first issue concerns the general nature of our scientific theories, which predominantly fall under what Lee Smolin (2013) has called the Newtonian paradigm, or "physics in a box," and which Rosen (1991) identifies with mechanistic reductionism. Theories that adhere to this paradigm are *object theories*: they describe and explain the dynamics of a set of objects in a predefined space of possibilities (the bounded "box" or *configuration space* of the system). There is *transcendence*: the behavior of the objects is determined entirely by principles (forces, laws, etc.) that are outside and beyond themselves. There is an *explanatory asymmetry*: these principles determine the properties of the objects, but the objects do not explain the principles. The traditional teleonomic account of "agency" outlined above corresponds to an object theory. It must avoid invoking actions generated from within the organism at all cost; otherwise, it would no longer comply with the rules that define an object theory. This is why it fundamentally fails to capture the nature of true agency in the first place.

An *agential or agent theory* of evolution is a fundamentally different kind of theory (Walsh, 2015). It does not conform to the Newtonian paradigm. Organisms become both the subject and the object of evolution (Levins & Lewontin, 1985). There is *immanence:* agents themselves cause changes in their own state and organization, through interactions with their perceived environments (see also Fontana & Buss 1994, 1996). There is *explanatory reciprocity*: agents both generate and respond to the conditions of their existence. There is no predefined "box" or configuration space. There is no list of prestatable possibilities. Agents generate their own rules internally, which is what enables their autonomy and, ultimately, their open-ended evolution (Kauffman, 2000, 2014; Ruiz-Mirazo et al.,2004; Longo et al.,2012). Some things in evolution happen *because organisms make them happen* (Walsh, 2015). This is the central conclusion we have to draw once we accept autonomous reproducers as the fundamental units of evolution.

The second issue concerns what we accept as a scientific explanation. Aristotle distinguished four ways of answering the question "why." His four causes—material, formal, efficient, and final—are not really causes in the modern sense, but rather aitia, denoting something (or someone) responsible for a given phenomenon. For simplicity, I will use the less technical (but also less precise) notion of *(be)causes* here. (Be)causes correspond to different categories of determinants that complement each other to yield a full understanding of a phenomenon. This does not imply that Aristotle had a non-factive notion of causation, even though our modern scientific notion of "cause" is much more restricted: it roughly corresponds to efficient (be)causes only. In addition, modern science implicitly takes the material (be)cause for granted, although it no longer considers it a proper cause.

This is illustrated by current accounts of mechanistic explanation (e.g., Nicholson, 2012; Craver & Tabery, 2019). Evo-devo, for instance, relies on dynamic mechanisms as explanations, which are formulated in terms of structures "that perform a function in virtue of [their] component parts, component operations, and their organization," and whose "orchestrated functioning" is "manifested in patterns of change over time in properties of its parts and operations" (Bechtel & Abrahamsen, 2010, p. 323). Developmental evolution is characterized by plausible transformation sequences for such dynamic mechanisms (DiFrisco & Jaeger, 2019). Such explanations (called *lineage explanations*; Calcott, 2009) rely entirely on material and efficient (be)causes, that is, on changes in the components of the mechanism and the operations between them.

Robert Rosen, in "Life Itself" (1991), formalizes Aristotelian (be)causation, explicitly distinguishing between material and efficient causes in his relational characterization of organizational closure. Formal (be)causes can also be integrated into his account. Very roughly speaking, formal (be)cause relates to the *kind of causal organization* that implements closure—for instance, the specification of particular functional relationships within the system (Hofmeyr, 2018). Similarly, process structuralism distinguishes between different kinds of morphogenetic fields based on the functional relations that determine their structure (Webster & Goodwin, 1982, 1996). In both cases, different systems (organisms or morphogenetic fields) are categorized by the relational properties that characterize their organization. Such *relational explanations* are perfectly scientific; but they are not mechanistic. They do not explain the behavior of a system in terms of cause and effect, but rather tell us what kind of a system it is in terms of its relational properties.

The organizational account of organismic agency relies on material, efficient, *and* formal (be)causes—mechanistic *and* relational explanations—which complement each other. Organizational closure, achieved through the closure of constraints, is the defining relational property of living systems. It is a formal (be)cause. However, it is not simply imposed on the material flows constituting the organism. Instead, it is continually regenerated, constantly (re)emerging over time through the dialectic dynamic interactions of material processes and the constraints they generate (DiFrisco, 2014; DiFrisco & Mossio, 2020). These processes represent the material and efficient (be)causes of the organism.

In contrast to all the above, population-level evolutionary genetics relies on *statistical explanations* that are neither structural nor mechanistic, accounting for their phenomena in terms of statistical relevance or conditional dependence instead (see Woodward, 2019). The agential perspective adds *a fourth kind of explanation* to evolutionary theory—*naturalistic teleological explanation* (Walsh, 2015)—thus completing the Aristotelian repertoire of (be)causes.

Natural teleological explanation does not describe any large-scale trends or tendencies in evolution. It applies exclusively at the level of the evolutionary individual. While mechanistic explanations show how specific causes *produce* their effects (answering the question "how" something happens), teleological explanations account for the means that are *conducive to* the attainment of the organism's goals (answering the question "why" something happens). The latter kind of questions are often the most relevant in evolutionary biology, but have been considered philosophically troublesome for a very long time. This does not have to be case (Walsh, 2015). Goal-oriented behavior is empirically observable. The goals of an organism do not exert any pull on it from the future, but naturally emerge from its interactions with its experienced environment (the individual's struggle for existence).

For these reasons, naturalistic teleological explanation does not suffer from any of the problems that usually render teleological explanations problematic. First, it does not imply non-actual (e.g., future) causes of present effects. Second, it does not imply any intentionality (or even cognitive abilities). Third, it is based on a naturalized notion of normativity (see Sect. 8.2, and Mossio et al., 2009a). Goal-oriented behaviors arise because the organism strives to maintain organizational closure and continuity in order to continue living, to reproduce, and to evolve. It does this by autonomously selecting actions from its behavioral repertoire in response to opportunities and obstacles in its experienced environment (see Sect. 8.6). Put simply, naturalistic teleological explanation is a necessary part of any agential theory of evolution, because of the immanence of *rules which are generated by the agents themselves* (Walsh, 2015).

This leads us to the third and last issue, which consists of a number of methodological challenges concerning the mathematical and conceptual tools we use to study evolving systems. These tools are often borrowed from physics (as discussed, e.g., in Fontana & Buss, 1996, or Knuuttila & Loettgers, 2016), and most of them were originally developed within a strictly Newtonian paradigm. Let us take dynamical systems theory as an example, which is used to support dynamic mechanistic explanations in evo-devo (see Brigandt, 2015; DiFrisco & Jaeger, 2019). In this framework, we first prestate the space of possible trajectories of a system (its configuration space) before homing in on those that are actually realized in specific circumstances through validation of the model with empirical data (see, e.g., Jaeger & Crombach, 2012; Jaeger et al., 2012; Jaeger & Monk, 2014; Crombach & Jaeger, 2021). This is classical "physics in a box." It is a very powerful approach for simulating developmental processes, but breaks down at the level of whole-cell or whole-organism models, since traditional dynamical systems models cannot deal with systems based on organizational closure. In fact, it cannot deal with self-constructing systems in general (Fontana & Buss, 1994, 1996).

Organizational closure, considered in a dynamic context, leads to the continuous (re)generation of the rules and constraints that determine the behavior of the system. Therefore, systems with organizational closure require models that rewrite their own equations and boundary conditions based on principles generated from *within* themselves. This recursiveness lies at the heart of Rosen's (1991) conjecture that organisms cannot be completely captured by any finite algorithm. Although recursive formalisms (such as Lambda calculus), which allow for operations on operators, can be used for simulating organized systems with closure (Fontana & Buss, 1994, 1996; Mossio et al.,2009b), this still falls short of capturing the full potential of an evolvable living system. The reason for this is because the different processes and constraints that constitute an organism not only recursively influence but mutually depend on each other for their very existence (see Sect.8. 2). Organisms *embody* their self-generated rules in a way which is impossible to fully implement within any predefined computational environment with its externally specified hardware and syntactic rules (Rosen, 1991). It may be the case that to fully capture an organism with its capabilities of survival, reproduction, and open-ended evolution, we'd have to actualize a synthetic, evolvable reproducer with a complete life cycle in the laboratory.

8.8   Conclusion

What I have presented here remains a very tentative outline of an agential approach to evolution (cf. Walsh, 2015). It rests on the organizational account of reproduction and organismal agency (Moreno & Mossio, 2015; Mossio & Pontarotti, 2020; Mossio, this volume; Pontarotti, this volume), as well as the perspective of the reproducer and its life cycle as the fundamental unit of evolution by natural selection (Griesemer, 2006). The major implication of the theory is that all evolving systems are agents (or involve agents among their parts), which implies that understanding organismic agency is absolutely fundamental for understanding evolution. These are strong claims that must be supported by strong evidence. Unfortunately, we have barely begun to study agency and its role in evolution, and many obstacles remain on the road to a more mature, robust, and empirically supported theory.

Some of these challenges are philosophical: they concern the nature of the new account, its mathematical methodology, and the kind of explanations we need to understand the role of organismic agency in evolution. What should be clear by now is that an agential theory will look very different from what we are used to calling a scientific theory within the traditional Newtonian paradigm. Instead of providing a mechanistic explanation of agency and its evolution in terms of efficient causation only, this new theory will rely on all Aristotelian aitia or (be)causes: material, efficient, formal, and final. Final (be)causes will be incorporated in the form of naturalistic teleological explanations for the behavior of individual evolutionary units (reproducers) engaged in their struggle for existence. I must emphasize again: this kind of teleology does *not* imply any large-scale goal-directedness in evolution. Whether macroevolutionary trends exist or not is not at issue here. Naturalistic teleological explanation strictly only applies to the goal-directedness of an individual's behavior.

Some of the challenges are empirical. These apply at two different levels. To fully understand Darwinian evolution—and its underlying struggle for existence—we need a naturalistic account of organismic agency. Organizational closure through the closure of constraints, and organizational continuity enabling the continual (re)emergence of organization and the completion of the life cycle, provides the most detailed and convincing explanation for agency and its role in evolution that we have today. Yet, they still remain largely disconnected from the empirical study of regulatory processes in systems and synthetic biology. The first challenge will be to cross this divide in order to test the organizational account empirically. The second challenge resides at the level of the organism's goal-oriented behavior, which is itself an empirical observable (Walsh, 2015). Now that we have a philosophical justification to do so, we can apply naturalistic teleological explanations to account for the means that are conducive to the pursuit of an organism's goals. In a way, this is already common practice in ecological research. What remains to be done is to embed this existing practice as a research program within the agential perspective on evolution.

Despite all the remaining challenges—and the speculative nature of the argument—I hope to have convinced the reader that it is worthwhile to take process and agency more seriously in the study of evolution. Most current research remains restricted to traditional structural and functional approaches, with genetic replicators as their focus. Processual and even more so agential perspectives remain severely understudied, mainly because of theoretical objections and prejudices that do not hold up under closer scrutiny. This unnecessarily limits the scope and depth of evolutionary research today. Efforts to provide extended synthetic accounts do not really solve this issue, because the problem is the attempt at synthesis itself. Evolution is a process that generates diversity. Why not embrace an equally diverse approach to evolutionary explanation? An agential theory properly contextualizes and enriches existing structural, functional, and processual approaches. It provides *a fourth perspective on evolution, a truly organismic angle*. All four illuminate each other's limitations and domains of applicability, and each provides its own epistemic approach (Griesemer, 2006; Wimsatt, 2007). Together, they address a much greater range of evolutionary phenomena than any single perspective could ever cover on its own.

Acknowledgments James Griesemer, Andrea Loettgers, James DiFrisco, Rick Gawne, Jannie Hofmeyr, and Denis Walsh provided extremely useful comments on earlier versions of this manuscript. This chapter ties together strands of thinking that have deeply fascinated me since the earliest days of my career. I thank the late Brian Goodwin for igniting many of those sparks. Eva Neumann-Held and Christoph Rehmann-Sutter not only published the volume in which Griesemer's reproducer perspective was published but also organized the symposium which originally set me on my journey. I thank Nick Monk for his long-term collaboration on our own processual perspective. I thank my collaborators Dan Brooks and James DiFrisco for countless illuminating discussions and for introducing me to Bill Wimsatt and his perspectival philosophy. I thank Denis Walsh for his essential moral and intellectual support last year in Paris and beyond. I thank Jannie Hofmeyr for the all too brief time we had discussing the nature of living organization last year in Stellenbosch. I thank my collaborator Andrea Loettgers for taking my emergent view of life to the next level. Last but not least, I thank Matteo Mossio for introducing me to the organizational account, for inviting me to contribute to this volume, and for his helpful criticisms of my arguments. I truly stand on the shoulders of giants.

References

Amundson, R. (2005). *The changing role of the embryo in evolutionary thought*. Cambridge University Press.

Barandiaran, X., & Moreno, A. (2008). Adaptivity: From metabolism to behavior. *Adaptive Behavior, 14*, 171–185.

Barandiaran, X., Di Paolo, E., & Rohde, M. (2009). Defining agency. Individuality, normativity, asymmetry and spatio-temporality in action. *Journal of Adaptive Behavior, 17*, 367–386.

Bechtel, W., & Abrahamsen, A. (2010). Dynamic mechanistic explanation: Computational modeling of circadian rhythms as an exemplar for cognitive science. *Studies in History and Philosophy of Science, 41*, 321–333.

Bich, L., & Skillings, D. (this volume). There are no intermediate stages: An organizational view on development. In M. Mossio (Ed.), *Organization in biology*. Springer.

Bickhard, M. H. (2000). Autonomy, function, and representation. *Communication and Cognition – Artificial Intelligence, 17*, 111–131.

Brandon, R. (1990). *Adaptation and environment*. Princeton University Press.

Brigandt, I. (2015). Evolutionary developmental biology and the limits of philosophical accounts of mechanistic explanation. In P.-A. Braillard & C. Malaterre (Eds.), *Explanation in biology* (pp. 135–173). Springer.

Buskell, A. (2019). Reciprocal causation and the extended evolutionary synthesis. *Biological Theory, 14*, 267–279.

Calcott, B. (2009). Lineage explanations: Explaining how biological mechanisms change. *British Journal for the Philosophy of Science, 60*, 51–78.

Christensen, W. D., & Bickhard, M. H. (2002). The process dynamics of normative function. *The Monist, 85*, 3–28.

Craver, C., & Tabery, J. (2019). Mechanisms in science. In: E. N. Zalta, (Ed.), *The Stanford encyclopedia of philosophy*. https://plato.stanford.edu/archives/sum2019/entries/science-mechanisms

Crombach, A., & Jaeger, J. (2021). Life's attractors continued: Progress in understanding developmental systems through reverse engineering and *in silico* evolution. In A. Crombach (Ed.), *Evolutionary systems biology: Advances, questions, and opportunities*. Springer.

Darwin, C. (1859). *On the origin of species by means of natural selection*. John Murray.

Dawkins, R. (1976). *The selfish gene*. Oxford University Press.

Dawkins, R. (1982). *The extended phenotype*. Oxford University Press.

DiFrisco, J. (2014). Hylomorphism and the metabolic closure conception of life. *Acta Biotheoretica, 62*, 499–525.

DiFrisco, J., & Jaeger, J. (2019). Beyond networks: Mechanism and process in evo-devo. *Biology and Philosophy, 34*, 54.

DiFrisco, J., & Jaeger, J. (2020). Genetic causation in complex regulatory systems: An integrative dynamic perspective. *BioEssays, 42*, 1900226.

DiFrisco, J., & Mossio, M. (2020). Diachronic identity in complex life cycles: An organizational perspective. In A. S. Meincke & J. Dupré (Eds.), *Biological identity: Perspectives from metaphysics and the philosophy of biology* (pp. 177–199). Routledge.

DiFrisco, J., Love, A. C., & Wagner, G. P. (2020). Character identity mechanisms: A conceptual model for comparative-mechanistic biology. *Biology and Philosophy, 35*, 44.

Eigen, M., & Schuster, P. (1977). The hypercycle: A principle of natural self-organization – Part A: Emergence of the hypercycle. *Naturwissenschaften, 64*, 541–565.

Eigen, M., & Schuster, P. (1979). *The hypercycle – A principle of natural self-organization*. Springer.

Fontana, W., & Buss, L. W. (1994). 'The arrival of the fittest': Toward a theory of biological organization. *Bulletin of Mathematical Biology, 56*, 1–64.

Fontana, W., & Buss, L. W. (1996). The barrier of objects: From dynamical systems to bounded organizations. In J. Casti & A. Karlqvist (Eds.), *Boundaries and barriers* (pp. 56–116). Addison-Wesley.

Fulda, F. C. (2017). Natural agency: The case of bacterial cognition. *Journal of the American Philosophical Association, 3*, 69–90.

Giere, R. N. (2006). *Scientific Perspectivism*. University of Chicago Press.

Godfrey-Smith, P. (2007). Conditions for evolution by natural selection. *Journal of Philosophy, 104*, 489–516.

Godfrey-Smith, P. (2009). *Darwinian populations and natural selection*. Oxford University Press.

Goodwin, B. C. (1982a). Biology without Darwinian spectacles. *Biologist, 29*, 108–112.

Goodwin, B. C. (1982b). Development and evolution. *Journal of Theoretical Biology, 97*, 43–55.

Grey, D., Hutson, V., & Szathmáry, E. (1995). A re-examination of the stochastic corrector model. *Proceedings of the Royal Society of London B, 262*, 29–39.

Griesemer, J. (2000a). Development, culture, and the units of inheritance. *Philosophy of Science, 67*, S348–S368.

Griesemer, J. (2000b). Reproduction and the reduction of genetics. In P. Beurton, R. Falk & H.-J. Rheinberger (Eds.), *The concept of the gene in development and evolution: Historical and epistemological perspectives* (pp. 240–285). Cambridge University Press.

Griesemer, J. (2000c). The units of evolutionary transition. *Selection, 1*, 67–80.

Griesemer, J. (2005). The informational gene and the substantial body: On the generalization of evolutionary theory by abstraction. In M. R. Jones & N. Cartwright (Eds.), *Idealization XII: Correcting the model – Idealization and abstraction in the sciences* (pp. 59–115). Brill.

Griesemer, J. (2006). Genetics from an evolutionary process perspective. In E. M. Neumann & C. Rehmann-Sutter (Eds.), *Genes in development* (pp. 199–237). Duke University Press.

Griffiths, P. E. (1994). Developmental systems and evolutionary explanation. *Journal of Philosophy, 91*, 277–304.

Griffiths, P. E. (1996). Darwinism, process structuralism, and natural kinds. *Philosophy of Science, 63*, S1–S9.

Hofmeyr, J.-H. S. (2017). Basic biological anticipation. In R. Poli (Ed.), *Handbook of anticipation.* Springer.

Hofmeyr, J.-H. S. (2018). Causation, constructors and codes. *Biosystems, 164*, 121–127.

Hofmeyr, J.-H. S., & Cornish-Bowden, A. (2000). Regulating the cellular economy of supply and demand. *FEBS Letters, 476*, 47–51.

Hull, D. L. (1980). Individuality and selection. *Annual Review of Ecology and Systematics, 11*, 311–332.

Hull, D. L. (1981). The units of evolution: A metaphysical essay. In U. Jensen & R. Harré (Eds.), *The philosophy of evolution* (pp. 23–44). The Harvester Press.

Hull, D. L. (1988). *Science as a process.* University of Chicago Press.

Jaeger, J., & Crombach, A. (2012). Life's attractors: Understanding developmental systems through reverse engineering and *in silico* evolution. In O. Soyer (Ed.), *Evolutionary systems biology* (pp. 93–120). Springer.

Jaeger, J., & Monk, N. (2014). Bioattractors: Dynamical systems theory and the evolution of regulatory processes. *Journal of Physiology, 592*, 2267–2281.

Jaeger, J., Irons, D., & Monk, N. (2012). The inheritance of process: A dynamical systems approach. *Journal of Experimental Zoology (Molecular and Developmental Evolution), 318B*, 591–612.

Jonas, H. (1966). *The phenomenon of life – Towards a philosophical biology.* Northwestern University Press.

Kauffman, S. A. (1971). Cellular homeostasis, epigenesis, and replication in randomly aggregated macromolecular systems. *Journal of Cybernetics, 1*, 71–96.

Kauffman, S. A. (1986). Autocatalytic sets of proteins. *Journal of Theoretical Biology, 119*, 1–24.

Kauffman, S. A. (1993). *The origins of order: Self-organization and selection in evolution.* Oxford University Press.

Kauffman, S. A. (2000). *Investigations.* Oxford University Press.

Kauffman, S. A. (2014). Prolegomenon to patterns in evolution. *Biosystems, 123*, 3–8.

Knuuttila, T., & Loettgers, A. (2016). Model templates within and between disciplines: From magnets to gases – and socio-economic systems. *European Journal for Philosophy of Science, 6*, 377–400.

Laland, K. N., Uller, T., Feldman, M. W., Sterelny, K., Müller, G. B., Moczek, A., Jablonka, E., & Odling-Smee, J. (2015). The extended evolutionary synthesis: Its structure, assumptions and predictions. *Proceedings of the Royal Society London B, 282*, 20151019.

Levins, R., & Lewontin, R. (1985). *The dialectical biologist.* Harvard University Press.

Levis, N. A., & Pfennig, D. W. (2016). Evaluating 'plasticity-first' evolution in nature: Key criteria and empirical approaches. *Trends in Ecology & Evolution, 31*, 563–574.

Lewontin, R. C. (1970). The units of selection. *Annual Review of Ecology and Systematics, 1*, 1–18.

Lewontin, R. C. (1978). Adaptation. *Scientific American, 239*, 212–231.

Longo, G., Montévil, M., and S. Kauffman (2012). No entailing laws, but enablement in the evolution of the biosphere." In: T. Soule (ed.), *GECCO '12: Proceedings of the 14th annual conference on genetic and evolutionary computation,* Association for Computing Machinery (ACM), New York, pp. 1379–1392.

Massimi, M. (2016). Four kinds of perspectival truth. *Philosophy and Phenomenological Research, 96*, 342–359.

Maturana, H. (1980). Autopoiesis: Reproduction, heredity and evolution. In M. Zelený (Ed.), *Autopoiesis, dissipative structures, and spontaneous social orders* (pp. 45–79). Westview.

Mayr, E. (1961). Cause and effect in biology. *Science, 134*, 1501–1506.

Moczek, A. P., Sultan, S., Foster, S., Ledón-Rettig, C., Dworkin, I., Nijhout, H. F., Abouheif, E., & Pfennig, D. W. (2011). The role of developmental plasticity in evolutionary innovation. *Proceedings of the Royal Society B: Biological Sciences, 278*, 2705–2713.

Montévil, M., & Mossio, M. (2015). Biological organisation as closure of constraints. *Journal of Theoretical Biology, 372*, 179–191.

Montévil, M., & Soto, A. M. (this volume). Modeling organogenesis from biological first principles. In M. Mossio (Ed.), *Organization in biology*. Springer.

Montévil, M., Mossio, M., Pocheville, A., & Longo, G. (2016). Theoretical principles for biology: Variation. *Progress in Biophysics and Molecular Biology, 122*, 36–50.

Moreno, A., & Etxeberria, A. (2005). Agency in natural and artificial systems. *Artificial Life, 11*, 161–175.

Moreno, A., & Mossio, M. (2015). *Biological autonomy*. Springer.

Mossio, M. (this volume). Introduction: Organization as a scientific blind spot. In M. Mossio (Ed.), *Organization in biology*. Springer.

Mossio, M., & Bich, L. (2017). What makes biological organisation teleological? *Synthese, 194*, 1089–1114.

Mossio, M., & Pontarotti, G. (2020). Conserving functions across generations: Heredity in light of biological organization. *British Journal for the Philosophy of Science*, axz031.

Mossio, M., Saborido, C., & Moreno, A. (2009a). An organizational account of biological functions. *British Journal for the Philosophy of Science, 60*, 813–841.

Mossio, M., Longo, G., & Stewart, J. (2009b). A computable expression of closure to efficient causation. *Journal of Theoretical Biology, 257*, 489–498.

Mossio, M., Montévil, M., & Longo, G. (2016). Theoretical principles for biology: Organization. *Progress in Biophysics and Molecular Biology, 122*, 24–35.

Nicholson, D. J. (2012). The concept of mechanism in biology. *Studies in History and Philosophy of Biological and Biomedical Sciences, 43*, 152–163.

Nicholson, D. J. (2018). Reconceptualizing the organism – From complex machine to flowing stream. In D. J. Nicholson & J. Dupré (Eds.), *Everything flows – Towards a processual philosophy of biology* (pp. 139–166). Oxford University Press.

Odling-Smee, F. J., Laland, K. N., & Feldman, M. W. (2003). *Niche construction – The neglected process in evolution*. Princeton University Press.

Oyama, S. (1986). *The ontogeny of information*. Duke University Press.

Oyama, S., Griffiths, P. E., & Gray, R. D. (2001). *Cycles of contingency – Developmental systems and evolution*. MIT Press.

Piaget, J. (1967). *Biologie et Connaissance*. Éditions Gallmard.

Pontarotti, G. (this volume). Organization and inheritance in twenty-first-century evolutionary biology. In M. Mossio (Ed.), *Organization in biology*. Springer.

Rosen, R. (1991). *Life itself*. Columbia University Press.

Ruiz-Mirazo, K., Peretó, J., & Moreno, A. (2004). A universal definition of life: Autonomy and open-ended evolution. *Origins of Life and Evolution of the Biosphere, 34*, 323–346.

Saborido, C., Mossio, M., & Moreno, A. (2011). Biological organization and cross-generation functions. *British Journal for the Philosophy of Science, 62*, 583–606.

Scott-Phillips, T. C., Laland, K. N., Shuker, D. M., Dickins, T. E., & West, S. A. (2014). The niche construction perspective: A critical appraisal. *Evolution, 68*, 1231–1243.

Shadlen, M. N., & Gold, J. I. (2004). The neurophysiology of decision-making as a window on cognition. In M. S. Gazzaniga (Ed.), *The cognitive neurosciences*. MIT Press.

Smocovitis, V. B. (1996). *Unifying biology: The evolutionary synthesis and modern biology*. Princeton University Press.

Smolin, L. (2013). *Time reborn*. Boston, MA.

Sober, E. (1984). *The nature of selection: Evolutionary theory in philosophical focus*. University of Chicago Press.

Svensson, E. I. (2018). On reciprocal causation in the evolutionary process. *Evolutionary Biology, 45*, 1–14.

Szathmáry, E. (1986). The eukaryotic cell as an information integrator. *Endocytobiosis and Cell Research, 3*, 113–132.

Szathmáry, E., & Demeter, L. (1987). Group selection of early replicators and the origin of life. *Journal of Theoretical Biology, 128*, 463–486.

Szathmáry, E., & Maynard Smith, J. (1993). The origin of genetic systems. *Abstracta Botanica, 17*, 197–206.

Toepfer, G. (this volume). "Organization": Its conceptual history and its relationship to other fundamental biological concepts. In M. Mossio (Ed.), *Organization in biology*. Springer.

Uller, T., & Helanterä, H. (2019). Niche construction and conceptual change in evolutionary biology. *The British Journal for the Philosophy of Science, 70*, 351–375.

Uller, T., Feiner, N., Radersma, R., Jackson, I. S. C., & Rago, A. (2019). Developmental plasticity and evolutionary explanations. *Evolution & Development, 22*, 47–55.

Varela, F. G., Maturana, H. R., & Uribe, R. (1974). Autopoiesis: The organization of living systems, its characterization and a model. *Biosystems, 5*, 187–196.

Waddington, C. H. (1957). *The strategy of the genes*. Routledge.

Wagner, G. P., & Altenberg, L. (1996). Complex adaptations and the evolution of evolvability. *Evolution, 50*, 967–976.

Wagner, G. P., Chiu, C.-H., & Laubichler, M. (2000). Developmental evolution as a mechanistic science: The inference from developmental mechanism to evolutionary processes. *American Zoologist, 40*, 819–831.

Walsh, D. (2015). *Organisms, agency, and evolution*. Cambridge University Press.

Webster, G., & Goodwin, B. C. (1982). The origin of species: A structuralist approach. *Journal of Social and Biological Structures, 5*, 15–47.

Webster, G., & Goodwin, B. C. (1996). *Form and transformation: Generative and relational principles in biology*. Cambridge University Press.

West-Eberhard, M.-J. (2003). *Developmental plasticity and evolution*. Oxford University Press.

Wimsatt W. (1994). "The ontology of complex systems: levels of organization, perspectives, and causal thickets." *Canadian Journal of Philosophy* 20 (Supp.): 207–74.

Wimsatt, W. (2007). *Re-engineering philosophy for limited beings*. Harvard University Press.

Woodward, J. (2019). Scientific explanation. In: E. N. Zalta (Ed.), *The Stanford encyclopedia of philosophy*. https://plato.stanford.edu/archives/win2019/entries/scientific-explanation

Zintzaras, E., Santos, M., & Szathmáry, E. (2002). 'Living' under the challenge of information decay: The stochastic corrector model vs. hypercycles. *Journal of Theoretical Biology, 217*, 167–181.

Chapter 9
On the Evolutionary Development of Biological Organization from Complex Prebiotic Chemistry

Kepa Ruiz-Mirazo and Alvaro Moreno

Abstract In this chapter we offer a critical analysis of organizational models about the process of origins of life and, thereby, a reflection about life itself (understood in a general, minimal sense). We begin by demarcating the idea of organization as an explanatory construct, linking it to the complex relationships and transformations that the material parts of (proto-)biological systems establish to maintain themselves under non-equilibrium dynamic conditions. The diverse ways in which this basic idea has been applied within the prebiotic field are then reviewed in relative detail. We distinguish between "network" and "protocell" approaches, discussing their specific implications and explaining the greater relevance of the latter in the current state of affairs. Despite the key role that such organizational approaches play (and should keep playing) to advance on the problem of primordial biogenesis, the second half of our contribution is devoted to argue that they must be combined with other explanatory accounts, which go beyond the physiology of any single (proto-)organism. With that aim, we underline the fundamental differences between the autonomous, metabolic dynamics that individual (proto-)cells perform and the evolutionary and ecological dynamics that take place in a collective and transgenerational dimension. Apart from obvious gaps in the characteristic temporal and spatial scales involved, the corresponding causal and interactive regimes also reveal themselves as neatly distinct, what is reflected in the unpaired functional integration and the agent behavior displayed by biological individuals. Nevertheless, any living organism (and life in a wider, general sense) derives from the deep interweaving of those two phenomenological domains: namely, the "individual-metabolic" and the "collective-evolutionary" domains. At the end of the chapter, we propose the principle of dynamical decoupling as the core idea to develop a more comprehensive

K. Ruiz-Mirazo (✉)
Department of Philosophy, University of the Basque Country, Bilbao, Spain

Biofisika Institute (CSIC, UPV-EHU), Leioa, Spain
e-mail: kepa.ruiz-mirazo@ehu.eus

A. Moreno
Department of Philosophy, University of the Basque Country, Bilbao, Spain

Donostia International Physics Centre (DIPC), San Sebastián, Spain

© The Author(s) 2024   187
M. Mossio (ed.), *Organization in Biology*, History, Philosophy and Theory of
the Life Sciences 33, https://doi.org/10.1007/978-3-031-38968-9_9

theoretical framework to understand how this intricate, causally asymmetric connection must be articulated during the actual process of biogenesis (as it happened here on Earth or anywhere else in the universe), so that life's minimal complexity threshold is reached.

Keywords Primordial biogenesis · Prebiotic transitions · Molecular reaction networks · Protocell models · Organizational integration · Minimal metabolism · Regulation · Origins of agency · Functional domain · Reproduction · Material/trans-generational constraints · Pre-Darwinian evolution · Sedimentation · Protophylogenies · Dynamical decoupling · Informational records · Genetic code · Biological organization

9.1 Introduction: Organization as an Explanatory Construct in Origins-of-Life Research

Most of the research work in the field of prebiotic chemistry has been focused, so far, on discovering reaction mechanisms and transformation pathways for the abiotic synthesis of biopolymers, their monomers, or some other biologically relevant molecules. In addition, the replicative and catalytic properties of those molecules have been explored in considerable detail (for an extensive review, see Ruiz-Mirazo et al., 2014). However, all that body of empirical and theoretical knowledge tells us very little, unfortunately, about the main transitions during the process of primordial biogenesis. Somehow, Miller's (1953) famous experiment was a turning point in the field that has been transformed, over the last decades, into a wider and more solid platform to approach the problem of origins of life, but further progress has been quite modest, really: like Sutherland (2017) says, all what we have achieved so far (including his own investigations) is just «the end of the beginning» in terms of solving the question.

Many authors consider that *natural selection*, combined with long enough time periods, up to the geological scale, could lead all the way, from populations of biomolecules (in particular, if the latter developed the capacity for multiplication, variation, and heredity (Maynard Smith, 1986; Szathmáry & Maynard Smith, 1997)) to living cells. Yet, this involves a huge assumption, based on the premise that the principles of Darwinian evolution can be readily applied to bare sets of molecular replicators. There is ample evidence, indeed, to support that molecular structures, like RNA strands, undergo artificial evolution in vitro, being able to reach pre-established target motifs (e.g., ribozymes with specific features (Bartel & Szostak, 1993; Johnston et al., 2001; Tjhung et al., 2020)) or even follow a potentially endless process (Lincoln & Joyce, 2009). Nevertheless, there are no obvious results showing evolutionary dynamics that bring about a relevant increase in the complexity of the individuals that constitute those populations, even when replicators are used in combination with protocellular structures (e.g., Chen et al., 2005; Mansy et al., 2008 – see also the more recent review: (Joyce & Szostak, 2018)). Therefore, although hopes still remain for an "RNA world" (the hypothesis that all started from RNA molecules (Crick, 1968; Orgel, 1968; Gilbert, 1986)) which could turn into a full-fledged biological world (Higgs & Lehman, 2015; Joyce & Szostak, 2018; Krishnamurthy, 2020), more and more skeptical voices are rising, advocating the need to conceive alternative scenarios (Ruiz-Mirazo et al., 2017; Le Vay & Mutschler, 2019; Kroiss et al., 2019; Preiner et al., 2020). Among other reasons, it could well be the case that a molecularly richer, more varied, and heterogeneous prebiotic milieu is required, right from the beginning, to trigger off those evolutionary processes that may lead to an open-ended increase in functional/phenotypic diversity, as it was argued more extensively in (Wicken, 1987; Moreno & Ruiz-Mirazo, 2009; Ruiz-Mirazo et al., 2008, 2020).

The main alternative to the "RNA world" has traditionally been the "metabolism-first" hypothesis (De Duve, 1991; Dyson, 1999; Morowitz, 1999; Shapiro, 2000), which defends a completely different plan of attack and explanatory framework for life's origin. The key question, according to this approach, would be discovering the combination of energy inputs, material components, and chemical transformation processes that put together a self-maintaining system in non-equilibrium, precarious conditions, in transition toward minimal (unicellular) organisms. It is in this context, precisely, where the idea of *organization* comes to the center of stage, as a fundamental mereological construct through which complex systems, like living cells (or their precursors, "protocells" or "proto-metabolisms"), demand a proper characterization. The work of a number of classical authors in theoretical biology (Rosen, 1971, 1991; Varela et al., 1974; Maturana & Varela, 1980; Ganti, 1975, 2003; Eigen & Schuster, 1979; Kauffman, 1986, 1993; Fontana & Buss, 1994, 1996), trying to determine the idiosyncratic nature of biological organization, in its most elementary and general sense, in contrast to other types of organization one may find in the natural world, also helped to elaborate and support this view. A view that has been reinforced in recent times, as well, with the advent of new research programs like "systems biology" (Kitano, 2002; Westerhoff & Palsson, 2004) and "systems chemistry" (von Kiedrowski, 2005; Ludlow & Otto, 2008) that insist on the irreducible complexity of biological and proto-biological entities (de la Escosura et al., 2015; Kroiss et al., 2019). This does not always mean defending metabolism as the first or most important landmark in the prebiotic process, but it stands much closer to that way of framing the question, as an investigation into the intricate material and energetic couplings that enable non-equilibrium, dynamic systems whose emergence and maintenance rely on the strong functional integration of a variety of molecular components in continuous transformation (Ruiz-Mirazo et al., 2017; Lauber et al., 2021).

In this chapter we will briefly review the different organizational approaches that have been pursued within the field of origins of life and their relative success, classifying them in two main groups: *network* models and *protocell* models. These two types of model can be both experimental (in vitro) and theoretical/computational (in silico): the fundamental feature that distinguishes them relates to the degree of diversity and interdependence required among the various processes, material components, and constraints that constitute the system. Although protocellular models tend to be more demanding and encompassing in that sense (and, therefore, more interesting in order to develop a complete theory of biogenesis), they are also more complicated to handle and analyze. In any case, we will try to show how the concept of organization, in its various meanings, can be used both as *explanans* (i.e., to describe intermediate hypothetical model-systems, taken as necessary conjectures or stages to make sense of such a long and complex transition) and *explanandum* (i.e., to account for the end result, prokaryotic cell organization, taken as factual "minimal life").

However, the most important message of our chapter is to remark that the *organizational* framework, being of primary and central importance, is not sufficient to elucidate the nature of the living phenomenon nor the way it came to be, from physics and chemistry. Although the inert and the living worlds are of course linked, in various ways and diverse planes, the jump between their corresponding phenomenologies is too big, or too high, to be taken in just a few steps. Even if the first stages of primordial biogenesis should already involve organized individuals (for the reasons suggested above, see again, in particular Moreno & Ruiz-Mirazo, 2009), additional explanatory principles, beyond the sphere of individual organizations, must be at work for prebiotic systems to overcome the bottlenecks that were surely present throughout such an intricate process. More precisely, increases in complexity (including the internal complexity of the individual organizations leading the process) require dynamics and interactions that take place at the level of population dynamics, as we will see, to ensure minimal robustness at the intermediate phases, paying back for the energetic and material costs involved.[1]

Regulation, for instance, understood in a biologically relevant sense (i.e., not simply as chemical feedback but as a hierarchy of controls operating in minimally adaptive systems, so they can select among a diversity of metabolic/behavioral regimes in response to changes in internal/external variables (Bich et al., 2016)), appears as a key property to be developed by protocells, on their way toward more autonomous, efficient, and sophisticated cells. Although the presence of regulatory mechanisms has also deep implications in terms of how each individual is organized, the appearance and stabilization of such mechanisms cannot be conceived but as the result of an *evolutionary* process. Namely, a process in which a large population of similar, precarious but proliferating systems, in a variable and challenging environment, try various (constitutive or behavioral) options, and come out with "a solution" that spreads and eventually becomes built-in, hard-wired in each system

---

[1] This makes our position quite different from other authors', like Varela's (1979) or Rosen's (1991) who defended that the living phenomenon could be fully captured in terms of the organization that each organism continuously realizes. In more pragmatic terms, the conception that we will defend here, based on our previous work on the nature of life (Ruiz-Mirazo et al., 2004, 2010), highlights that there are yet many hard issues to address in the field of origins (as well as in theoretical biology) related to the intricate link between the physiological, ecological, and evolutionary spheres of the living. That move also makes more understandable the huge gap in complexity that still lies between our current prebiotic (or "bottom-up") models/systems and any real cell.

(i.e., it is adopted as a reliable mechanism by all subsequent individuals in the population). Similarly, genetic mechanisms are incardinated within individual organizations, where they play their fundamental physiological roles (as a guide for protein synthesis, more prominently), but their raison d'être and functional contribution do not make full sense unless it is considered in the context of a wider and longer evolutionary pathway, an open pathway that transcends any of those particular individuals (Ruiz-Mirazo et al., 2020).

Therefore, we will put forward the thesis that complex chemical systems (self-producing and self-reproducing protocells) progressively transform into hypercomplex biological organisms (living cells) thanks to a combination of factors that operate not only at different spatial/temporal scales and with different weights but also following intrinsically different dynamic principles. Some of these principles have to do with the composition, architecture, and necessarily interactive self-maintaining dynamics of the individuals involved, whereas some others have to do with their reproduction, inheritance, diversification, and open, collective dynamics. Accordingly, *organizational* aspects will be primarily associated to the development – and adequate coupling – of basic control mechanisms at the molecular and physiological description levels (Ruiz-Mirazo et al., 2017), while *evolutionary* and *ecological* aspects (Moreno, 2016) will rather cover the "propagation" (Kauffman, 2000) and "sedimentation" (Walsh, 2018) processes working at the level of the population, or the whole ecosystem/biosphere. We will argue that, even if they seem quite orthogonal to each other, these two phenomenological domains must actually get tied up during the process of origins of life, establishing a mutual – though causally asymmetric – connection that is further reinforced once biological evolution takes off. Our discussion will reveal, in any case, how much ground science must still cover in order to solve the problem of primordial biogenesis.

9.2 Organizational Accounts at the Onset of Prebiotic Evolution: Network Versus Protocell Models

As we just advanced in the introduction, tackling the problem of origins of life from an organizational perspective implies a theoretical scheme according to which different molecular components and transformation processes come together to constitute prebiotic systems that maintain themselves and proliferate in non-equilibrium conditions, on their way toward living organisms. Depending on the diversity and complexity of the material components/transformations involved, as well as on their relationships and interactive properties, one can propose a variety of architectures to characterize such systems. A complete organizational theory of biogenesis should provide a plausible sequence of transitions, starting from relatively simple systems toward increasingly complex ones (both in terms of molecular ingredients and architectural/interactive features), with the aim to bridge the gap currently observed in nature between physical-chemical and biological phenomena.

Living beings, as pointed out so wisely by Kant, long before the development of modern biology, have a very special kind of organization, in which even the mere existence of many fundamental system parts cannot be taken for granted, since they result from the collective transformation dynamics of the whole. This idea is, in our view, of central importance to understand biological phenomena, and, thus, it must also play a key role in any explanation of the process of biogenesis. Yet, what Kant did not anticipate is that the roots of this complex dynamic behavior could actually be found in the domain of physics and chemistry: in other words, that matter is inherently active, given the adequate conditions, as it became apparent in the study of non-equilibrium self-organization and self-assembly processes last century (Nicolis & Prigogine, 1977; Lehn, 1995) and has been, thereafter, reinforced (Showalter & Epstein, 2015; Semenov et al., 2016). These processes, being necessary to understand the workings of any cell (Karsenti, 2008), are nevertheless not sufficient. The capacity for self-maintenance characteristic of biological organisms, their surprising endurance as non-equilibrium, dissipative systems, involves not only the organization of already existing parts into a whole but a proper *metabolism*: continuous constructive and reconstructive transformations, which actually synthesize the key ingredients that rule its complex behavior (i.e., a diversity of material constraints that operate on those same transformations (Ruiz-Mirazo & Moreno, 2004; Lauber et al., 2021)).

This circularity or cyclic/self-referential, collective dynamics was the theoretical target of a number of classical models of minimal biological organization (Rosen, 1971; Varela et al., 1974; Ganti, 1975; Pattee, 1977; Kauffman, 1986). Although the direct impact of such abstract models on the research field of prebiotic chemistry has been relatively modest, we consider that they can still be helpful to draw a conceptual distinction between two general approaches to the problem of origins of life, when this is envisaged in terms of the emergence of metabolic organizations. On the one hand, one can identify the *network* approach, in which the dynamics of a population of reacting molecules in homogenous – typically aqueous solution – conditions is explored, assuming a more or less concentrated "organic chemistry soup" where the potential couplings/interactions among those molecular components can be captured through mathematical mappings or graphs.[2] On the other hand, we find the *protocell* approach, in which both physical and chemical transformations take place in heterogeneous conditions – typically a mixture of aqueous and organic domains, like a lipid vesicle suspension – where the couplings/interactions among the system components must be analyzed making use of additional tools, since they are also influenced by spatial constraints on their free movement/diffusion. With variations (which we will not go into here – see Moreno & Ruiz-Mirazo, 1999;

---

[2] Some "network" models/theories would not work, strictly speaking, in homogeneous 3D conditions, but on surfaces. For instance, the classical proposal of Wächtershäuser (1988, 1990) would constitute a two-dimensional metabolism working on the surface of pyrite. However, we are not going to pay special attention to this type of scenario here. It could be interesting in terms of finding synthetic pathways to generate some organic compounds, but they are severely limited for any further organizational developments (Ruiz-Mirazo et al., 2020; Lauber et al., 2021).

Hofmeyr, 2007; Cornish-Bowden & Cárdenas, 2020 for more detailed reviews), Rosen's *M-R systems* or Kauffman's *autocatalytic sets* would represent the former (i.e., network approaches) and Maturana and Varela's *autopoiesis* or Ganti's *chemoton* the latter (protocell approaches). Let us briefly review, with a critical eye, the effective progress made in the prebiotic research camp, over the years, by following these two general organizational schemes.

9.2.1 Network Models

Most empirical approximations to the problem of origins of life have been championed by chemists (organic synthetic chemists, in particular), whose main interest is deciphering abiotic reaction pathways that could lead to various, specific biomolecules (Ruiz-Mirazo et al., 2014). This, although important to address, does not lead very far in terms of understanding the first biologically relevant *organizations*, as we said above. Nevertheless, in recent years, with the advent of systems biology and systems chemistry, an increased awareness in the community about the importance of dealing with complex mixtures in a prebiotic context has brought about a much more compelling research scene (Ashkenasy et al., 2017; Kroiss et al., 2019; Wolos et al., 2020), in which strongly reductionist approximations to the problem (e.g., working with one type of molecule, or one type of chemistry – even if this is claimed to be fundamental for life) are no longer valid.

There were some remarkable achievements associated with the idea of autocatalytic networks in the past (e.g., von Kiedrowski's (1986) self-replicating oligonucleotides, or the analogous oligopeptide systems developed by Ghadiri's group (Lee et al., 1996)), but these were just networks of oligomer-pairs with a template that coupled through a single, potentially autocatalytic recognition mechanism, rather than a collectively or reflexively autocatalytic network. Subsequent expansions toward more complex systems (employing combinations of more diverse components and higher-order catalytic and cross-catalytic mechanisms operating in parallel, within the same pot) led to interesting, emergent properties at the collective level (for a review, see Dadon et al., 2008). Similar investigations have also been carried out with populations of different RNA molecules designed to build cooperative relationships among them in order to achieve some collective autocatalytic behavior (see, e.g., Vaidya et al. (2012) or, more recently, Ameta et al. (2021)). Nevertheless, despite their obvious interest, we consider that these emergent phenomena are not so relevant, prebiotically speaking. Although they do reflect a complex global dynamic behavior that could not be predicted from the pieces of the puzzle (like in self-organizing phenomena, in which the more pieces you mix, the more difficult it becomes inferring, from individual molecular/mechanistic properties, what will happen at the overall, network level), their potential to build minimally robust, integrated material organizations is still unclear.[3] Many similarities can be drawn with the general case of dynamic combinatorial libraries (DCLs), and more so if they are under non-equilibrium conditions: the higher compositional diversity one introduces, the more interesting (and difficult-to-predict/analyze) phenomena one obtains (Corbett et al., 2006; Reek & Otto, 2010) but, without more demanding *systems* requirements (in particular, without the development of spatial and energetic control mechanisms) where do those phenomena lead to?

Many of these works, in addition, are not so concerned about the prebiotic plausibility of the components used. They are just demonstrating that chemistry is much wider than biology, in terms of molecular structures and nontrivial combinations thereby. Nevertheless, complex dynamic behavior does not immediately lead to molecules and transformation processes that establish and develop *functional* relationships, like those so characteristic of living organisms (see Sect. 9.3.1). The key question does not seem to be molecular and interactive diversity per se but playing with the biologically relevant type of diversity, with the aim to open a new window of dynamic behavior, performed by more *complexly organized* systems. Yet, this combination of material ingredients and conditions for viability does not come for free: they need to be physically constructed and maintained. Unlike self-organizing phenomena, which often run spontaneously (given some initial/boundary conditions), biological organization involves a thermodynamic effort right from the beginning (that is probably the reason why proto-metabolisms are not so easy to implement).

A good number of labs and researchers are actually focusing on how fundamental metabolic cycles and synthetic pathways (as they are realized in biochemistry, or in similar versions) could run under prebiotic conditions (i.e., in the absence of enzymes, making use of alternative catalysts) (Keller et al., 2014, 2016; Coggins & Powner, 2017; Muchowska et al., 2017, 2019; Springsteen et al., 2018; Stubbs et al., 2020). These groups are opening the origins-of-life field in a really interesting direction, demonstrating that there could be a natural bridge (or several bridges, right now under exploration) between organic chemistry and biochemistry, to be then reinforced through the development of proteins and enzymes, but not necessarily dependent on the latter at the very beginning. The importance of considering metabolism as the central problem in biogenesis is that the material and thermodynamic hurdles involved become apparent: they actually turn to be the main focus of research. However, the results obtained on these lines, though highly promising, are still far from "minimal metabolisms" because they have not managed to couple the

---

[3] By an "integrated material organization," we mean a molecular system where all components are strongly interdependent and constitute a coherent, operational unit that self-maintains. Minimal robustness, in this context, requires the combination of different physical and chemical factors. More precisely, compositional and interactive diversity, along with phase heterogeneity (the coupling of chemistries taking place in various reaction domains) seems critical to achieve this kind of collective and operational molecular interdependences (Ruiz-Mirazo et al., 2017; Lauber et al., 2021).

reactions with adequate, endogenously synthesized material constraints that should act as first-order control mechanisms on those same reactions. This is crucial, as we will expand below, for any material organization to be able to *construct* itself autonomously (see Ruiz-Mirazo & Moreno, 2004, 2012 and also Lauber et al., 2021).

9.2.2 Protocell Models

Network approaches are mostly concerned with chemical reactions (in particular, their stoichiometry and kinetics). Yet, the problem of origins of life is not only chemical: physics also plays a fundamental, complementary role in it. In support of that claim, one can always bring to the fore the fact that all biological systems heavily rely upon boundaries and compartments, as their universal cellular character indicates, which has deep energetic and thermodynamic implications (Harold, 1986, 2001). Following this premise (more explicitly stated in footnote 3), the protocell research camp has flourished in the last couple of decades. There were, of course, remarkable pioneers earlier on, starting from one of the founders of the origins-of-life field, Oparin, but also including other key figures, like Deamer or Luisi, who defended the prebiotic importance of lipid compartments in times when it was still a rather marginal line of work (for a nice review on the history of the field, see: Hanczyc, 2009). The situation changed with the turn of the century, when the "lipid world" hypothesis was introduced (Segré et al., 2001) and highly influential researchers, like Jack Szostak, coming from the RNA-camp, started investigating protocellular systems in depth (Szostak et al., 2001).

This contributed to widen the field of origins of life, embracing in the same move some of the non-reductionist postulates coming from the field of systems biology (Ruiz-Mirazo et al., 2014, 2017). Indeed, the assumption that protocellularity is central in the early stages of biogenesis brings forward a concept of prebiotic individual that goes definitely beyond the molecular level: rather than populations of molecules as such, what one should consider is populations of molecular *organizations* constructed within compartments. Or, more accurately expressed, one should consider molecular organizations that also build their own boundaries and constantly traffic with matter and energy through them to achieve a precarious self-maintenance, with potential to propagate through reproduction and evolve as a protocell population. Taking seriously into account a global constraint, like a vesicle membrane, that derives *from* and exerts spatial control *on* a set of encapsulated chemical species/transformations (introducing new rules for dynamic behavior that need not be strictly stoichiometric – e.g., osmotic and volume effects, generation/management of electrochemical gradients) has far-reaching implications, both in a proto-metabolic and in a proto-evolutionary sense. Unfortunately, many chemists feel out of their "comfort zone" working with colloidal systems (like lipid vesicle suspensions), hence their traditional reluctance to investigate this domain. But postponing the problem of compartmentalization to later stages in biogenesis only makes it worse (Piedrafita et al., 2012; Szostak, 2012), and the community is beginning to realize this.

Thus, during the last two decades, there has been a remarkable increase in the scientific exploration of protocellular systems, in diverse directions, and taking up both bottom-up and top-down approaches. The development of synthetic biology, in particular the "synthetic cell" research program, has also contributed to this expansion (de la Escosura et al., 2015), even if most of that work is far from being prebiotic, and often just recreates biochemical processes under well-controlled, artificial conditions (e.g., through the use of synthetic liposomes). Nevertheless, understanding the principles of organization underlying real, prokaryotic cells (the *end result*, from an origins perspective) or simpler, hypothetical versions of them (the *intermediate steps*) also requires making use of material components and conditions that are alternative to the standard, biological ones. In this vein, we will briefly review here experimental work that is especially interesting from a particular theoretical perspective on biology and primordial biogenesis, the "autonomy perspective" (that we embrace and have contributed to develop (Ruiz-Mirazo & Moreno, 2004, 2012; Moreno & Mossio, 2015)),[4] but without paying so much attention on whether the material aspects involved in those (proto-)cellular systems exactly match the actual biochemistry and biophysics that we know on planet Earth.[5] The autonomy view, when focused on the process of biogenesis, is particularly interested in finding paths toward prebiotic systems whose internal complexity (i.e., the diversity of components and interrelations among them) is organized in such a way as to achieve their own sustainability. Namely, systems that can build – at least, part of – the boundary conditions that allow for their existence as precarious organizations in far from equilibrium conditions.

Maturana and Varela (1980) and their theory of *autopoiesis*, forerunners in this way of thinking, were more concerned about capturing the "organizational core" of the living phenomenon than to understand its origins. There were others, like Luisi, who took up the job of trying to implement those ideas in an empirical research program that could illuminate biogenesis (Luisi & Varela, 1989; Walde et al., 1994; Luisi, 2006; Bich & Green, 2018). That research program, established 30 years ago, is still active and giving interesting results, e.g., Hardy et al. (2015); Post and Fletcher (2020). The main motivation that articulates this type of investigation (which blends very nicely with our conception of the origins of life as the evolutionary development of autonomous, protocellular systems (Ruiz-Mirazo & Moreno, 2004; Shirt-Ediss et al., 2017; Ruiz-Mirazo et al., 2020)) is the exploration

---

[4] The idea that biological organisms are autonomous systems has deep historical roots, although the modern explicit use of it can be attributed to the Chilean biologist Francisco Varela (Varela, 1979). The general claim is that the property of autonomy can be naturalized and applied to molecular systems with an organization that produces and maintains itself.

[5] This is often taken as a criterion for prebiotic plausibility, but we consider it is somewhat narrow-minded (too Earth-chauvinist, as it is commonly expressed), especially from the wider perspective that fields like astrobiology, artificial life, and synthetic biology have given to the problem of origins.

of how the generative power of chemistry (typically, autocatalysis) can be coupled to the (self-assembly) dynamics of the compartment, so that relatively simple protocells stay in non-equilibrium conditions, through that mutual reinforcement, becoming *active* and, potentially, *reproductive* systems.[6] This (the capacity to make thermodynamically viable the synthesis, growth, and reproduction of a system) is of fundamental importance not only to understand how chemistry may get organized *biologically* but, furthermore, to realize how such an achievement actually requires the unfolding an evolutionary dimension – an aspect that was utterly disregarded by the autopoietic school.

Other "bottom-up" approaches, like the one pursued by the Szostak's lab, have provided key insights into protocell growth and division processes, usually in the context of a population of vesicles competing for the available lipid monomer, either through osmotic effects (Chen et al., 2004), differences in the membrane lipid composition (Budin & Szostak, 2011), or internal synthesis of a hydrophobic compound (e.g., a dipeptide) that could spontaneously join the membrane (Adamala & Szostak, 2013). However, despite some interesting excursions into aspects like vesicle homeostasis (Engelhart et al., 2016) or membrane functionalization, combining lipids with peptides and RNA (Izgu et al., 2016), this group has not focused on the development of autonomous protocell behavior, as such, but on finding an adequate companion for RNA evolution, so that natural selection starts operating at a supramolecular level. Yet, as we already argued above, when an evolutionary scenario is advocated as necessary to tackle the origins-of-life problem, this should be done taking into account the organizational complexity of the primitive individuals involved (like it is shown, for instance, in Piedrafita et al., 2017). On those lines, a former researcher of Szostak's lab, Sheref Mansy, has recently established an independent line of research that is more directly tackling the issue of how complex should the "original protocells" be. In other words, is there a minimal threshold of complexity, like we suggest in Ruiz-Mirazo et al. (2017) and Lauber et al. (2021) for prebiotic evolution to get started? How many different constraints (i.e., material controls) must be put together to reach the platform for taking off? The Mansy group are pushing quite promisingly in this direction, in an effort to combine compartments, catalysts, energy currencies within the same experimental system (Bonfio et al., 2017, 2018), keeping also an eye on how chemical diversity can contribute to protocell growth and division processes (Toparlak et al., 2021).

In addition to these "bottom-up" strategies that start from scratch, so to speak (i.e., from physics and chemistry), other researchers take minimal-life exemplars (microorganisms or parasites) and try to simplify or deconstruct them. From such a "top-down" perspective (which should show us the finish line for the process of primordial biogenesis), there have been very interesting results in the last years, as well. The new Craig Venter *Mycoplasma* construct (Hutchison et al., 2016) was of course a landmark, in that regard: it has provided plenty of opportunities for further

---

[6] That coupling between chemistry and compartment may actually be considered as the key feature to define what a "protocell" is (Ruiz-Mirazo, 2011).

exploration, not only about its physiology and metabolism (a highly complex, genetically instructed metabolism, as one could expect (Breuer et al., 2019)), but also about its reproductive potential or reliability (Pelletier et al., 2021). Although these minimalist approaches push in a direction in which both the autonomy of the cells (i.e., their actual capacity to survive in "free-living," changeful environmental conditions) and their reliable reproduction (i.e., their ability to generate "normal offspring") are taken to the limit, their study is critical to discern, precisely, the boundaries of biology. In a similar vein, "semisynthetic" constructs, like the bioreactors developed by Noireaux et al. (2011) or, more recently, Blanken et al. (2020), are also very informative. These involve biomolecules and other parts/subsystems of biological organisms under compartmentalized (*in vesiculo*) artificial conditions, with the aim to investigate the complementary relationship between membrane and endogenous reaction pathways, specifically focusing on the implications for autonomous behavior – an illuminating and very interesting line of work for the future.

Nevertheless, in order to conclude this section, we must acknowledge that the empirical evidence available to date is still clearly insufficient to elaborate a minimally consistent and complete *organizational* account for the origins of life, in the sense of establishing a plausible sequence of transitions that cover all the ground from complex, non-equilibrium chemical systems to the simplest biological ones. There are theoretical models (in particular, protocell models – from the classical (Varela et al., 1974; Ganti, 1975; Dyson, 1982) to much more recent and refined ones (Ono & Ikegami, 1999; Castellanos et al., 2004; Macía & Solé, 2007; Mavelli & Ruiz-Mirazo, 2007; Ruiz-Mirazo & Mavelli, 2008; Van Segbroek et al., 2009; Mavelli, 2012; Shirt-Ediss et al., 2015; Piedrafita et al., 2017; Pechuan et al., 2018; Attal & Schwartz, 2021) that try to fill in the current holes and open new avenues of research. The advantage of the latter (as compared to strict molecular simulations of prebiotic chemistry, usually linked to the network models reviewed above – or to other protocell models that tackle evolutionary dynamics but simplify so much organizational aspects that cannot be called properly "protocellular," e.g., Kamimura and Kaneko (2010, 2019) is that they offer a richer picture in terms of "constraint-based" or "rule-based" modeling techniques (see Lauber et al., 2021 and references therein). Thus, they are probably much closer to the complex reality of the first protocells that were involved in the process of biogenesis. Yet, without more solid, ample, and informative experimental results, it is very difficult to move forward through theoretical-computational approaches.

Furthermore, by focusing on the organization of individual protocells, like most of the previous works do, we may be limiting ourselves, "hitting a wall" that is there, but not so easy to see. In other words, we may be overlooking a fundamental bottleneck that needs to be addressed at the population level, as will be discussed in the remaining of the chapter. The crux of the matter will be finding the adequate balance between the two perspectives, organizational and evolutionary, and how they actually get intermingled. The discussion that we open here, in any case, points at one of the most difficult issues in the problem of origins of life, for which scientific insights and methods are still to be developed, so our aim is just to pose it in conceptual terms and draw some implications for future research.

9.3   The Interweaving of Organizational and Evolutionary Processes in Biogenesis: A Complementary but Causally Asymmetric Relationship

The historical dimension of life is a commonplace, a recurrent theme. While life is manifested in the form of organisms (and associations of organisms, whose spatial borders are often not so trivial to determine), the complexity of their material components and organization seems inexplicable unless appealing to a long process of evolution, beyond the time span of each of those individuals. «Nothing in biology makes sense, except in the light of evolution», as Dobzhansky famously remarked. This means that we must consider systems that, before disintegration, reproduce the essential features of their organization (what is usually understood, in general terms, as "heredity") and generate, in this way, a set of causal entailments that propagates in time and space, transcending the limits of such an organization (i.e., of the actual organization that constitutes each individual). The collection of temporally similar systems brought about through reproduction across "successive generations" constitutes a "lineage" (or a "phylogeny" – when genetic mechanisms are under focus). In that context, where the analysis must obviously scale up to a "population" level, variability also tends to be assumed (linked to some inevitable, random modifications) in the reproductive success of the individuals (i.e., their "fitness"), which leads (through a combination of selective pressures and cooperative dynamics) to a highly complex phenomenon that shows both long-term maintenance, in a basic sense, but also continuous change and diversification along the way.

In contrast to the intricate molecular and energetic couplings that constitute the organizational core (i.e., the metabolism/physiology) of organismic processes, as we discussed in the first part of this chapter, evolutionary processes cover a completely different dimension of the phenomenon of life, where causal connections extend across much larger temporal and spatial scales. In fact, the interesting point is not just that evolutionary processes are spatially and temporally wider than organismic processes: they are also ontologically different. They concern population dynamics, in which remarkably looser "organism-environment" and "organism-organism" causal interactions (i.e., less demanding or stringent than the molecular interactions *within* each organism) are the key. An additional peculiarity is that the relevant effects of these interactions can only be adequately analyzed statistically and, even more importantly, through a very long time window: a time window during which most of the causally responsible entities or agents (the actual organisms, the "tokens") have already disappeared, after participating in a *sedimentation* process of the most successful lineages (i.e., the "types," which are conserved).[7] Thus,

---

[7] We will use here the term "sedimentation" (Walsh, 2018, personal communication) as a generalization of the idea of "selection." Evolution does not only result from competition dynamics but also from cooperative relationships among the individuals/agents of a population, which play an active role in the process (Walsh, 2015). The idea of sedimentation conveys long temporal scales, in which different types of hereditary mechanisms, with different degrees of reliability (i.e., different "trans-generational depth," genetic and nongenetic) could be operating in parallel (Danchin et al., 2019).

the organismic and the evolutionary dimensions of life, despite being deeply entangled (and necessarily so, as we will expand below), hold an essential asymmetry: the former relies on molecular components, processes, and interactions that continuously sustain each other in a tightly cyclic, self-constructing, and self-referential manner, whereas the latter is the result of an open, long-term, and much wider process of sorting out that takes place in populations of reproducing agents, across many successive generations (Ruiz-Mirazo et al., 2020).

How can all this get started? And in what sense does the origin of such a complex, asymmetric entanglement help us understand the unfolding of a biological domain? Well, a central issue that must be highlighted straightaway (in line with what we just described in the previous section) is that the first chemical systems with potential to start turning biological were relatively complex but still precarious, given their far-from-equilibrium nature. How were these systems, then, capable of increasing their stability and robustness? We should realize that this is not a trivial task, especially if it requires an effort of synthesis of progressively more complex molecular ancillary. Fortunately, steady self-maintenance in this context would be the exception, rather than the rule: vesicles in heterogeneous, changing conditions naturally tend to undergo fission and fusion processes and more so if they are coupled with physical gradients and chemical reactions (see, e.g., Carrara et al., 2012; Oglêcka et al., 2014; Toparlak et al., 2021) for an experimental survey of this type of scenario). In other words, it is more realistic to consider that the large majority of such primitive protocells were very dynamic (favoring either growth or shrinkage, potential division, intermingling, decay, etc.) not organized for the stabilization of a steady state – like it is often assumed in theoretical models about minimally autonomous (autopoietic) protocells. Therefore, one should imagine this setting as a mess of diverse "populations" of organizationally similar systems, i.e., groups of growing and dividing protocells with their own suite of dynamic and plastic behaviors, which also brought about many processes of merging and content reshuffling (e.g., through vesicle fusion).

The advantage of such a scenario is twofold: (i) on the one hand, protocells would have an intrinsic tendency to grow and divide, to reproduce and propagate;[8] (ii) this intense activity would be an obvious source of novelties, which could eventually be kept in the system if they contributed to the far-from-equilibrium maintenance of a given type of protocell (including here the first point, too – i.e., *maintenance through reproduction*). Nevertheless, these mechanisms for preservation through statistical reproduction and generation of molecular novelties (that could be recruited for the protocellular organization) were probably quite poor during the initial stages. Under such conditions, the main "driving forces" for prebiotic complexification would depend on some specific boundary conditions (a range of temperatures, osmolarity, pH values, gradients, etc.) that could sustain protocell

---

[8] By "reproduction" (or "propagation" (Kauffman, 2000)) of an organization, we mean here the process through which a complex system (in this case, a protocell) generates physically detached similar systems (i.e., other protocells with a similar material composition and organization).

synthesis and dynamics, rather than on the robustness of their internal self-constructing organization or on their agency.

Anyhow, that incipient capacity for propagation of an organization could explain that, at a given stage, growth and fission led to the generation of protocells capable to reproduce through some primitive (still statistical, stochastic) mechanisms of transmission of their compositional and organizational identity (as suggested, for instance, by Segre and Lancet (2000) through their "composome" idea).[9] The iteration of self-reproducing cycles would generate a somewhat longer-term continuity of a specific *type* – the incipient lineage – constituted by populations of similar self-reproducing protocells. At the level of each protocell – as a particular *token* – the mechanisms involved in its reliable reproduction would trigger a diachronic succession of similar self-reproducing organizations, and in this way the innovations may have been retained beyond the particular fate of each individual protocell (Ruiz-Mirazo et al., 2020).

Hence reproduction (viz., growth and fission ending up in at least one new, physically separated, similar entity) would become the way in which the system displays its own far-from-equilibrium self-maintaining dynamics, and, at the same time, the consequence of these dynamics is the maintenance of a similar type of protocells (a particular "protocell lineage") through the continuity of generations. Interestingly, some proto-organismic innovations could be thus stabilized, and, even more importantly, this trans-generational continuity (type preservation) would also allow organizational changes, because it may have involved the accumulation of variations across long time periods. All this, of course, would be enhanced if molecular mechanisms to *record* (at least, to some extent) the increasing complexity of the protocell were developed, in parallel, at the molecular level (e.g., template mechanisms) giving way to progressively more reliable "hereditary" transmission of various features – even if the evolutionary (trans-generational) depth of these mechanisms would be rather small at those early stages (nothing comparable to later, genetic mechanisms).

The central issue here, in a situation in which nature must have faced a huge bottleneck (perhaps the biggest bottleneck it has ever faced), would be to develop material constraints that would enable these systems to solve two fundamental problems at once: (i) increase the robustness of the precarious individuals/agents and (ii) preserve the level of complexity they reach, in a way that is both operational for each individual, during its existence as a protocell (its "proto-ontogeny"), and for the collection of individuals it may bring about (its "proto-phylogeny" or "proto-lineage"). Solving these two problems requires obviously higher metabolic efficiency of the protocells, which is necessary both to ensure maintenance against perturbations and reliable reproduction. But the key is to realize that the solution is not at reach for any kind of metabolism: a remarkable threshold of molecular and organizational complexity must be reached, and systems below that threshold will

---

[9] This proposal was made in the wider context of a "lipid world" (Segré et al., 2001), some of whose assumptions we share, but some others we don't (in particular, the open-ended character of the evolution that such systems could implement – see: (Ruiz-Mirazo et al., 2008)).

naturally tend to decay. Von Neumann's idea of the "universal constructor," of course, resonates with force in this context (i.e., the problem of determining the logic of a system, the architecture of relationships among its operational modules, so that it builds itself, avoiding disintegration, across generations – see McMullin (2000) and Ruiz-Mirazo et al. (2008) for a more extended discussion). Yet the way this issue was originally posed avoided many aspects that had to do with the physical/material implementation of the systems that could become universal constructors, which might be crucial, as von Neumann (1966 [1948]) himself acknowledged. By focusing on the question of primordial biogenesis, we are precisely trying to naturalize the problem, going all the way back to its primary roots, and taking up a conceptual but unmistakably nonabstract standpoint.

Thus, as we were saying, evolutionary changes are the consequence of a long, historical series of causal actions performed by particular protocells belonging to a "proto-population" (or a family of protocells). Most of these changes (generated through a large number of reproductive cycles in a pool of similar systems) will be lost; but some variations in certain protocells will contribute to an increase in organizational integration and adaptive potential, generating more stable and somewhat deeper lineages (proto-phylogenies). In this way, trans-generational continuity may afford the maintenance and slow transformation of protocell lineages, facilitating the appearance of new protocellular types, whose organization is metabolically more efficient and has more and more control over external conditions. This is why reproduction with heredity is so important for the progressive complexification of such proto-organisms, allowing for their transition toward full-fledged biological organisms. Therefore, the evolutionary dimension indirectly (but with an increasing weight) affects the composition and organization of new generations of protocells, and more so as their reproductive capacities become more reliable, enabling higher and higher levels of sustainable organizational complexity.

In Sect. 9.4, below, we will give a rationale about the actual transition from these initial stages toward a situation in which the interweaving between the organismic and evolutionary dynamics becomes really profound, inextricable, as it is necessary for the unfolding of biological phenomena. But, once the complementarity between these two dimensions of life has been brought to the fore, let us say a few more words on the asymmetry involved in their relationship. Metabolic organization (the core of the individual dimension) is run and maintained in each (proto-)organism through a set of "rate-dependent" causal connections (Pattee 1977): namely, causal connections that crucially depend on specific conditions of distance, velocity, and energy requirements. In Ruiz-Mirazo et al. (2017), following Ruiz-Mirazo and Moreno (2004), we propose a minimal set of (first-order) control mechanisms that would be necessary to keep basic autonomous systems, like these protocells, in far-from-equilibrium conditions (kinetic, spatial, and energetic control mechanisms, more specifically). In any case, what is important to highlight for the discussion here is that this kind of organization can only stand robustly and efficiently on its own feet if, and only if, its constituent parts are highly *integrated*. The idea of organizational integration (as it was discussed in the first part of the chapter) expresses the fact that the different parts and processes of a system are highly interdependent:

there is a need to coordinate the distances, times, rates, and energies involved in all of them. And when the system's complexity increases, the need to introduce regulatory mechanisms that reorganize some parts in differentiated levels (constraints on top of constraints) also becomes apparent (again, see Sect. 9.4, below, for further explanations).

Thus, the pressure for integration is inherent in any system whose identity is based on a far-from-equilibrium, cyclic set of synthetic processes (always coupled to matter and energy sources from the environment), namely, on a logic of self-construction that depends on the specific energy requests and the actual rates of their (always precarious) constitutive/interactive dynamics. That is why such systems cannot increase in complexity unless they enlarge the web of endogenous (higher-order) constraints and their assorted integration – including mechanisms to control the relationship with the environment (which will lead to the development of minimal forms of agency). In sharp contrast with this, the maintenance of an evolutionary process, per se, is much less demanding. Or, rather, it is demanding but in a completely different way: what matters there is the reliability in the transmission of constraints across generations, within the dynamics of populations of reproducing systems, all of which is averaged out in a very long and complex sedimentation process. In this context, part of the causal connections operate as if they were "rate-independent" (Pattee, 1977), even if they must be continuously supported by the set of (rate-dependent) cyclic causal connections that constitute, maintain, and reproduce each protocell in far-from-equilibrium conditions. As the mechanisms of heredity (or "control on variability" (Ruiz-Mirazo et al., 2017)) become more and more reliable, the relevant historical series (i.e., the "trans-generational depth" or the average number of generations through which those constraints do no suffer relevant changes (Danchin et al., 2019)) becomes longer, more relevant, and profound in evolutionary terms. This has very important implications for "open-ended evolution" (Ruiz-Mirazo et al., 2008), as we will recall in the next section, but a fundamental related issue must be explicitly addressed first: functional expansion and diversification.

9.3.1   Trans-generational Constraints and the Expansion of Functional Space

The emergence of a functional domain (a world where material systems exist by virtue of what they do – i.e., by virtue of their dynamic causal effects (Mossio et al., 2009; Moreno & Mossio, 2015)) is important in this prebiotic context precisely because it is behind the key fact that during biogenesis chemical diversity gets reduced and narrows down to a relatively small subspace of "the molecularly possible." As the rich mess of prebiotic processes and material transforms into more elaborate chemical organizations, a progressive selection takes place, favoring those molecular components capable of putting together cohesive far-from-equilibrium systems. Regardless of the time this may take, only those components that have allowed further complexification will be retained, and that has some important implications. In particular, it means that the chemical diversity will suffer a significant decrease, as this is a condition for systemic and highly integrated material organizations. The development of the necessary mechanisms of control (spatial, catalytic, energetic) actually requires fixing some of the molecular rules and components operating in these systems. More specifically, a subset of chemicals and reaction processes must be chosen both to generate components of control (internal constraints) and to be amenable to that autonomous control.

This, in our account, coincides with the emergence of "minimal metabolisms" (Lauber et al., 2021) and is actually the first moment in natural history where one can begin to speak properly in terms of *functions*, the claim being that these don't emerge "one-by-one" (Ruiz-Mirazo et al., 2017): a combination of endogenously produced and tightly coupled constraints (operating as first-order control mechanisms on the underlying, far-from-equilibrium reaction network) must come together, from the very beginning, so as to constitute a minimally robust chemical system, similar to the protocell systems that we described at the end of Sect. 9.2. Therefore, the basic idea of "functional organization" (as an enduring form of self-maintenance) is deeply linked to that of material control and organizational integration. In this context, we should remark that the appearance of self-reproducing protocells already requires, as a precondition, the existence of populations of protocells with – still strongly limited but – nontrivial functional domains. The reason is that the reproduction of a protocellular and minimal metabolic organization involves managing quite a number of processes, like the duplication of certain structures of the system, coordinated with surface increase (and other modifications) in the compartment, as well as with an adequate temporal and spatial allocation of the components during growth (so as to ensure that, when fission actually occurs, the new entity is able to repeat a similar self-productive cycle). In other words, reproduction requires a fair degree of control of the proto-metabolic processes, since growth and fission are the specific expression of the self-production regime of these protocells (Mavelli & Ruiz-Mirazo, 2007). One could say that spatial, kinetic, and energy control mechanisms, including the suitable coordination among them, constitute the necessary functional basis for any reliable trans-generational propagation of protocell organization (Moreno, 2019).

One should also recognize that the very idea of reproduction (see footnote 8) implies, right from the start, a minimal degree of reliability or "inheritance," namely, the new, spatially separated entity has to be molecularly and organizationally similar to the parental entity. The first forms of reproduction would have been statistical, which means that similarity among the different members of the progeny was ensured only partially – at some percentage, so to speak. As it is argued in (Danchin et al., 2019), an increase in the reliability of reproduction was most probably a consequence of a stronger degree of functional integration, and this, in turn, through reproductive steps, resulted in the selection – or sedimentation – of the most efficiently integrated protocells, which nicely illustrates how the aforementioned link between evolutionary and physiological (proto-metabolic/protocellular) processes can be, in practice, coherently articulated.

Conversely, the incipient connection between these two phenomenological dimensions (that will develop and get reinforced throughout primordial biogenesis) has really interesting and far-reaching consequences with regard to the functional domain itself, which can expand through novel ways of contributing to maintenance that become available to those protocellular populations. Indeed, such a connection opens the door to a completely new set of functionalities, which lie beyond the strictly physiological sphere of each protocell. A function in this extended functional domain can acquire "temporal/historical depth," in so far as some feature/property of the system is linked to the new ways of ensuring organizational maintenance across generations. This allows to establish a natural conceptual bridge between the *organizational* and the *evolutionary* interpretations of function (Saborido et al., 2011). Indeed, from this stage onward, it makes sense to say that a trait (a component, a mechanism, a property of the organization) X in a population is there because it has been selected/sedimented through a complex evolutionary pathway. In other words, in our prebiotic context, X would be there because those protocells that bear X – and were capable to transmit it to their offspring – have a (relatively long-term) history of reproductive success through which X remains in the population. In this sense, we must open ourselves to the possibility that there are functions whose contribution to the current individual organization of the system is not so obvious, and they should be analyzed in a wider time frame, i.e., there could be functional traits that contribute to the maintenance of the *type* and, thus, only indirectly to the maintenance of any particular *token*.

Let us explain how, in just a couple of paragraphs, before moving on. As we have discussed, the possibility that some protocells managed to achieve relatively reliable reproduction cycles would depend critically on the synthesis of a number of material constraints controlling the processes of growth and fission. Certainly, there would be an organizational and material continuity between the initial, "mother protocell" and its subsequent offspring and, in this sense, the functional role of the constraints more specifically involved in the reproductive processes would not be distinguishable, in principle, from the nonreproductive functions. However, more and more reliable self-reproductive systems require additional control mechanisms: in particular, hereditary mechanisms that should be focused in managing the variability generated in these protocells, preserving the level of complexity reached and making "statistical numbers," so to speak, "no-longer-statistical." But this, in turn, requires, as we will expand in the next section, a *dynamic decoupling* with regard to the current organization and the specific times, rates, and energies required by each individual metabolism. In a concurrent way, the organizational architecture of the protocells must be profoundly modified, through a *hierarchical coupling* with these new mechanisms that can no longer be considered, simply, as the result of the constructive power of each individual but rather as the result of a much more complex evolutionary process in which whole populations are involved.

Somehow, we are facing a scenario in which the organization of individual entities transforms the way in which evolution occurs and evolution also transforms, more and more profoundly, those individuals. But what should be especially underlined here is that all this takes place in the context of – and thanks to – the capacity of these protocellular systems to enlarge and diversify, enormously, the space of possible functions though which they are realized (starting from that initial, minimal set that we mentioned above). Although a good part of such a space will be filled by strictly physiological control mechanisms, some other regions will not simply belong anymore to individual "tokens" but to the "types," the lineages, that consolidate through longer-and-longer-term population dynamics. Therefore, the main problem at this stage (and from this stage onward) is not dealing with chemical diversity, heterogeneity, and messiness (what is classically regarded as the combinatorial explosion of molecular interactions and transformations) but dealing with an increasingly rich space of functionalities, expanding in different – though interconnected – directions. A fundamental issue will be addressed next, in Sect. 9.4.

9.4 "Dynamical Decoupling": A Key Principle to Understand the Evolutionary Development of Complex Material Organizations

Taming complexity in systems that develop numerous functionalities and thus, a large space of possible dynamic states/behaviors is not a trivial task. These systems can realize in multiple ways their basic constitutive regime, as self-constructing protocells (minimal metabolisms) in constant interaction with a variable environment, shifting from one stationary state to another, depending on the conditions that they meet at any given time. More precisely, dynamic multistability poses a remarkable organizational challenge in an evolutionary setting like the one we just described above: the challenge of how to navigate efficiently that space without wasting time and resources that could be critical for the persistence of the individuals involved. This is a problem that cannot be taken for granted, nor assumed to be spontaneously solved by nature: it requires work, literally (viz., in a thermodynamic sense), and time, plenty of time, to develop the necessary mechanisms. In line with other authors that have previously addressed it (in particular Christensen, 2007), we consider that simple feedback mechanisms, or even combinations of positive and negative feedbacks, if they work "online" (at the same rates/conditions in which metabolic processes take place), are not sufficient to deal with it. Let us try to explain, briefly, why.

In principle, when facing perturbations, a functional system can restore its constitutive and behavioral coherence through *self-organization*[10], namely, through parallel local interactions that generate emergent outcomes (without making use of specifically devoted mechanisms of regulation or higher-order controls). Yet, this solution only works when the number of different functions to be coordinated is not very high. As the complexity of a system increases, these dynamically coupled

---

[10] In the context of our discussion here, this would apply to *minimal metabolisms* (i.e., *basic "self-construction"*).

("online") mechanisms become clearly insufficient. As Christensen (2007) rightly points out (reasoning in a cognitive context but using arguments that are perfectly applicable at a much more basic level), self-organization has intrinsic limitations to achieve functional coordination. The reason is that the process of reaching a certain global state, in such a case, depends on the reliable concatenation of state changes through local interactions, and those cascades of events add a delay as the functional diversity of the system increases. Nevertheless, robust global coherence/behavior against variations requires selecting very precisely a given dynamical attractor and maintaining the system there during a given period of time. Therefore, if the organization of the system stays "flat" and "online," it faces an obvious dilemma: either its capacity to generate multiple finely differentiated global states is limited, or, instead, the system will have to sacrifice the reliability of attaining a specific state out of that multiple choice. In Christensen's own words, «slow action and poor targeting capacity severely limit the capacity of self-organization to achieve the kind of coherence that functional complexity requires (...) Consequently, the most effective means for achieving the type of global coherence required for functional complexity is through regulation, including feedback mechanisms and instructive signals operating at both local and larger scales. The key feature that distinguishes regulation from self-organization is the presence of a functionally specialized system that differentially specifies one or a restricted set of states from the range of possible states the regulated system might take, based on the sensing of system conditions and the production of control signals that induce changes in functional state» (Christensen, 2007, pp. 265–266).

In Bich et al. (2016) we argued, precisely, that (biological) regulation involves second-order control hierarchies that necessarily work "offline" in a relevant sense, or to a relevant extent. In other words, achieving effective control when the complexity of a system is very high requires a subsystem that is endogenously synthesized but operationally decoupled from the dynamics of the controlled processes, so that it can be modified without disrupting those underlying synthetic processes (Bechtel, 2007; Bich et al., 2016). Minimal metabolisms, as generally characterized in Lauber et al. (2021), do not constitute completely "flat" organizations, in the sense that they do require first-order controls (i.e., a set of elementary constraints) to operate. But regulation involves constraints on constraints, which make decoupling mechanisms effectively feasible in an autonomous organization. Basic, first-order controls are required to put the system together, but they are too closely engaged in the metabolic dynamics to be able to work "offline." The question that we must address here, in any case, is why minimal forms of regulation, interpreted precisely in this vein (i.e., already implying a *dynamically decoupled* but *hierarchically coupled* individual system organization), were necessary during primordial biogenesis and how they were actually implemented in the (pre-Darwinian) evolutionary context of protocell populations described above.

As for the first point, we concluded the previous section highlighting that self-reproducing protocell systems demand, right from the beginning, a rather elaborate set of basic functions (those first-order, material constraints acting as "process controllers": catalysts, compartments, etc.) just to realize themselves and that the prebiotic evolutionary dynamics they bring about would contribute to expand their potentially available space for functionalities (including trans-generational constraints – such as hereditary mechanisms of various kinds). We consider that this hypothetical but plausible protocellular scenario is, indeed, complex enough to defend the need for second-order control mechanisms that help those systems navigate an internal dynamic space with multiple stationary states. The reason why such mechanisms should be considered as "second order" is because they must operate *on top of* the basic set of functions that already put together the constitutive regime of the system, with a variety of accessible dynamic attractors. In brief, the new controllers must be constituted by material constraints operating on other material constraints: there is no other way for nature to do it. And the reason why this action is "offline" has to do with the second point, which we must address now: how were such *hierarchical* autonomous organizations (Pattee, 1973) actually implemented for the first time?

In concrete operational terms, regulation is commonly understood as the harnessing of a system according to a set of rules (e.g., «in case of situation X, do Y»). Contrary to what occurs in artificially designed systems, where the rules are a collection of external norms, in natural (biological or infra-biological) systems, the idea of regulation points to an internal set of constraints that *functionally select* some specific dynamical configuration of the system, among several possibilities, as we expressed above (and as some other authors have also argued, to distinguish this estate of affairs from strict or minimal autopoiesis (Di Paolo, 2005)). Yet, what kind of "function" is this? In principle, it looks physiological, difficult to distinguish from the other, elementary ones – since it is exerted in ontogenic time scales, as an adaptive response of the individual, here and now, to a given environmental challenge: e.g., «if this nutrient is detected, swim up its gradient» or «if this toxin is found, do not absorb it». Nevertheless, these behavioral *shortcuts* are quite more complex than direct controls on a process. In fact, when one thinks carefully about their emergence, they cannot be easily understood outside an evolutionary perspective: regulatory mechanisms definitely seem to require a different time scale to appear and get stabilized in the population. They look anticipatory, when they are analyzed at the scale of a single individual – who "seems to know," in advance, the outcome of its actions. Instead, these self-imposed instructions most probably come from a history of interactions that have taken place in the population and are linked to the persistence of those individuals, but throughout many generations.

As we explain in more detail in Bich et al. (2016), the nature of regulatory mechanisms is not straightforward: they involve material gears to shift from one constitutive regime to another depending on circumstances that must be associated to internal/external variables but without responding directly (through online mechanisms) to those actual variables (e.g., the concentrations of metabolites in the system). The system must "detect" internal/external circumstances selectively, which means distinguishing some inputs as "signals" that will trigger a rapid shift (the adaptive shortcut) to a given behavior (an alternative stationary state). Thus, it is not easy to describe in detail how these regulatory meta-constraints (including the molecular machinery that determines accurately when and how their action should be executed) could have appeared. Further empirical and theoretical research needs to be carried out on this topic, within a pre-Darwinian evolutionary setting where different stages are distinguished and compared to the pre-regulatory (i.e., minimal metabolic protocell) phase. Yet, it seems quite reasonable to conjecture that regulatory mechanisms should be the result of a long series of "trials and errors," in the context of protocell population dynamics in which subsequent generations of prebiotic individuals were developing, competing for resources, probing their local environments, etc. How it actually happened, putting all the pieces of the mechanism together, avoiding potential disintegration pathways, overcoming external perturbations, and keeping internal coherence, will not be obvious, but if it came about, the regulatory device would for sure be retained, because of its immediate contribution to the persistence of those protocellular systems that integrate it in their organization.

Interestingly, regulatory mechanisms may constitute one of the most prominent pieces of evidence to demonstrate that a sedimentation process is taking place at larger and longer scales, with very important implications at the level of the individual, here and now. The history of interactions of a population of similar protocells with their environment (including the interactions among them) gets eventually distilled or condensed into a relatively complex, built-in mechanism that ensures higher robustness and better adaptivity (quicker responses) by the members of the population to certain variations in the medium. In other words, regulatory meta-controls somehow reflect, also due to the intrinsic dynamical decoupling they involve, the interweaving between the physiological and the evolutionary dimensions of biological (in this case, proto-biological) phenomena. This interweaving is asymmetric, as we discussed in Sect. 9.3, because the physiological sphere always has causal priority (real self-constructing individuals are the material agents performing all relevant interactions, after all) even if the evolutionary sedimentation process, working at larger and longer scales, has a deep impact on the physiological mechanisms and organization of the resulting individuals.

However, regulation by itself is not enough to ensure reliability in the transmission of increasingly complex molecular and organization features to the offspring (including the regulatory apparatus itself, which could also face the risk of getting lost on the way). Hereditary mechanisms must be specifically developed for such a fundamental task. The conservation of system features across generations can be implemented through different means (and then interpreted according to different theoretical frameworks – e.g., Bonduriansky and Day (2018) or Mossio and Pontarotti (2020)), but we are particularly referring here to *molecular records* (Pattee, 1969, 1977) that, through their template properties, are capable of replication with conservation of their monomeric sequences. These hereditary mechanisms would be completely futile if they were not linked to concrete functionalities of the protocells in evolution, in either metabolic or global reproductive terms. Under the hypothesis of an "RNA world," the same kind of molecule could be carrying catalytic power ("proto-phenotype") and replicative potential ("proto-genotype"), but such a reductionist interpretation does not fit in our account. From a more encompassing organizational perspective, like the one we embrace here, the phenotype-genotype mapping would be quite more complex, right from the beginning.

In such a context, as primordial biogenesis proceeded forward, protocells would inherit, among many other components, molecular records whose functionality would be the result of a longer and longer evolutionary process, beyond the "life span" of each of them. At the same time, each hereditary record would play a key causal role in the metabolic organization of the protocell where it exists. In other words, a trans-generational constraint of this kind also embodies two different temporal scales: one that corresponds to its causal activity in the current physiological processes of the protocell and another one that corresponds to the long evolutionary history that has shaped the specificity of its functional sequence. Thus, hereditary mechanisms bring some other kind of dynamical decoupling into these systems. Records are *dynamically decoupled* from (but *functionally connected* to) the metabolic organization because (like regulatory mechanisms) they have been shaped in a different temporal and spatial domain. But in contrast to the rest of the system components, which functionally depend on each other (in the sense that the effects of some components generate and transform the others), hereditary records are not strictly generated within the metabolic organization of each protocell (although they are physically constructed, repaired, and replicated by it). More exactly, they are materially regenerated, preserved, and used within that metabolic organization, but they are not *informationally* generated within each protocell. And yet, it is precisely for this reason that the specific sequence of those hereditary components – what will come to be their "informational content" – allows for a much more robust and efficient mechanism of reproduction, even if the complexity of the metabolism would be much higher at these later stages. Eventually, "genetically instructed metabolisms" would introduce a completely different way of exploring innovations and variation in time: "open-ended evolution" (Ruiz-Mirazo et al., 2008).

As hereditary mechanisms (and phenotype-genotype mappings, in general) develop in protocell populations, regulation can also be applied, in turn, to all the processes in which those material records (which are meta-constraints, too, of course – but with their own specificities (Pattee, 1977, 1982)) are involved. In fact, the constructive and transformative power of combining these two different modes of dynamical decoupling, as it is reflected in the basic organizational architecture shared by all living beings (prokaryotic cell metabolisms, already endowed with a translation apparatus and a common genetic code), was surely crucial to reach the "hypercomplexity" that life required to maintain itself on the surface of the Earth in the long run (a situation that, we guess, should be similar anywhere in the universe, since the problems addressed here would apply to any material organization dwelling close to the von Neumann threshold).[11]

---

[11] A more extended analysis and conceptual reflection on these issues lie beyond the scope of this contribution but should be an interesting topic for future work.

9.5 Concluding Remarks

The aim of this chapter has been, as indicated in the title, to provide a theoretical framework, a plausible and reasonable account to understand how a biological domain could unfold from complex, nonliving matter. We are convinced that such an intricate transition must be a very long process, involving myriads of molecular systems that generate a great diversity of reaction networks which, over time, lead to increasingly complex material organizations. As this process of biogenesis proceeds, something quite intriguing happens: the phenomena taking place at a given stage, being based on the previous, somehow manage to redefine the conditions, the rules of the game, bringing about systems/organizations that overcome in efficiency and performance the preceding ones and ruthlessly eradicate the latter, leaving no traces behind. However, at a given stage, things radically change: systems sharing an organization with a set of fundamental features similar to what we call nowadays "prokaryotic life" come about and that evolutionary dynamics of "continuous substitution of the old by the new" stops. Not only because there is an unprecedented explosion of diversity and proliferation of these systems (probably all over the surface of the planet) but also because, from that moment onward, subsequent organizational innovations do not (perhaps, cannot) erase this basic type of organization. Instead, all novel biological complexifications become dependent and supported by prokaryotic life – they become, so to speak, curlicues, "convoluted redefinitions" of that same type of phenomenon. Hence, the target of any theory of primordial biogenesis should be to explain how such a fundamental but far-from-trivial material organization (genetically instructed metabolic cells) could naturally emerge.

Within this general context, we have focused the discussion on several key issues. The scientific work reviewed in Sect. 9.2 was mostly related to the early stages of the process: in particular, we collected evidence on how under favorable environmental conditions catalytically driven sets of reactions could turn into self-sustaining protocellular systems, which probably constitute, at those first steps, just a mess of growing and shrinking individuals, only later leading to more sequential fission and fusion events. Then, in the following sections, we explained how, over time, some of such protocells could manage to reproduce their characteristic type of organization, opening in this way a completely new scenario, which has not been explored empirically yet. On the one hand, more and more integrated functional systems (protocellular *individuals* of higher complexity) should start developing. But, on the other hand, lineages of different families of protocells (evolutionary *populations*) would also begin to form. These two apparently orthogonal dimensions of the phenomenon, unfolding in very different scales (both spatially and temporally), get nevertheless deeply entangled. And, through that entanglement, a really powerful driving force is generated that overcomes the apparent physical and material bottlenecks present at those stages, bringing about much more integrated protocells. In turn, these new protocells would not only be more robust but also capable of more reliable reproduction, which would then increase the weight of evolutionary aspects in the process.

Finally, we also discussed the importance of having protocells that develop hierarchical relationships within their organization, namely, complex functional mechanisms that operate on top of the (first-order) controllers of metabolic processes (i.e., only indirectly on metabolism), at rates significantly different from the ones involved in those basic transformation processes, and thus, look as if they were working "offline." Embodied in two very different modes, regulation and heredity, this *dynamic decoupling principle* also seems to play two complementary roles in prebiotic evolution: in the first case, enhancing individual (i.e., ontogenetic) adaptiveness, and in the second, increasing lineage (i.e., phylogenetic) fidelity. Nevertheless, both modes (an effective combination of the two, more precisely speaking) are apparently crucial to complete the process of primordial biogenesis, leading eventually to complex material organizations similar to prokaryotic cells. The physiological plasticity of these cells, together with their capacity for open-ended evolution, lies at the heart of the impressive robustness and long-term sustainability of the phenomenon of life. Modeling all these prebiotic transitions, from initial families of minimal metabolic protocells, to full-fledged living organisms (individuals with a translation apparatus, a complex, code-mediated, phenotype-genotype mapping, etc.) is still a great challenge for science. But making the challenge conceivable, under realistic assumptions, is a first, necessary step to tackle it.

Acknowledgments This research work has been supported with grants from the Basque Government (ref.: IT 1228-19, recently extended to: IT1668-22) and the Spanish Ministry of Science and Innovation (PID2019-104576GB-I00), as well as from the John Templeton Foundation (Project Title: "Integration and Individuation in the Origin of Agency and Cognition" – Grant 62220). KR-M also takes part in a Horizon 2020 Marie Curie ITN ("ProtoMet" – Grant Agreement no. 813873) from the European Commission. We would like to thank Cliff Hooker and Juli Peretó, as well as the editor of the book, Matteo Mossio, for carefully reading a first manuscript of this contribution and providing critical feedback that certainly helped us improve it.

References

Adamala, K., & Szostak, J. W. (2013). Competition between model protocells driven by an encapsulated catalyst. *Nature Chemistry, 5*, 495–501.

Ameta, S., Arsène, S., Foulon, S., Saudemont, B., Clifton, B. E., Griffiths, A. D., & Nghe, P. (2021). Darwinian properties and their trade-offs in autocatalytic RNA reaction networks. *Nature Communications, 12*(1), 842.

Ashkenasy, G., Hermans, T. M., Otto, S., & Taylor, A. F. (2017). Systems chemistry. *Chemical Society Reviews, 46*, 2543–2554.

Attal, R., & Schwartz, L. (2021). Thermally driven fission of protocells. *Biophysical Journal, 120*(18), 3937–3959.

Bartel, D. P., & Szostak, J. W. (1993). Isolation of new ribozymes from a large pool of random sequences. *Science, 61*(5127), 1411–1418.

Bechtel, W. (2007). Biological mechanisms: Organized to maintain autonomy. In F. Boogerd, F. Bruggerman, J. H. Hofmeyr, & H. V. Westerhoff (Eds.), *Systems biology: Philosophical foundations* (pp. 269–302). Elsevier.

Bich, L., & Green, S. (2018). Is defining life pointless? Operational definitions at the frontiers of biology. *Synthese, 195*, 3919–3946.

Bich, L., Mossio, M., Ruiz-Mirazo, K., & Moreno, A. (2016). Biological regulation: Controlling the system from within. *Biology and Philosophy, 31*, 237–265.

Blanken, D., Foschepoth, D., Serrão, A. C., & Danelon, C. (2020). Genetically controlled membrane synthesis in liposomes. *Nature Communications, 11*(1), 4317.

Bonduriansky, R., & Day, T. (2018). *Extended heredity: A new understanding of inheritance and evolution*. Princeton University Press.

Bonfio, C., Valer, L., Scintilla, S., Shah, S., Evans, D. J., Jin, L., Szostak, J. W., Sasselov, D. D., Sutherland, J. D., & Mansy, S. S. (2017). UV-light-driven prebiotic synthesis of iron-sulfur clusters. *Nature Chemistry, 9*(12), 1229–1234.

Bonfio, C., Godino, E., Corsini, M., Fabrizi de Biani, F., Guella, G., & Mansy, S. S. (2018). Prebiotic iron–sulfur peptide catalysts generate a ph gradient across model membranes of late protocells. *Nature Catalysis, 1*(8), 616–623.

Breuer, M., Earnest, T. M., Merryman, C., Wise, K. S., Sun, L., Lynott, M. R., Hutchison, C. A., Smith, H. O., Lapek, J. D., Gonzalez, D. J., de Crecy-Lagard, V., Haas, D., Hanson, A. D., Labhsetwar, P., Glass, J. I., & Luthey-Schulten, Z. (2019). Essential metabolism for a minimal cell. *eLife, 8*, e36842.

Budin, I., & Szostak, J. W. (2011). Physical effects underlying the transition from primitive to modern cell membranes. *Proceedings of the National Academy of Sciences of the United States of America, 108*, 5249–5254.

Carrara, P., Stano, P., & Luisi, P. L. (2012). Giant vesicles "colonies": A model for primitive cell communities. *Chembiochem: A European Journal of Chemical Biology, 13*(10), 1497–1502.

Castellanos, M., Wilson, D. B., & Shuler, M. L. (2004). A modular minimal cell model: Purine and pyrimidine transport and metabolism. *Proceedings of the National Academy of Sciences of the United States of America, 17*, 6681–6686.

Chen, I. A., Roberts, R., & Szostak, J. W. (2004). The emergence of competition between model protocells. *Science, 305*, 1474–1476.

Chen, I. A., Salehi-Ashtiani, K., & Szostak, J. W. (2005). RNA catalysis in model protocell vesicles. *Journal of the American Chemical Society, 127*(38), 13213–13219.

Christensen, W. (2007). The evolutionary origins of volition. In D. Spurrett, H. Kincaid, D. Ross, & L. Stephens (Eds.), *Distributed cognition and the will: Individual volition and social context* (pp. 255–287). MIT Press.

Coggins, A. J., & Powner, M. W. (2017). Prebiotic synthesis of phosphoenol pyruvate by α-phosphorylation-controlled triose glycolysis. *Nature Chemistry, 9*(4), 310–317.

Corbett, P. T., Leclaire, J., Vial, L., West, K. R., Wietor, J.-L., Sanders, J. K. M., & Otto, S. (2006). Dynamic combinatorial chemistry. *Chemical Reviews, 106*, 3652–3711.

Cornish-Bowden, A., & Cárdenas, M. L. (2020). Contrasting theories of life: Historical context, current theories. In search of an ideal theory. *BioSystems, 188*, 104063.

Crick, F. H. (1968). The origin of the genetic code. *Journal of Molecular Biology, 38*(3), 367–379.

Dadon, Z., Wagner, N., & Ashkenasy, G. (2008). The road to non-enzymatic molecular networks. *Angewandte Chemie, 47*, 6128–6136.

Danchin, É., Pocheville, A., & Huneman, P. (2019). Early in life effects and heredity: Reconciling neo-Darwinism with neo-Lamarckism under the banner of the inclusive evolutionary synthesis. *Philosophical Transactions of the Royal Society of London. Series B, Biological Sciences, 374*(1770), 20180113.

De Duve, C. (1991). *Blueprint for a cell: The nature and origin of life*. Neil Patterson Publishers.

de la Escosura, A., Briones, C., & Ruiz-Mirazo, K. (2015). The systems perspective at the crossroads between chemistry and biology. *Journal of Theoretical Biology, 381*, 11–22.

Di Paolo, E. (2005). Autopoiesis, adaptivity, teleology, agency. *Phenomenology and the Cognitive Sciences, 4*(4), 429–452.

Dyson, F. J. (1982). A model for the origin of life. *Journal of Molecular Evolution, 18*(5), 344–350.

Dyson, F. J. (1999). *Origins of life* (2nd ed.). Cambridge University Press.

Eigen, M., & Schuster, P. (1979). *The hypercycle: A principle of natural self-organization*. Springer.

Engelhart, A. E., Adamala, K. P., & Szostak, J. W. (2016). A simple physical mechanism enables homeostasis in primitive cells. *Nature Chemistry, 8*(5), 448–453.

Fontana, W., & Buss, L. W. (1994). The arrival of the fittest: Toward a theory of biological organization. *Bulletin of Mathematical Biology, 56*(1), 1–64.

Fontana, W., & Buss, L. W. (1996). The barrier of objects: From dynamical systems to bounded organizations. In J. Casti & A. Karlqvist (Eds.), *Boundaries and barriers* (pp. 56–116). Addison-Wesley.

Ganti, T. (1975). Organization of chemical reactions into dividing and metabolizing units: The chemotons. *Biosystems, 7*, 15–21.

Ganti, T. (2003). *The principle of life*. Oxford University Press.

Gilbert, W. (1986). Origin of life: The RNA world. *Nature, 319*, 618.

Hanczyc, M. (2009). The early history of protocells. In S. Rasmussen (Ed.), *Protocells: Bridging nonliving and living matter* (pp. 3–17). MIT Press.

Hardy, M. D., Yang, J., Selimkhanov, J., Cole, C. M., Tsimring, L. S., & Devaraj, N. K. (2015). Self-reproducing catalyst drives repeated phospholipid synthesis and membrane growth. *Proceedings of the National Academy of Sciences of the United States of America, 112*, 8187–8192.

Harold, F. M. (1986). *The vital force: A study of bioenergetics*. Freeman.

Harold, F. M. (2001). *The way of the cell: Molecules, organisms and the order of life*. Oxford University Press.

Higgs, P. G., & Lehman, N. (2015). The RNA world: Molecular cooperation at the origins of life. *Nature Reviews. Genetics, 16*(1), 7–17.

Hofmeyr, J.-H. S. (2007). The biochemical factory that autonomously fabricates itself: A systems-biological view of the living cell. In F. C. Boogerd, F. J. Bruggeman, J.-H. S. Hofmeyr, & H. V. Westerhoff (Eds.), *Towards a philosophy of systems biology* (pp. 217–242). Elsevier.

Hutchison, C. A., Chuang, R.-Y., Noskov, V. N., Assad-Garcia, N., Deerinck, T. J., Ellisman, M. H., Gill, J., Kannan, K., Karas, B. J., Ma, L., Pelletier, J. F., Qi, Z.-Q., Alexander Richter, R., Strychalski, E. A., Sun, L., Suzuki, Y., Tsvetanova, B., Wise, K. S., Smith, H. O., et al. (2016). Design and synthesis of a minimal bacterial genome. *Science, 351*(6280), aad6253.

Izgu, E. C., Björkbom, A., Kamat, N. P., Lelyveld, V. S., Zhang, W., Jia, T. Z., & Szostak, J. W. (2016). N-carboxyanhydride-mediated fatty acylation of amino acids and peptides for functionalization of protocell membranes. *Journal of the American Chemical Society, 138*(51), 16669–16676.

Johnston, W. K., Unrau, P. J., Lawrence, M. S., Glasner, M. E., & Bartel, D. P. (2001). RNA-catalyzed RNA polymerization: Accurate and general RNA-templated prime extension. *Science, 292*, 1319–1325.

Joyce, G. F., & Szostak, J. W. (2018). Protocells and RNA self-replication. *Cold Spring Harbor Perspectives in Biology, 10*(9), a034801.

Kamimura, A., & Kaneko, K. (2010). Reproduction of a protocell by replication of a minority molecule in a catalytic reaction network. *Physical Review Letters, 105*(26), 268103.

Kamimura, A., & Kaneko, K. (2019). Molecular diversity and network complexity in growing protocells. *Life (Basel, Switzerland), 9*(2), 53.

Karsenti, E. (2008). Self-organization in cell biology. A brief history. *Nature Reviews, 9*, 255–262.

Kauffman, S. (1986). Autocatalytic sets of proteins. *Journal of Theoretical Biology, 119*, 1–24.

Kauffman, S. (1993). *The origins of order: Self-organization and selection in evolution*. Oxford University Press.

Kauffman, S. (2000). *Investigations*. Oxford University Press.

Keller, M. A., Turchyn, A. V., & Ralser, M. (2014). Non-enzymatic glycolysis and pentose phosphate pathway-like reactions in a plausible Archean ocean. *Molecular Systems Biology, 10*, 725.

Keller, M. A., et al. (2016). Conditional iron and pH-dependent activity of a non-enzymatic glycolysis and pentose phosphate pathway. *Science Advances, 2*, e1501235.

Kitano, H. (2002). Systems biology: A brief overview. *Science, 295*, 1662–1664.

Krishnamurthy, R. (2020). Systems chemistry in the chemical origins of life: The 18th camel paradigm. *Journal of Systems Chemistry, 8,* 40–62.

Kroiss, D., Ashkenasy, G., Braunschweig, A. B., Tuttle, T., & Ulijn, R. V. (2019). Catalyst: can systems chemistry unravel the mysteries of the chemical origins of life? *Chem, 5*(8), 1917–1920.

Lauber, N., Flamm, C., & Ruiz-Mirazo, K. (2021). 'Minimal metabolism': A key concept to investigate the origins and nature of biological systems. *BioEssays, 43,* 2100103.

Le Vay, K., & Mutschler, H. (2019). The difficult case of an RNA-only origin of life. *Emerging Topics in Life Sciences, 3*(5), 469–475.

Lee, D. H., Granja, J. R., Martinez, J. A., Severin, K., & Ghadiri, M. R. (1996). A self-replicating peptide. *Nature, 382,* 525–528.

Lehn, J.-M. (1995). *Supramolecular chemistry: Concepts and perspectives.* Wiley.

Lincoln, T. A., & Joyce, G. F. (2009). Self-sustained replication of an RNA enzyme. *Science, 323*(5918), 1229–1232.

Ludlow, R. F., & Otto, S. (2008). Systems chemistry. *Chemical Society Reviews, 37,* 101–108.

Luisi, P. L. (2006). *The emergence of life: From chemical origins to synthetic biology.* Cambridge University Press.

Luisi, P. L., & Varela, F. J. (1989). Self-replicating micelles – A chemical version of a minimal autopoietic system. *Origins of Life and Evolution of the Biosphere, 19,* 633–643.

Macía, J., & Solé, R. V. (2007). Protocell self-reproduction in a spatially extended metabolism-vesicle system. *Journal of Theoretical Biology, 245*(3), 400–410.

Mansy, S. S., Schrum, J. P., Krishnamurthy, M., Tobé, S., Treco, D. A., & Szostak, J. W. (2008). Template-directed synthesis of a genetic polymer in a model protocell. *Nature, 454*(7200), 122–125.

Maturana, H., & Varela, F. J. (1980). *Autopoiesis and cognition. The realization of the living.* D. Riedel Publishing Company.

Mavelli, F. (2012). Stochastic simulations of minimal cells: The Ribocell model. *BMC Bioinformatics, 13*(Suppl 4), S10.

Mavelli, F., & Ruiz-Mirazo, K. (2007). Stochastic simulations of minimal self-reproducing cellular systems. *Philosophical Transactions of the Royal Society B, 362,* 1789–1802.

Maynard Smith, J. (1986). *The problems of biology.* Oxford University Press.

McMullin, B. (2000). John von Neumann and the evolutionary growth of complexity: Looking backward, looking forward.... *Artificial Life, 6,* 347–361.

Miller, S. L. (1953). A production of amino acids under possible primitive Earth conditions. *Science, 117,* 528–529.

Moreno, A. (2016). Some conceptual issues in the transition from chemistry to biology. *History & Philosophy of the Life Sciences, 38*(4), 16.

Moreno, A. (2019). The origin of a trans-generational organization in the phenomenon of biogenesis. *Frontiers in Physiology, 10,* 1222.

Moreno, A., & Mossio, M. (2015). *Biological autonomy: A philosophical and theoretical enquiry.* Springer.

Moreno, A., & Ruiz-Mirazo, K. (1999). Metabolism and the problem of its universalization. *Biosystems, 49*(1), 45–61.

Moreno, A., & Ruiz-Mirazo, K. (2009). The problem of the emergence of functional diversity in prebiotic evolution. *Biology and Philosophy, 24,* 585–605.

Morowitz, H. J. (1999). A theory of biochemical organization, metabolic pathways, and evolution. *Complexity, 4*(6), 39–53.

Mossio, M., & Pontarotti, G. (2020). Conserving functions across generations: Heredity in light of biological organization. *British Journal for the Philosophy of Science* (accepted – ahead of print).

Mossio, M., Saborido, C., & Moreno, A. (2009). An organizational account of biological functions. *The British Journal for the Philosophy of Science, 60,* 813–841.

Muchowska, K. B., Varma, S. J., Chevallot-Beroux, E., Lethuillier-Karl, L., Li, G., & Moran, J. (2017). Metals promote sequences of the reverse Krebs cycle. *Nature Ecology & Evolution, 1*(11), 1716–1721.

Muchowska, K. B., Varma, S. J., & Moran, J. (2019). Synthesis and breakdown of universal metabolic precursors promoted by iron. *Nature, 569*(7754), 104–107.

Nicolis, G., & Prigogine, Y. (1977). *Self-organization in non-equilibrium Systems*. Wiley.

Noireaux, V., Maeda, Y. T., & Libchaber, A. (2011). Development of an artificial cell, from self-organization to computation and self-reproduction. *Proceedings of the National Academy of Sciences of the United States of America, 108*(9), 3473–3480.

Oglêcka, K., Rangamani, P., Liedberg, B., Kraut, R. S., & Parikh, A. N. (2014). Oscillatory phase separation in giant lipid vesicles induced by transmembrane osmotic differentials. *eLife, 3*, e03695. https://doi.org/10.7554/eLife.03695

Ono, N., & Ikegami, T. (1999). Model of self-replicating cell capable of self-maintenance. In D. Floreano (Ed.), *Proceedings of the 5th European conference on artificial life* (pp. 399–406). Springer.

Orgel, L. E. (1968). Evolution of the genetic apparatus. *Journal of Molecular Biology, 38*(3), 381–393.

Pattee, H. H. (1969). How does a molecule become a message? *Developmental Biology Supplement, 3*, 1–16.

Pattee, H. H. (1973). The physical basis and origin of hierarchical control. In H. H. Pattee (Ed.), *Hierarchy theory* (pp. 73–108). Braziller.

Pattee, H. H. (1977). Dynamic and linguistic modes of complex systems. *International Journal of General Systems, 3*, 259–266.

Pattee, H. H. (1982). Cell psychology: An evolutionary approach to the symbol-matter problem. *Cognition and Brain Theory, 4*, 325–341.

Pechuan, X., Puzio, R., & Bergman, A. (2018). The evolutionary dynamics of metabolic protocells. *PLoS Computational Biology, 14*(7), e1006265.

Pelletier, J. F., Sun, L., Wise, K. S., Assad-Garcia, N., Karas, B. J., Deerinck, T. J., Ellisman, M. H., Mershin, A., Gershenfeld, N., Chuang, R. Y., Glass, J. I., & Strychalski, E. A. (2021). Genetic requirements for cell division in a genomically minimal cell. *Cell, 184*(9), 2430–2440.e16.

Piedrafita, G., Ruiz-Mirazo, K., Monnard, P.-A., Cornish-Bowden, A., & Montero, F. (2012). Viability conditions for a compartmentalized proto-metabolic system: a semi-empirical approach. *PLoS One, 7*(6), e39480.

Piedrafita, G., Monnard, P.-A., Mavelli, F., & Ruiz-Mirazo, K. (2017). Permeability-driven selection in a semi-empirical protocell model: The roots of prebiotic *systems* evolution. *Scientific Reports, 7*, 3141.

Post, E. A. J., & Fletcher, S. P. (2020). Dissipative self-assembly, competition and inhibition in a self-reproducing protocell model. *Chemical Science, 11*, 9434–9442.

Preiner, M., Asche, S., Becker, S., Betts, H. C., Boniface, A., Camprubi, E., Chandru, K., Erastova, V., Garg, S. G., Khawaja, N., Kostyrka, G., Machné, R., Moggioli, G., Muchowska, K. B., Neukirchen, S., Peter, B., Pichlhöfer, E., Radványi, Á., Rossetto, D., Salditt, A., Schmelling, N. M., Sousa, F. L., Tria, F. D. K., Vörös, D., & Xavier, J. C. (2020). The future of origin of life research: Bridging decades-old divisions. *Life, 10*, 20.

Reek, J. H. R., & Otto, S. (2010). *Dynamic combinatorial chemistry*. Wiley-VCH.

Rosen, R. (1971). Some realizations of (M, R)-systems and their interpretation. *Bulletin of Mathematical Biophysics, 33*, 303–319.

Rosen, R. (1991). *Life itself: A comprehensive inquiry into the nature, origin and fabrication of life*. Columbia University Press.

Ruiz-Mirazo, K. (2011). Protocell. In M. Gargaud, R. Amils, J. Cernicharo Quintanilla, H. J. Cleaves, W. M. Irvine, D. Pinti, & M. Viso (Eds.), *Encyclopedia of astrobiology* (Vol. 3, pp. 1353–1354). Springer.

Ruiz-Mirazo, K., & Mavelli, F. (2008). On the way towards 'basic autonomous agents': Stochastic simulations of minimal lipid-peptide cells. *Biosystems, 91*(2), 374–387.

Ruiz-Mirazo, K., & Moreno, A. (2004). Basic autonomy as a fundamental step in the synthesis of life. *Artificial Life, 10*(3), 235–259.

Ruiz-Mirazo, K., & Moreno, A. (2012). Autonomy in evolution: from minimal to complex life. *Synthese, 185*(1), 21–52.

Ruiz-Mirazo, K., Peretó, J., & Moreno, A. (2004). A universal definition of life: autonomy and open-ended evolution. *Origins of Life and Evolution of the Biosphere, 34*, 323–346.

Ruiz-Mirazo, K., Peretó, J., & Moreno, A. (2010). Defining life or bringing biology to life. *Origins of Life and Evolution of the Biosphere, 40*, 203–213.

Ruiz-Mirazo, K., Umerez, J., & Moreno, A. (2008). Enabling conditions for 'open-ended evolution'. *Biology and Philosophy, 23*(1), 67–85.

Ruiz-Mirazo, K., Briones, C., & de la Escosura, A. (2014). Prebiotic system chemistry: New perspectives for the origins of life. *Chemical Reviews, 114*(1), 285–366.

Ruiz-Mirazo, K., Briones, C., & de la Escosura, A. (2017). Chemical roots of biological evolution: The origins of life as a process of development of autonomous functional systems. *Open Biology, 7*, 170050.

Ruiz-Mirazo, K., Shirt-Ediss, B., Escribano-Cabeza, M., & Moreno, A. (2020). The construction of biological 'inter-identity' as the outcome of a complex process of protocell development in prebiotic evolution. *Frontiers in Physiology – Systems Biology, 11*, 530.

Saborido, C., Mossio, M., & Moreno, A. (2011). Biological organization and cross-generation functions. *The British Journal for the Philosophy of Science, 62*(3), 583–606.

Segré, D., & Lancet, D. (2000). Composing life. *EMBO Reports, 1*(3), 217–222.

Segré, D., Ben-Eli, D., Deamer, D. W., & Lancet, D. (2001). The lipid world. *Origins of life & Evolution of Biospheres., 31*(1–2), 119–145.

Semenov, S. N., Kraft, L. J., Ainla, A., Zhao, M., Baghbanzadeh, M., Campbell, V. E., Kang, K., Fox, J. M., & Whitesides, G. M. (2016). Autocatalytic, bistable, oscillatory networks of biologically relevant organic reactions. *Nature, 537*(7622), 656–660.

Shapiro, R. (2000). A replicator was not involved in the origin of life. *IUBMB Life, 49*(3), 173–176.

Shirt-Ediss, B., Solé, R. V., & Ruiz-Mirazo, K. (2015). Emergent chemical behavior in variable-volume protocells. *Life, 5*(1), 181–211.

Shirt-Ediss, B., Murillo-Sánchez, S., & Ruiz-Mirazo, K. (2017). Framing major prebiotic transitions as stages of protocell development: three challenges for origins-of-life research. *Beilstein Journal of Organic Chemistry, 13*, 1388–1395.

Showalter, K., & Epstein, I. R. (2015). From chemical systems to systems chemistry: Patterns in space and time. *Chaos, 25*(9), 097613.

Springsteen, G., Yerabolu, J. R., Nelson, J., Rhea, C. J., & Krishnamurthy, R. (2018). Linked cycles of oxidative decarboxylation of glyoxylate as protometabolic analogs of the citric acid cycle. *Nature Communications, 9*(91), 1–8.

Stubbs, R. T., Yadav, M., Krishnamurthy, R., & Springsteen, G. (2020). A plausible metal-free ancestral analogue of the Krebs cycle composed entirely of α-ketoacids. *Nature Chemistry, 12*(11), 1016–1022.

Sutherland, J. D. (2017). Opinion: Studies on the origin of life—The end of the beginning. *Nature Reviews Chemistry, 1*(2), 0012.

Szathmáry, E., & Maynard Smith, J. (1997). From replicators to reproducers: The first major transitions leading to life. *Journal of Theoretical Biology, 187*, 555–571.

Szostak, J. W., Bartel, D. P., & Luisi, P. L. (2001). Synthesizing life. *Nature, 409*(6818), 387–390.

Szosztak, J. W. (2012). The eightfold path to non-enzymatic RNA replication. *Journal of Systems Chemistry, 3*, Article No. 2.

Tjhung, K. F., Shokhirev, M. N., Horning, D. P., & Joyce, G. F. (2020). An RNA polymerase ribozyme that synthesizes its own ancestor. *PNAS, 117*(6), 2906–2913.

Toparlak, Ö. D., Wang, A., & Mansy, S. S. (2021). Population-level membrane diversity triggers growth and division of protocells. *Journal of the American Chemical Society, 1*(5), 560–568.

Vaidya, N., Manapat, M. L., Chen, I. A., Xulvi-Brunet, R., Hayden, E. J., & Lehman, N. (2012). Spontaneous network formation among cooperative RNA replicators. *Nature, 491*(7422), 72–77.

van Segbroeck, S., Nowé, A., & Lenaerts, T. (2009). Stochastic simulation of the chemoton. *Artificial Life, 15*(2), 213–226.

Varela, F. J. (1979). *Principles of biological autonomy*. Elsevier.

Varela, F. J., Maturana, H., & Uribe, R. (1974). Autopoiesis: The organization of living systems, its characterization and a model. *Biosystems, 5*, 187–196.

von Kiedrowski, G. (1986). A self-replicating hexadeoxy nucleotide. *Angewandte Chemie (International Ed. in English), 25*, 932–935.

von Kiedrowski, G. (2005). *Public communication in systems chemistry workshop* (pp. 3–4). Venice International University.

von Neumann, J. (1966 [1948]). *Theory of self-reproducing automata* (A. W. Burks, Ed.). University of Illinois.

Wächtershäuser, W. (1988). Before enzymes and templates: Theory of surface metabolism. *Microbiological Reviews, 52*, 452–484.

Wächtershäuser, W. (1990). Evolution of the first metabolic cycles. *Proceedings of the National Academy of Science USA, 87*, 200–204.

Walde, P., Wick, R. F., Fresta, M., Mangone, A., & Luisi, P. L. (1994). Autopoietic Self-Reproduction of Fatty Acid Vesicles. *Journal of the American Chemical Society, 116*, 11649–11654.

Walsh, D. M. (2015). *Organisms, agency, and evolution*. Cambridge University Press.

Walsh, D. M. (2018). *Personal Communication – December 11, 'IAS-Research Talk': "Summoning and Sedimentation: Concepts for an agent-centred evolutionary biology"*.

Westerhoff, H. V., & Palsson, B. O. (2004). The evolution of molecular biology into systems biology. *Nature Biotechnology, 22*(10), 1249–1252.

Wicken, J. S. (1987). *Evolution, thermodynamics and information. Extending the Darwinian program*. Oxford University Press.

Wołos, A., Roszak, R., Żądło-Dobrowolska, A., Beker, W., Mikulak-Klucznik, B., Spólnik, G., Dygas, M., Szymkuć, S., & Grzybowski, B. A. (2020). Synthetic connectivity, emergence, and self-regeneration in the network of prebiotic chemistry. *Science, 369*(6511), eaaw1955.

Chapter 10
Organization and Inheritance in Twenty-First-Century Evolutionary Biology

Gaëlle Pontarotti

Abstract During the last few years, various authors have called for the elaboration of a theoretical framework that would better take into account the role of organisms in evolutionary dynamics. In this paper, I argue that an organism-centered evolutionary theory, which implies the rehabilitation of an organizational thinking in evolutionary biology and should be associated with what I will call a heuristic of collaboration, may be completed by an organizational perspective of biological inheritance. I sketch this organizational perspective – which allows going beyond gene-centrism –, show how it grounds a systemic concept of heritable variation suited to the new evolutionary framework, and highlight some of its explanatory value and theoretical implications for evolutionary thinking.

10.1 Introduction

The gene-centered theory of evolution is sometimes presented as obsolete. Associated with twentieth century's modern synthesis, it is accused to outlook the role of organisms and of their properties in evolutionary dynamics (Walsh, 2006; Nicholson, 2014). Many authors have therefore recently called for the elaboration of a more organism-centered evolutionary biology (Walsh, 2010; Laland et al., 2015), notably in the context of an extended evolutionary synthesis (Pigliucci & Müller, 2010). Such biology is notably expected to integrate non-genetic channels of inheritance in its models but also to make some room to the concept of agency (Walsh, 2015) and biological organization (Müller, 2017) insofar as organisms – at the center of its preoccupations – are generally considered as paragons of organized and purposive biological systems. The objective of this paper is to argue that an organism-centered evolutionary biology may be enriched by a not only extended but also organizational perspective of biological inheritance, to sketch this perspective and to highlight its theoretical implications for evolutionary thinking.

G. Pontarotti (✉)
Institut d'Histoire et de Philosophie des Sciences et des Techniques (IHPST, CNRS/Paris1), Paris, France

© The Author(s) 2024
M. Mossio (ed.), *Organization in Biology*, History, Philosophy and Theory of the Life Sciences 33, https://doi.org/10.1007/978-3-031-38968-9_10

The argument is structured as follows. In Sect. 10.2, I briefly present the contemporary literature which invites departing from a gene-centered evolutionary theory and embracing a more organism-centered framework. I further suggest that an organizational perspective of biological inheritance appears as a missing ingredient in this theoretical movement that not only involves the return of an organizational thinking in evolutionary biology but that also follows a more global perspective shift, from a heuristic of replication – in which evolution is thought as a competition among self-replicating objects endowed with their own adaptive value – toward a heuristic of collaboration – in which biological objects are necessarily considered as parts of integrated wholes and cannot replicate independently. In Sect. 10.3, I rest on earlier studies (Pontarotti, 2015; Mossio & Pontarotti, 2019) to sketch an organizational perspective of biological inheritance suited to an organism-centered evolutionary biology, and I notably highlight that this perspective grounds a systemic concept of heritable variation appropriate to the new evolutionary biology's framework. In Sect. 10.4, I evoke some theoretical implications of an organizational account of inheritance for evolutionary thinking. I show how this account allows making sense of the evolution of "non-standard" biological systems[1] and how it induces a change of perspective, in the wake of earlier contributions, as far as lineages, fitness, selection, and evolution are concerned.

10.2 Toward a More Organization-Centered Framework for Twenty-First-Century Evolutionary Biology

In this Section, I briefly present the literature announcing a perspective shift, from gene-centrism toward organism-centrism, in evolutionary biology. I then highlight that an organism-centered evolutionary biology is expected to make important room for the concept of organization in its explanations. Consequently, I argue that it may be completed by an organizational perspective of biological inheritance.

10.2.1 An Extended Evolutionary Synthesis to Fill in the Explanatory Gaps of the Gene-Centered Framework

The theoretical framework of evolutionary biology has been seriously challenged for the last few years. Many authors have indeed advocated the necessity to adopt an extended evolutionary synthesis (EES) in order to overcome some of the theoretical and explanatory limitations of modern synthesis (MS) (Pigliucci & Müller, 2010).

---

[1] The concept of non-standard biological systems usually refers to symbiotic associations or to insect colonies including abiotic parts (mounds). Here, it will designate all biological systems whose parts cannot simply be accounted by classical interactionist accounts (gene/environment). For more details, see Sect. 10.4.

EES is described as a movement of conceptual and disciplinary extension (Pigliucci & Müller, 2010) but also as an alternative ecological-developmental perspective to evolution (Laland et al., 2015). In this respect, EES is not just an extension of MS but rather a "distinctively different framework for understanding evolution" (Laland et al., 2015). EES is meant to be more inclusive than MS. Indeed, while the latter makes sense of evolutionary phenomena through the articulation of Neo-Darwinism, Mendelism, and population genetics,[2] the former is willing to include new elements in evolutionary thinking, notably concepts of evolutionary-developmental biology (e.g., plasticity), an extended vision of inheritance, as well as ideas about evolvability (Pigliucci, 2009, p. 218).

Let us go into more details. While MS ignores developmental processes, EES intends to shed light on the developmental origin of organismal variations. It stresses on the role of developmental constraints regarding the diversification of forms[3] (Müller, 2017) and that of plasticity – "the capacity of organisms to develop altered phenotypes in reaction to different environmental conditions" (Müller, 2017, p. 5) – on evolutionary dynamics. Besides, while MS is based on a genetic account of inheritance according to which the trans-generational reoccurrence of features is exclusively underpinned by the replication of genes, EES integrates data about so-called non-genetic inheritance, for example, epigenetic and behavioral transmission. The framework also takes into account niche construction (Laland et al., 2015), namely, the fact that organisms modify their surroundings in such a way that they alter the selection pressure exerted on their offspring (Odling-Smee et al., 2003). This inclusion stresses on the "reciprocal causality" (Müller, 2017) at play in evolution, which means that organisms are not only submitted to independent selective forces but that they also define the selective pressures exerted on them and their offspring.

More generally, EES is meant to go beyond some "basic restrictions and methodological commitments" of MS (Pigliucci & Müller, 2010, p. 13). According to MS, evolution is a gradual process mainly driven by the selection of small and random genetic variations correlated with phenotypic differences (Mayr, 1998). ESS, as for it, intends to overcome gradualism (Pigliucci & Müller, 2010; Laland et al., 2015; Müller, 2017) in highlighting that evolutionary change can follow various paths (Pigliucci & Müller, 2010). As mentioned above, EES also aims at going beyond externalism, the hypothesis according to which independent selection

---

[2] Beyond the articulation of Mendelian genetics and Neo-Darwinian evolutionary theory through the mediation of population genetics, MS refers to the agreement of various disciplines – systematics, zoology, botany, paleontology, and natural history – on a set of core hypothesis (e.g., gradualism, creativity of natural selection, etc.).

[3] It is important to make a clear distinction between the hypothesis of organismal origin of variation, according to which variation is originated and constrained by organisms themselves (and their developmental processes), and trade-off adaptationism, which states that organisms are trade-offs of adapted traits (e.g., trades between traits enhancing survival and traits enhancing reproduction). In the first case, organisms (and developmental processes) impose constraints on variation and have a key explanatory value in evolutionary theory. In the second case, natural selection is still the main explanans of organismal characteristics (for a detailed analysis, see Huneman, 2017).

pressures are the main drivers of evolutionary change (Pigliucci & Müller, 2010; Müller, 2017). Finally, while MS is based on statistical analysis, EES appears as a causal-mechanistic framework (Pigliucci & Müller, 2010). In this perspective, evolution is not primarily portrayed as a change in gene frequencies mainly caused by natural selection (Dobzhansky, 1937) but as a change in phenotypes partly driven by developmental processes (Helanterä & Uller, 2010). These processes are thought to "share responsibility" with natural selection in the determination of evolutionary trajectories (direction and rate of evolution, origin of variation, etc.) (Laland et al., 2015). Genes, as for them, are sometimes described as followers (West-Eberhard, 2003; Pigliucci, 2009).

10.2.2 Focus on Organisms and Introduction of an Organizational Thinking

In this context, the focus of evolutionary biology changes radically. Evolution is not anymore thought as a matter of genetic dynamics but rather of organismal changes. Organisms – which are often described as developmental systems – appear as key causal agents in evolution. As summarized by Laland and colleagues (2015), EES is "characterized by the central role of the organism in the evolutionary process and by the view that the direction of evolution does not depend on selection alone and need not start with mutation." While MS explains biological evolution by focusing on the scale of genes, ESS is grounded on the assumption that "the organisms themselves represent the determinants of selectable variation and innovation" (Pigliucci & Müller, 2010, p. 13). In brief, EES represents a "different way of thinking about evolution, historically rooted in the organicist tradition" (Müller, 2017). As a result, it is meant to better take into account the role of organisms' properties in the determination evolutionary trajectories.

On this specific point, the literature about EES meets other studies dedicated to the return of organisms in evolutionary biology (Bateson, 2005; Walsh, 2006, 2015).[4] For example, Walsh (2006) analyzes that contemporary evolutionary biology has forgotten organisms in asking how supra-organismal entities (populations) change under the effect of sub-organismal entities (genes, replicators). He calls for the development of a Kantian-flavored biology which would take into account organismal properties in its explanations. Inspired by West-Eberhard's contribution (2003), Walsh also suggests that phenotypic accommodation can sometimes precede genotypic one, and that genes can thus be followers in evolution (Walsh, 2006, p. 778). In a way, all these contributions follow considerations early on made by Mayr (1963, p. 184), who claimed that changes in gene frequencies is an effect and not a cause of evolution. To him, describing evolution as a change in gene

---

[4] For a general appraisal regarding the return of organisms in evolutionary biology, see Huneman (2010) and Nicholson (2014).

frequencies amounts to neglect the mechanisms that cause organisms and populations' transformations.

To sum up, "the emerging view of evolution" presents organisms as "the primary agents of evolutionary change" (Nicholson, 2014). In this perspective, organisms are thought as a major *explanans* – and not only *explanandum* – of evolutionary processes (Huneman, 2010); organismal properties do not only appear as elements that should be explained, but they are also – and crucially – conceived as elements which contribute to the explanation of evolutionary phenomena. Now, if one considers, in line with an old tradition usually thought of as tracing back to Kant (1790), that the most fundamental and distinctive property of organisms is to be (self)organized, the emerging view of evolution should involve the rehabilitation of the concept of organization and the introduction of an organizational thinking in evolutionary biologists' toolkit.

10.2.3   From a Heuristic of Replication to a Heuristic of Collaboration

These elements, I argue, are part of a more global perspective shift that is more or less implicitly announced in the literature. Such shift takes its distance with what I will hereinafter call a *heuristic[5]of replication* (atomistic, gene-eye view) and embraces what I will name a *heuristic of collaboration* (systemic view).

The *heuristic of replication*, embodied by Dawkins's work on the selfish gene (1976, 1982), states that evolution can be conceived as a process mainly driven by the selection of virtually[6] atomized units endowed with intrinsic capacities of self-replication and with their own adaptive value. It corresponds to what Walsh (2015) calls the Replicator biology. The *heuristic of collaboration*, as for it, rests on the hypothesis that biological objects cannot be considered otherwise than as parts of integrated wholes: they cannot replicate, evolve, and have any adaptive value independently from these wholes. In other words, this heuristic implies that heritable variations cannot be considered as virtually atomized traits correlated to virtually atomized genes (or replicators) but rather as parts of systems including interdependent elements.

The push toward the heuristic of collaboration subtly emerged in various contributions. It notably appeared in Gould and Lewontin's (1979) critique of the adaptationist program which considers organisms as aggregates of virtually atomized

---

[5] A heuristic is not a faithful account of reality but rather a theoretical tool that is supposed to help scientists grasping something from the objects that they study.

[6] The adverb "virtually" should be understood in the context of a given heuristic. It denotes the fact that even if biologists have always been aware that evolution takes place among organisms and that heritable variations are necessarily located in integrated biological systems, these elements were overlooked in theoretical models (see, e.g., Dobzhansky, 1970, p. 65, who recognizes the limits of considering traits as independent entities).

traits. It was also expressed in Lewontin's argument according to which the only biological entities able to self-replicate are not genes but organisms as complex systems (1993) or in developmental systems theory, which states that full developmental systems, and not genes alone, can replicate (Griffiths & Gray, 1994). More recently and in a similar vein, Fox Keller noted that DNA is neither stable nor able to replicate independently from a full cellular machinery (2000, pp. 26–27). Turner (2000, 2004), as for him, proposed a physiological interpretation of Dawkins' extended phenotype (1982). He argued that selection does not target replicators but rather whole systems able to self-maintain through the collaboration of various internal and external physiological parts which specify how flows of matter and energy are channeled. Finally, one can mention Walsh's (2010) elaboration of an alternative Neo-Darwinism that would not focus on replicators but on organisms and the various calls that are made for a "shift toward a network thinking" in evolutionary biology (Bapteste & Huneman, 2018). Even if all these approaches should not be conflated, they all reflect some endeavor toward the elaboration of an evolutionary biology focusing on integrated wholes and on networks and not on atomized objects, be they genes, or traits associated with these genes.

10.2.4   A Missing Organizational Perspective of Biological Inheritance?

Finally, the integration of an organizational thinking in evolutionary biology cannot be limited to the inclusion of organismal properties (and developmental timescales) in theoretical models. It might also require the integration of organizational concerns at the level of inheritance, which is traditionally thought as a key element for evolution (Lewontin, 1970; Sterelny, 2001).

It is generally asserted that EES notably relies on an extended vision of inheritance (Pigliucci & Müller, 2010; Laland et al., 2015). This means that the framework takes into account, in addition to genetic replication and transmission, various channels involved in the reoccurrence of traits across generations (Jablonka & Lamb, 2005; Danchin et al., 2011). Epigenetic inheritance (through the maintenance of epigenetic marks such as DNA methylation) can underpin the return of phenotypic outcomes such as defense against predators and pathogens (Holeski et al., 2012) and floral symmetry (Cubas et al., 1999). Behavioral inheritance takes place when social interactions mediate the reoccurrence of behavioral traits (Galef & Laland, 2005), notably those involved in the determination of the feeding niche (Slagsvold & Wiebe, 2007). Symbiotic transmission, which can be considered as a second mode of genetic inheritance (Gilbert et al., 2012), is linked, in many insects, to the trans-generational maintenance of metabolic capacities (Douglas, 2009).

The integration, in the evolutionary framework, of multifarious channels of inheritance and therefore of multifarious heritable variations[7] could have major theoretical consequences. It could notably weaken the statement that heritable variation is always small and random (Jablonka & Lamb, 2005). In this respect, it would damage the MS's core hypothesis which states that natural selection of small and random heritable variation is the main determinant of evolutionary change.[8] However, the inclusion of extended inheritance into evolutionary theory comes with some requirements. First, it demands the *elaboration of a consistent theoretical framework* regarding inheritance. This framework should include more than genetic mechanisms, but it should not result from a mere cumulative approach (Merlin, 2017) which would basically consist in integrating, into biological legacies, anything that appears as a "good" transmitted across generations (developmental resources, developmental factor, source of information). Such approach would make sense from a metaphorical point of view,[9] but it would be theoretically unsatisfactory insofar as it would turn inheritance into a vague, ill-defined concept (Mossio & Pontarotti, 2019). Some of the accounts of extended inheritance outlined during the last few years have intended to establish this consistent framework (Griesemer, 2000; Jablonka, 2002; Pontarotti 2015; Mossio & Pontarotti, 2019).

Second, one could also consider that a framework suited to an evolutionary biology characterized by an organizational thinking should make some room to the concept of organization. While different extended perspectives of inheritance could be compatible with the emerging view of evolution, an organizational approach, beyond overcoming genocentrism, would be fully consistent with an organization-minded evolutionary biology. Besides, it would notably present the advantage of implying a theoretically fecund *systemic* concept of heritable variation, as explained in Sect. 10.3. In this respect, it would unambiguously participate in the perspective shift from a *heuristic of replication* (atomistic, gene-eye view) to a *heuristic of collaboration* (systemic view).

---

[7] Inheritance usually refers to the transmission of traits – eye color and liver metabolic capacities – across generations of organisms. When compared with other instances in the population, these traits can be considered as heritable *variations*. For example, we can say that inheritance is responsible for the recurrence of a *trait* like a specific eye color in a lineage but also that it is responsible for the recurrence of variation in eye color when the whole population is taken into account. This variation can be linked to differential adaptive value. On this topic, Mameli (2005, p. 367) makes a distinction between inheritance of features and inheritance of differences ("'trait' can be used to refer to a particular value (being 176 cm tall) as well as to sets of possible values (height)".)

[8] This consequence is made obvious by a famous historical episode: that of the temporary eclipse of Darwinism at the dawn of the twentieth century, caused by the mutationist vision of heritable variation adopted by the first Mendelians (Huxley, 2010 [1942], p. 22; Gayon, 1992a, p. 14).

[9] It is useful to remind that, according to historians, biological inheritance was initially a metaphorical concept (López-Beltrán, 1994; van der Lugt & de Miramon, 2008). It was indeed imported from the legal sphere into the medical vocabulary to refer to diseases that appeared to be transmitted like goods from parents to offspring.

10.3   An Organizational Perspective of Biological Inheritance

In this section, I present an organizational account of inheritance that has been elaborated in earlier studies (Pontarotti, 2015; Pontarotti 2017; Mossio & Pontarotti, 2019). I show that, beyond overcoming genocentrism, opening a way to explain organisms' stability and bounding the phenomenon of biological (extended) inheritance, this account appears as a key ingredient for an organism- and organization-centered evolutionary biology grounded on a heuristic of collaboration.

10.3.1   Principles of an Organizational Perspective of Biological Inheritance

The organizational account of biological inheritance is grounded on recent theoretical studies dedicated to biological autonomy and putting emphasis on the concept of biological organization (Mossio & Moreno, 2010; Montévil & Mossio, 2015; Moreno & Mossio, 2015). These studies place themselves in the wake of earlier contributions which conceive of biological systems as organized beings (Kant, 1790; Bichat, 1801; Bernard, 1885; Bertalanffy, 1968; Kauffman, 1995).[10] According to them, biological systems include differentiated parts (cells, tissues, organs) that collaborate in order to maintain the system to which they belong.

In this view, biological systems are more precisely conceptualized as far from equilibrium open thermodynamical systems which maintain themselves through exchanges of matter and energy with their environment. They display differentiated parts that are interdependent[11] for their maintenance and that collectively channel flows of matter and energy so as to maintain themselves and the system to which they belong. These parts are called functional constraints insofar as they are said to perform biological function (Mossio et al., 2009). They notably display stability with respect to the process they harness in a given system (Montévil & Mossio, 2015). For example, the cardiovascular system can be depicted as an organized (or functional) constraint given that (1) it contributes to channel flows of matter an energy in the organism, (2) it is dependent on the organisms' other parts (e.g., the digestive system, the respiratory system, etc.) to maintain the organism (and, thereby, itself) as a whole, and (3) it exhibits stability with respect to the process that it harnesses, namely, blood circulation.

---

[10] It is important to specify that these contributions are not equivalent even if they all conceive of living beings as organized ones. To Kant, for example, purposiveness and "self-organization" are regulative concepts necessary to make sense of the movement observed in some natural objects. This transcendental consideration is not endorsed by the other authors. In addition, while Bichat considers that living beings are animated by vital forces, Bernard rejects this concept. Bertalanffy's approach, as for it, is characterized by thermodynamics considerations.

[11] See Sect. 10.4 for an analysis of the concept of "interdependence".

The constraints that constitute an organized system are involved in a Kantian-flavored circular causality: they produce each other in the system that they contribute to maintain and which reciprocally contributes to their maintenance. The loop of interdependencies among a set of constraints is referred to as "organizational closure". Organization, in this context, is defined as closure of constraints and is associated with intrinsic teleology (Mossio & Bich, 2014).

Organization as closure of constraints is a theoretical principle that allows explaining how a biological system maintains itself in an environment with which it exchanges flows of matter and energy. In this respect, it plays the role of *explanans* for the stability of biological systems and their constitutive parts. The theoretical principle of organization as closure of constraints can therefore be used to conceive of biological inheritance, which traditionally refers to the like-begets-like phenomenon (Darwin, 1859) and more globally to the idea of trans-generational stability.

From an organizational point of view, biological inheritance refers to the continuity, across generation breaks, of organizational patterns displayed by biological systems (Mossio & Pontarotti, 2019). It does not primarily designate the reoccurrence of – virtually atomized – traits underpinned by the replication of, virtually atomized, genetic factors but rather the trans-generational *conservation of functional networks*.[12] Because organized constraints collectively channel flows of matter and energy in such a way that they maintain themselves and the system to which they belong, inheritance can also be described, in this context, as the cross-generation conservation of specific regimes of flow of matter and energy channeling (Pontarotti, 2017).

10.3.2   Inheritance and Organization: Toward the Conception of Multifarious Heritable Variations

The organizational perspective of inheritance is compatible with the hypothesis according to which inherited objects can be multifarious: traditional organs such as hearts but also epigenetics marks, external artifacts produced by organisms or symbionts. As argued elsewhere (Pontarotti, 2016; Mossio & Pontarotti, 2019), the concept of (inherited) organized constraints is abstract enough to be applied to objects

---

[12] One may object that the concept of constraints is too narrow to include, in biological legacies, elements that appear as not functional but that are traditionally thought as heritable (e.g., short-sightedness). The question of inheritance of prima facie nonfunctional or dysfunctional traits has been addressed in another paper (Mossio & Pontarotti, 2019). To sum up, from an organizational point of view, heritable dysfunctional objects are still falling under the definition of constraint when they contribute to the maintenance of a given organizational regime, even if it is in a poorer way (e.g., short-sightedness refers to a poorer way of perceiving the environment). Besides, non-functional traits (e.g., eye color) are considered as "subordinary hereditary characteristic" when they are one aspect of an object (e.g., the eye) which is itself functional.

which are not traditional organs; similarly, the concept of biological organization as closure of constraints applies to natural systems which depart from traditional organisms. It more precisely permits going beyond the common and simplified vision, reminded by Dupré & O'Malley (2007, p. 834), according to which biological systems are free-living cells or coordinated groups of cells containing the same genome.

In other words, an organizational perspective of biological inheritance allows conceiving of non-standard biological systems exhibiting multifarious heritable variations. Non-standard biological systems usually refer to symbiotic associations or to insect colonies including abiotic parts. In this paper, it designates all biological systems whose parts cannot simply be accounted by classical interactionist accounts (gene/environment). In other words, it refers to biological systems – such as human beings – that are more than groups of coordinated cells containing the same genome and that possibly include symbiotic and/or behavioral parts (involving or not the use of artifacts).

For pragmatic reason, it can be argued that non-standard biological system can undergo two kinds of variation: *genetic mutations* (changes in a DNA sequences) and *non-genetic acquisitions* (development of a new behavior, recruitment of new microorganisms, etc.) The first are conserved through *genetic inheritance* and the second through *non-genetic inheritance*. In this view, a functional variation appearing in a biological lineage is not necessarily due to a genetic mutation and can be conserved through non-genetic channels. For example, a heritable metabolic change in the capacity to degrade cellulose can be due to a mutation in some DNA sequences, to the acquisition of a cooking technique, to the acquisition of some microorganisms, etc. It can be conserved through various genetic and non-genetic mechanisms.

10.3.3 Inheritance and Organization: An Approach Suited to the Heuristic of Collaboration

When compared to other accounts that widen the scope of inheritance beyond geno-centrism (Jablonka, 2002; Bonduriansky, 2012; Griffiths & Stotz, 2013), the organizational perspective presents some important and specific characteristics that make it more suited to the emerging organisms-centered view of evolution.

- *Extension without dilution.* First, it extends inheritance beyond genetics while keeping it clearly bounded and thereby avoids its dilution into the vague concept of biological stability.[13] Indeed, it offers tools to distinguish inheritance – conservation

---

[13] While the concept of extended inheritance first appears as theoretically and explanatorily fecund (insofar as it promises to overcome the limitations of gene-centrism), it can lead to consider as inherited any elements being stable across generations and having some causal influence on the reoccurrence of traits. In other words, it can lead to think about inheritance as a concept synonymous with stability (Mossio & Pontarotti, 2019). As a result, the extension of inheritance can

of functional patterns across generations – from stability of environment, stability of ecosystems, etc. The organizational perspective only grants the status of heritable objects to those elements which can fall under the definition of functional constraints (e.g., hearts, nests, etc.) at a given scale. It regards stable flows of matter and energy (e.g., nutrients) and stable functional elements whose (physical) persistence is not primarily explained by their being part of a networks of interdependent constraints (e.g., persisting caves used as shelters), as part of stable environments (Mossio & Pontarotti, 2019). In this respect, the organizational account specifically defines inheritance and environmental stability as two different phenomena. It clarifies that stable biological (or biologically relevant) objects are either part of a *heritable* organization or part of a *stable* environment but not both at the same time (Pontarotti, 2022).

- *Collective stability*. Second, the organizational perspective invites to invalidate the classical distinction between hereditary factors (genes), understood as the causes of heredity, and hereditary traits (observable features), conceived as effects of the former. This distinction has been expressed through the opposition between the genotype and the phenotype since Johannsen's (1911) seminal contribution and has been a structuring one in the twentieth century. However, it appears as irrelevant in the presented theoretical framework, where genes and other biological constraints belong to a network of interdependent objects involved in a circular causality (Pontarotti et al., 2022). In an organizational view, distinguishing supposedly causal factors from supposedly epiphenomenal traits makes no sense. Inheritance is not a matter of (selfish) replication but rather of systemic, and therefore collective, stability.

While the genetic theory explains the stability of organismal traits by that of DNA sequences[14] supposedly endowed with self-replicative and causal properties (Dawkins, 1976), an organizational perspective distributes the explanation for traits stability to various interdependent parts conceived as functional constraints: DNA sequences involved in the production of proteins, cells, socially learned behaviors, microorganisms performing some metabolic work in the system they constitute with their hosts, etc. On this point, the organizational perspective appears in line with the previously mentioned heuristic of collaboration. It also somehow echoes older "dynamical" or "energetist" conceptions of inheritance rejecting atomistic approach (Gayon, 1992b, pp. 432–433). Among them, Thompson's vision (1942) is critical about the fact of attributing to individual particles something that is due to the "energy of their collocation", while Nanney's conception (1957) suggests that inheritance can refer to the behavior of a full system.

---

conceal the fact that the concept initially designates a specific phenomenon – the stability of organismal traits – which is explained by the presence of some specific causal factors (López-Beltrán, 1994). When thinking about extended inheritance, the challenge is therefore to include more than genes in inheritance while avoiding to turn it into an all-inclusive concept.

[14] Mendelian genes are theoretical units but genes have been conceived of as DNA sequences since the middle of the twentieth century.

Other extended accounts of inheritance evoke the replication of elements belonging to organized biological systems (Jablonka, 2002) and state that replication is a collective matter (Griffiths & Gray, 2004). However, these accounts do not rest on a clear concept of biological organization and/or do not put the concept of organization at their core. This prevents them from clearly bounding extended inheritance and from proposing a theoretically informed explanation of biological transgenerational stability.

- *Systemic heritable variation.* The organizational perspective further and more importantly has the specificity of implying a *systemic* concept of heritable variation, which is also consistent with a heuristic of collaboration. Conceiving of inheritance as the conservation of functional patterns indeed implies that heritable variations cannot be considered per se but should first and foremost be conceptualized as changes affecting complete biological networks. Insofar as parts of biological systems are thought as constraints which are interdependent for their maintenance within and across generations, new heritable variations should be considered as events that modify organizational regimes and that ground new organizational deals (Pontarotti, 2017).

This implies that a new heritable variation theoretically limits or enables further variations in the considered organizational pattern, within and across generations, following whether it turns out to increase or reduce the cost of a function. For example, the acquisition of a new metabolic capacity to degrade cellulose could be energetically costly for a system and therefore limit the possibly for further variations, but it could on the contrary reduce the energetic cost of the function and therefore leave some room for further changes. The termites that come to rely on fungi to digest cellulose become free from constraints on the digestion rates faced by the termites that rely only on intestinal digestion (Turner, 2004, p. 335). They can mobilize energy at much higher rates than their competitors (Turner, 2004, p. 339). The fact of counting on symbionts to perform part of the digestive process – a phenomena that some authors call functional "outsourcing" (Turner, 2004, p. 335; Bouchard, 2013, p. 261) – can enable or limit further modifications in the considered systems. If the acquisition of the symbionts reduces the cost of nutrition or if it provides the system with more energy, this system may undergo other costly variations.

These considerations notably require admitting, in accordance with the life-history theory, that there is a trade-off for the allocation of resources among the parts of biological systems involved in survival and reproduction (Stearns, 1992, Fabian & Flatt, 2012). They also somehow appear in line with the thesis according to which biological systems are integrated wholes in which parts cannot be individually optimized (Gould & Lewontin, 1979), even if Gould and Lewontin's conception should not be conflated with the trade-off adaptationism endorsed by the life-history theory (see footnote 3).

10.4   Explanatory Value and Theoretical Implications of an Organizational Perspective on Biological Inheritance for Evolutionary Thinking

In this last section, I show that an organizational account of biological inheritance, beyond being a key ingredient for the elaboration of an organization-centered evolutionary biology, can modify perspectives and shed new light on various evolutionary phenomena.

10.4.1   Stabilization of Non-genetic Acquisitions and Evolution of Non-standard Biological Systems

It has been argued that non-genetic inheritance could have an impact on evolutionary trajectories (Jablonka & Lamb, 2005; Bonduriansky & Day, 2018). It has also been said that it is "crucial to make sense of the evolution of complex biological individuals" such as symbiotic associations or insects colonies including mounds and fungi (Bouchard, 2013, p. 259). However, non-genetic elements are generally thought as relatively labile when compared with genes (see Richards et al., 2010 for epigenetic marks), and non-genetic inheritance is sometimes referred to as transgenerational plasticity (Mesoudi et al., 2013). This seems to prevent non-genetic *acquisitions* from having any impact on evolutionary dynamics. Actually, such conclusion relies on a theoretical commitment toward a gene-centered evolutionary biology, based on a heuristic of replication where biological stability is thought as a property of virtually atomized objects able to make faithful copies of their structure and thought on the model of Dawkins's replicators (1976). Endorsing this view, Sterelny (2001) argues that, to have an effect on cumulative evolution mainly driven by natural selection, non-genetic inheritance should present the same properties as genetic inheritance and should notably ensure the reconstruction of highly variable replicators exhibiting stability and having a common evolutionary fate.

An organization-centered biology, based on a heuristic of collaboration where heritable variations are not conceived as virtually atomized and self-replicating elements, leads to a very different conclusion. More precisely, an organizational perspective of biological inheritance, grounded on the idea of collective stability, opens a way to make sense of the stabilization of prima facie labile non-genetic *acquisitions* in the course of evolution and, thereby, of the evolution of non-standard biological systems (as defined in Sect. 10.3). As explained above, an organizational account of biological inheritance implies that heritable variation – be it a genetic mutation or a (plastic) non-genetic acquisition – grounds a new organizational deal and can have systemic consequences. In modifying a system's access to flows of matter and energy (access to new resources, increased or decreased cost of a

function, functional redundancy[15]), it determines the possibility for further variations in this system. When these further variations occur, the other original parts of
the system may not be able to survive without the earlier changes, even if these
changes are non-genetic acquisitions (e.g., epigenetic marks, socially acquired
behaviors). More generally, a systemic vision of heritable variation allows outlining
three conditions favoring the stabilization of non-genetic inherited elements and the
consequent evolution of non-standard biological systems. These conditions are
those which increase the *interdependence* of parts, namely, environmental changes,
random functional losses, and appearance of other costly functional variations.

For example, an insect can acquire microorganisms that perform cellulose degradation via so-called facultative symbiosis, where both host and symbionts car reproduce independently (Moran et al., 2008). But the host-symbiont association can
become irreversible in the case of an environmental change (food shortage that
would favor the systems that are more performant for digestion), in the case of a
functional loss (if the insect loses the capacity to digest cellulose) or if a costly
variation arises in the system.[16] Another speculative example is the acquisition of
sewing techniques and of clothing traditions in humans. Under some climates, these
heritable acquisitions can be considered as a functional innovation regarding ther-
moregulation. In some circumstances (loss of genetic capacity to perform thermo-
regulation, costly variation, environmental change), this acquisition can become
more crucial for the maintenance of other parts of the systems, such as hearts.
Finally, one can imagine a situation where the decreased cost of the digestive function, related to the acquisition of cooking techniques destroying toxins, leaves some
rooms for costly mutations linked to the development of brain. This is what is suggested by the tenants of the expensive tissue hypothesis (Aiello & Wheeler, 1995).

Before concluding, it is important to make some clarifications regarding the concept of *interdependence*. According to the recent literature dedicated to biological
autonomy and biological organization, the constitutive constraints of biological systems are, by definition, interdependent. However interdependence can take different
forms and meanings. According to a first meaning, two objects are interdependent
when they are conserved by producing each other (*reciprocal production and symmetrical dependence involving joined conservation*). For example, the liver of an
organism cannot be conserved without the activity of the heart within and across
generations, and vice versa: the liver and the heart are therefore produced by each
other, and if the former is destroyed, the latter is also destroyed. However, reciprocal
production can also come with asymmetrical dependence. For example, the

---

[15] Note that methylation marks seem to be involved in the silencing of redundant genetic elements
(Rapp & Wendel, 2005, p. 82).

[16] More generally, the case of symbiosis, which is a paradigmatic example to think about organizational inheritance (Pontarotti, 2016), provides many examples of conditions leading to the
increased interdependence of parts. For example, the loss of genes in vertically transmitted symbionts is said to be at the origin of plasts and mitochondria (Sachs, 2013, p. 632). Besides, a host can
come to tolerate a parasite if even more dangerous parasites are present in the environment (van
Baalen & Jansen, 2001).

conservation of hunting tools participating in the food channeling process, in a human lineage, can depend on the conservation of hearts, and vice versa, but in some conditions (food abundance), hearts can be maintained without these manufactured tools. According to a second meaning, two objects are interdependent when they are maintained in a joint way, notably at the trans-generational timescale, even if they do not necessarily produce each other (*mere joined conservation*). For example, in an organism, the kidneys can be conserved if eyes are destroyed during the life cycle, but the kidneys and eyes can be jointly maintained at the trans-generational timescale: if the former reoccurs, the latter will in principle also reoccur. In this case, the kidneys are not directly dependent on the eye for their production and conservation,[17] but the elements needed to rebuild the kidneys and to rebuild the eyes (notably DNA sequences) are conserved together.[18] According to a third meaning, two objects are interdependent when they need to interact for the maintenance of a given organizational regime (and therefore for their maintenance as organizational constraints in this given regime) even if they can otherwise be maintained without one another. For example, the heart of an organism involved in a facultative symbiosis can be dependent on some microorganisms for the maintenance of a given pattern of matter and energy channeling, but not for its maintenance within and across generations (*joined action for the maintenance of a given organizational regime*).

These distinctions are important if one wants to apply the organizational framework to think about biological inheritance and the evolution of non-standard biological systems. They contribute to clarify that parts of biological systems exhibiting multifarious variations are minimally interdependent according to the third meaning (*joined action for the maintenance of a given organizational regime*) but that they can become interdependent in the first and strongest meaning (*reciprocal production and symmetrical dependence involving joined conservation*) in the circumstances mentioned above (environmental changes; random functional losses; appearance of other costly functional variations). This strongest kind of interdependence involves the common fate of parts (not mediated by bottleneck), a property which is one of the main hallmarks of biological individuals (Bouchard, 2013; Godfrey-Smith, 2009).

---

[17] However, the kidneys are fully dependent on sensitive organs more globally. One can therefore consider that there is interdependence in the strong sense (*reciprocal production*) between kidneys and sensitive organs.

[18] This is what happens when objects that do not have any function (which do not comply with the definition of constraints) are conserved. These objects are conserved jointly with others which have, as for them, a clear function (e.g., DNA sequences used to build eye color are jointly conserved with sequences used to build pupil). They can be considered as "subordinate hereditary characteristics" (Mossio & Pontarotti, 2019).

10.4.2   Perspectives on Fitness, Natural Selection, and Evolution

As explained above, the organizational perspective on biological inheritance appears as a major ingredient for an evolutionary biology based on a heuristic of collaboration. Below, I sketch how its integration into evolutionary thinking could induce, in the wake of earlier contributions, a change of perspective with regards to lineages, fitness, selection, and evolution.

First and foremost, assuming that inheritance is a matter of conservation of functional patterns – regardless of the parts being involved – and not of replication of genetic elements, implies that evolutionary biology should track *functional lineages* and not genetic ones. In this view, specialists should more precisely track the fate of integrated networks, not of virtually atomized alleles correlated with phenotypic variations. The key units of the living world are not elements heuristically depicted as selfish individuals eager to self-replicate but rather as parts collaborating with others in the context of organized networks. The stability of these parts is not linked to their intrinsic capacity of making faithful copies of themselves: it is related to their being integrated in networks channeling flows of matter and energy.

In this context, *fitness* cannot be thought of as the property of atomized objects but must be attributed to full organized systems whose spatial boundaries are outlined by interdependent constraints. It cannot be conceptualized, like in twentieth-century evolutionary biology, as a matter of differential replication (of genes) or as a matter of differential reproduction (of genetically homogenous organisms), but it should rather be thought as the differential capacity of integrated networks to channel flows of matter and energy in order to maintain themselves within and across generations.[19] In accordance with earlier studies (van Valen, 1975, p. 267), fitness can therefore be said to rest on the differential quantity of energy controlled by a biological system. Put another way, it can refer to differential management of resources (Pontarotti, 2017). Finally, it can globally be envisioned as a matter of differential expansion (van Valen, 1989, p. 7), some systems being more capable than others to make more of themselves in space and time (through reproduction, growth, etc.), depending on their performance in resources channeling.

From this point of view, *natural selection* does not target genes but networks with differential performances as far as resources management – and therefore spatiotemporal maintenance and expansion – is concerned. It selects among networks exhibiting differential efficiency regarding the control of material and energetic flows. This line of argument is consistent with the idea according to which natural selection targets effects and not structures as such, in a given environment (Rosenberg, 1994). It also somehow meets the hypothesis of physiological selection advanced by Turner (2004) when thinking about the evolution of "extended organisms" such as termite-fungi-mound systems. It is important to no note, here, that

---

[19] For details about intergeneration breaks and therefore about temporal limits of organized systems involved in evolution, see Mossio & Pontarotti, 2019.

according to Turner, genes should not be primarily considered as replicators but rather as specifiers[20] of future functions, as elements, among others, specifying how flows of matter and energy are channeled (Turner, 2004).

The *evolution* of non-standard biological systems, as for it, can be thought on the model of the evolution which took place before the appearance of DNA. In a world inhabited by autocatalytic sets of molecules, what matters, for evolution, is not the differential replication of discrete entities but the variations impacting network's efficiency regarding maintenance (Kauffman, 1995). The point can be summarized as follows: "if the result (of a variation in an autocatalytic set of molecules) were a more efficient network – one better able to sustain itself amid a harsh environment – then these mutations would be rewarded, the altered web crowding out its weaker competitors" (Kauffman, 1995, p. 73). In this view, evolution can no more be defined as a change in gene frequencies (Dobzhansky, 1937) or as a change in developmental programs during phylogeny (Oster & Alberch, 1982, p. 444). It must rather be viewed as a process leading to changes in regimes of canalization of flows of matter and energy through time (Pontarotti, 2017), as a change in organizational regimes.

It should also be noted that an organizational account of biological inheritance makes it possible to articulate three elements that were conceived separately in modern synthesis but that an organism-centered evolutionary biology is willing to link again: development, inheritance, and evolution (Walsh, 2010; Nicholson, 2014). Indeed, such an account makes no theoretical distinction between hereditary factors (genotype) and developed traits (phenotype), considering all of them as inherited organized constraints. It also acknowledges a continuity regarding the processes involved in the conservation of biological systems within and across generations (Mossio & Pontarotti, 2019).

Finally, let us go back to the role of natural selection in this theoretical context. This role would be limited by theoretical models, based on a heuristic of collaboration, in which the dynamics of interdependence between parts of biological systems would be more important than multi-level selection of selfish elements in the determination of biological evolution. These models would, for example, take their distance with Szathmáry and Maynard Smith (1995) work on evolutionary transitions. The latter indeed clearly relies on a heuristic of replication and insists on the role of multilevel selection in the appearance of new kinds of individuals, an event in which elements which could initially replicate independently become interdependent for their own replication. The collaborative point of view associated with the integration of an organizational thinking in evolutionary biology would rather be in line with the literature about constructive neutral evolution (Lukes et al., 2011). The latter indeed offers a perspective in which evolution principally rests on games of interdependence: "In this conception, mutation is not a source of raw materials, but

---

[20] "Specifiers are the catalytic surfaces that specify particular types of chemical reactions. These can be affected both by translated information in replicators (genes) and by environmental conditions" (Turner, 2004, p. 342).

an agent that introduces novelty, while selection is not an agent that shapes features, but a stochastic sieve" (Stoltzfus, 2012). This reduced role for natural selection would be consistent with the main statements of EES which gives an important explanatory role to the internal dynamics of organized systems.

10.5 Conclusion

In this paper, I have argued that the emerging organism-centered evolutionary biology, which theoretically makes some important room to the concept of organization, is missing some organizational perspective of inheritance, the latter being known as a key ingredient for evolutionary processes. I have outlined an organizational account of biological inheritance, and I have detailed the systemic concept of heritable variation (genetic mutation and non-genetic acquisition) that it contributes to ground. Finally, I have sketched some implications of an organizational perspective of biological inheritance for an evolutionary theory which would be based on a heuristic of collaboration rather than on a heuristic of replication. The big picture set in this article deserves being developed in future contributions. For instance, it will be important to further analyze the link between organization and developmental mechanisms, the latter being at the center of extended evolutionary synthesis. Organization, as presented in this paper, makes abstraction of mechanistic and temporal details. But these details could be of great relevance for the elaboration of a theoretical framework which would better take into account the causal role of organized biological systems in evolutionary dynamics.

References

Aiello, L. C., & Wheeler, P. (1995). The Expensive-Tissue Hypothesis: The Brain and the Digestive System in Human and Primate Evolution. *Current Anthropology, 36*(2), 199–221.

Bapteste, E., & Huneman, P. (2018). Towards a dynamic interaction network of life to unify and expand the evolutionary theory. *BMC Biology, 16*, 56. https://doi.org/10.1186/s12915-018-0531-6

Bateson, P. (2005). The return of the whole organism. *Journal of Biosciences, 30*(1), 31–39.

Bernard, C. (1885 [1878]). *Leçons Sur les phénomènes de la vie communs aux animaux et aux végétaux*. Librairie J.-B. Baillière et Fils (source : Gallica.fr).

Bertalanffy, L. (1973 [1968]). *Théorie générale des systèmes*. Traduction par J.-B. Chabrol. Bordas.

Bichat, F.-X. (1801). *Anatomie générale appliquée à la physiologie et à la médecine*. Brosson, Gabon et Cie (source : Gallica.fr).

Bondurianversky, R. (2012). Rethinking heredity, again. *Trends in Ecology & Evolution, 27*(6), 330–336.

Bondurianversky, R., & Day, T. (2018). *Extended inheritance: A new understanding of inheritance and evolution*. Princeton University Press.

Bouchard, F. (2013). What is a symbiotic superindividual and how do you measure its fitness? In F. Bouchard & P. Huneman (Eds.), *From groups to individuals: Evolution and emerging individuality* (Vienna series in theoretical biology) (pp. 243–264). MIT Press.

Cubas, P., Vincent, C., & Coen, E. (1999). An epigenetic mutation responsible for natural variation in floral symmetry. *Nature, 401*(6749), 157–161.

Danchin, E., Charmantier, A., Champagne, F., Mesoudi, A., Pujol, B., & Blanchet, S. (2011). Beyond DNA: Integrating inclusive inheritance into an extended theory of evolution. *Nature Reviews Genetics, 12*(7), 475–486.

Darwin, C. (1859 [2008]). *L'origine des espèces au moyen de la sélection naturelle ou la préservation des races favorisées dans la lutte pour la vie*. Flammarion.

Dawkins, R. (1976). *The Selfish Gene*. Oxford University Press.

Dawkins, R. (2008 [1982]). *The extended phenotype*. Oxford University Press.

Dobzhansky, T. (1937). *Genetics and the origin of species*. Columbia University Press.

Dobzhansky, T. (1970). *Genetics of the evolutionary process*. Columbia University Press.

Douglas, A. E. (2009). The microbial dimension in insect nutritional ecology. *Functional Ecology, 23*(1), 38–47.

Dupré, J., & O'Malley, M. (2007). Metagenomics and biological ontology. *Studies in History and Philosophy of Biological and Biomedical Sciences., 38*(4), 834–846.

Fabian, D., & Flatt, T. (2012). Life history evolution. *Nature Education Knowledge, 3*(10), 24.

Fox Keller, E. (2000). *The century of the gene*. Harvard University Press.

Galef, B. G., Jr., & Laland, K. N. (2005). Social learning in animals: Empirical studies and theoretical models. *Bioscience, 55*(6), 489–499.

Gayon, J. (1992a). *Darwin et l'après-Darwin : Une histoire de l'hypothèse de sélection dans la théorie de l'évolution*. Kimé.

Gayon, J. (1992b). Animalité et végétalité dans les représentations de l'hérédité. *Revue de Synthèse, Springer Verlag/Lavoisier, 4*(3), 49–61.

Gilbert, S., Sapp, J., & Tauber, A. (2012). A symbiotic view of life: We have never been individuals. *The Quarterly Review of Biology, 87*(4), 325–341.

Godfrey-Smith, P. (2009). *Darwinian populations and natural selection*. Oxford University Press.

Gould, S. J., & Lewontin, R. C. (1979). The spandrels of San Marco and the Panglossian paradigm: A critique of the adaptationist programme. *Proceedings of the Royal Society of London. Series B, Biological Sciences., 205*(1161), 581–598.

Griesemer, J. (2000). Development, culture, and the units of inheritance. *Philosophy of Science, 67*, S348–S368.

Griffiths, P. E., & Gray, R. D. (1994). Developmental systems and evolutionary explanation. *The Journal of Philosophy, 91*(6), 277–304.

Griffiths, P., & Gray, R. (2004). The developmental systems perspective: Organism-environment systems as units of evolution. In K. Preston & M. Pigliucci (Eds.), *Phenotypic integration: Studying the ecology and evolution of complex phenotypes* (pp. 409–431). Oxford University Press.

Griffiths, P. E., & Stotz, K. (2013). *Genetics and philosophy: An introduction*. Cambridge: Cambridge University Press.

Helantera, H., & Uller, T. (2010). The Price equation and extended inheritance. *Philosophy & Theory in Biology, 2*(201306), 1–17.

Holeski, L. M., Jander, G., & Agrawal, A. A. (2012). Transgenerational defense induction and epigenetic inheritance in plants. *Trends in Ecology & Evolution, 27*(11), 618–626.

Huneman, P. (2010). Assessing the prospects for a return of organisms in evolutionary biology. *History and Philosophy of the Life Sciences, 32*(2–3), 341–371.

Huneman, P. (2017). Why would we call for a new evolutionary synthesis? The variation issue and the explanatory alternatives. In P. Huneman & D. Walsh (Eds.), *Challenging the modern synthesis: Adaptation, development, and inheritance*. Oxford University Press.

Huxley, J. (2010 [1942]). *Evolution, the modern synthesis*. MIT Press.

Jablonka, E. (2002). Information: Its interpretation, its inheritance and its sharing. *Philosophy of Science, 69*(4), 578–605.

Jablonka, E., & Lamb, M. (2005). *Evolution in four dimensions*. MIT Press.

Johannsen, W. (1911). The genotype conception of inheritance. *The American Naturalist, 45*(531), 129–159.

Kant, E. (2000 [1790]). *Critique de la faculté de juger*. Vrin.

Kauffman, S. (1995). *At home in the universe. The search for the laws of self-organization and complexity*. Oxford University Press.

Laland, K. N., Uller, T., Feldman, M. W., Sterelny, K., Müller, G. B., Moczek, M., Jablonka, E., & Odling-Smee, J. (2015). The extended evolutionary synthesis: Its structure, assumptions and predictions. *Proceedings of the Royal Society B, 282*(1813), 1019.

Lewontin, R. C. (1970). The units of selection. *Annual Review of Ecology and Systematics, 1*, 1–18.

Lewontin, R. C. (1993). *Biology as ideology : The doctrine of DNA*. Harper Collins.

López-Beltrán, C. (1994). Forging inheritance: From metaphor to cause, a reification story. *Studies in History and Philosophy of Science Part A, 25*(2), 221–235.

Lukes, J., Archibald, J. M., Keeling, P. J., Doolittle, W. F., & Gray, M. W. (2011). How a neutral evolutionary ratchet can build cellular complexity. *International Union of Biochemistry and Molecular Biology Life, 63*(7), 528–537.

Mameli, M. (2005). The inheritance of features. *Biology and Philosophy, 20*(2), 365–399.

Mayr, E. (1963). *Animal species and evolution*. Harvard University Press.

Mayr, E. (1998). Prologue: Some thoughts on the history of the evolutionary synthesis. In E. Mayr & W. B. Provine (Eds.), *The evolutionary synthesis: Perspectives on the unification of biology* (pp. 1–48). Harvard University Press.

Merlin, F. (2017). Limited extended inheritance. In D. Walsh & P. Huneman (Eds.), *Challenging the modern synthesis* (pp. 263–279). Oxford University Press.

Mesoudi, A., Blanchet, S., Charmantier, A., Danchin, E., Fogarty, L., Jabolonka, E., Laland, K. N., Morgan, T. J. H., Müller, G. B., Odling-Smee, J., & Pujol, B. (2013). Is non-genetic inheritance just a proximate mechanism? A corroboration of the extended evolutionary synthesis. *Biological Theory, 7*(3), 189–195.

Moran, N. A., McCutcheon, J. P., & Nakabashi, A. (2008). Genomics and Evolution of Heritable Bacterial Symbionts. *Annual Review of Genetics, 42*, 165–190.

Montévil, M., & Mossio, M. (2015). Biological organisation as closure of constraints. *Journal of Theoretical Biology, 372*, 179–191.

Moreno, A., & Mossio, M. (2015). *Biological autonomy. A philosophical and theoretical enquiry*. Springer.

Mossio, M., & Bich, L. (2014). What makes biological organisation teleological? *Synthese*, Springer. http://link.springer.com/article/10.1007/s11229-014-0594-z

Mossio, M., & Moreno, A. (2010). Organisational closure in biological organisms. *History and Philosophy of the Life Sciences, 32*(2–3), 269–288.

Mossio, M., & Pontarotti, G. (2019, 2020). Conserving functions across generations: Inheritance in light of biological organization. *British Journal for the Philosophy of Science*, 1–33.

Mossio, M., Saborido, C., & Moreno, A. (2009). An organizational account of biological functions. *The British Journal for the Philosophy of Science, 60*(4), 813–841.

Müller, G. B. (2017). Why an extended evolutionary synthesis is necessary. *Interface Focus, 7*, 20170015.

Nanney, D. (1957). The role of cytoplasm in heredity. In W. D. McElroy & B. Glass (Eds.), *A symposium on the chemical basis of heredity, 1956* (pp. 134–135). The Johns Hopkins Press.

Nicholson, D. (2014). The return of the organism as a fundamental explanatory concept in biology. *Philosophy Compass, 9*(5), 347–359.

Odling-Smee, J., Laland, K. N., & Feldman, M. W. (2003). *Niche construction: The neglected process in evolution*. Princeton University Press.

Oster, G., & Alberch, P. (1982). Evolution and bifurcation of developmental programs. *Evolution, 36*(3), 444–459.

Pigliucci, M. (2009). "An extended synthesis for evolutionary biology", the year in evolutionary biology 2009. *Annals of the New York Academy of Sciences, 1168*, 218–228.

Pigliucci, M., & Muller, G. B. (2010). Elements of an extended evolutionary synthesis. In *Evolution – The extended synthesis*. MIT Press.

Pontarotti, G. (2015). Extended inheritance from an organizational point of view. *History and Philosophy of the Life Sciences, 37*, 430–448.

Pontarotti, G. (2016). Extended inheritance as reconstruction of extended organization: The paradigmatic case of symbiosis. *Lato sensu, 3*(1), 93–102.

Pontarotti, G. (2017). *Au delà du tout génétique: une perspective organisationnelle sur l'hérédité biologique et ses implications en biologie de l'évolution.* PhD dissertation. Université Paris 1 Panthéon-Sorbonne.

Pontarotti, G. (2022). Environmental inheritance: Conceptual ambiguities and theoretical issues. *Biological Theory, 17*(1), 36–51.

Pontarotti, G., Mossio, M., & Pocheville, A. (2022). The genotype–phenotype distinction: From Mendelian genetics to 21st century biology. *Genetica, 150*(3–4), 223–234.

Rapp, R. A., & Wendel, J. F. (2005). Epigenetis and plant evolution. *New Phytologist., 168*(1), 81–91.

Richards, C., Bossdorf, O., & Pigliucci, M. (2010). What role does heritable epigenetic variation play in phenotypic evolution? *Bioscience, 60*(3), 232–237.

Rosenberg, A. (1994). *Instrumental biology or the disunity of science*. University of Chicago Press.

Sachs, J. L. (2013). Origins, evolution, and breakdown of bacterial Symbiosis. In S. A. Levin (Ed.), *Encyclopedia of biodiversity* (Vol. 5, 2nd ed., pp. 637–644). Academic.

Slagsvold, T., & Wiebe, K. (2007). Learning the ecological niche. *Proceedings of the Royal Society B, 274*, 19–23.

Stearns, S. C. (1992). *The evolution of life histories*. Oxford University Press.

Sterelny, K. (2001). Niche construction, developmental systems and the extended replicator. In S. Oyama, P. E. Griffiths, & R. Gray (Eds.), *Cycles of contingency: Developmental systems and evolution* (pp. 333–349). MIT Press.

Stoltzfus, A. (2012). Constructive neutral evolution: exploring evolutionary theory's curious disconnect. *Biology Direct, 7*(1), 35.

Szathmáry, E., & Maynard Smith, J. (1995). The major evolutionary transitions. *Nature, 374*, 227–232.

Thompson, D. A. W. (1942). *On growth and form* (nouv. ed.). Cambridge University Press.

Turner, J. S. (2000). *The extended organism, the physiology of animal-built structures*. Harvard University Press.

Turner, J. S. (2004). Extended phenotypes and extended organisms. *Biology and Philosophy, 19*(3), 327–352.

Van Baalen, M., & Jansen, V. A. A. (2001). Dangerous liaisons: The ecology of private interest and common good. *Oikos, 95*(2), 211–224.

Van der Lugt, M., & De Miramon, C. (2008). Penser l'hérédité au Moyen Âge : une introduction. In M. van der Lugt & C. de Miramon (Eds.), *L'hérédité entre Moyen Âge et Époque moderne. Perspectives historiques* (pp. 3–37). Edizioni del Galluzzo.

Van Valen, L. M. (1975). Life, death, and energy of a tree. *Biotropica, 7*(4), 259–269.

Van Valen, L. M. (1989). Three paradigms of evolution. *Evolutionary Theory, 9*, 1–17.

Walsh, D. M. (2006). Organisms as natural purposes: The contemporary evolutionary perspective. *Studies in History and Philosophy of Biological and Biomedical Sciences, 37*(4), 771–791.

Walsh, D. M. (2010). Two Neo-Darwinisms. *History and Philosophy of the Life Sciences., 32*(2–3), 317–339.

Walsh, D. (2015). *Organisms, agency, and evolution*. Cambridge University Press.

West-Eberhard, M. J. (2003). *Developmental plasticity and evolution*. Oxford University Press.

Chapter 11
There Are No Intermediate Stages: An Organizational View on Development

Leonardo Bich 🆔 **and Derek Skillings** 🆔

Abstract Theoretical accounts of development exhibit several internal tensions and face multiple challenges. They span from the problem of the identification of the temporal boundaries of development (beginning and end) to the characterization of the distinctive type of change involved compared to other biological processes. They include questions such as the role to ascribe to the environment or what types of biological systems can undergo development and whether they should include colonies or even ecosystems. In this chapter we discuss these conceptual issues, and we argue that adopting an organizational approach may help solve or clarify them.

While development is usually identified with the achievement of an adult form with the capability to reproduce and therefore maintain a lineage, adopting the organizational approach may provide a different strategy, which focuses also on the maintenance of the current organization of the organism. By doing so an organizational approach favors a switch in perspective which consists in analyzing how organisms maintain their viability at each moment of development rather than considering them as going through intermediate stages of a process directed toward a specific goal state. This developmental dimension of biological organization has yet to be given a general and detailed analysis within the organizational theoretical perspective, apart from some preliminary attempts. How a biological organization is maintained through a series of radical organizational changes and what these changes are issues that still require clarification. In this chapter we offer the beginnings of such an analysis of developmental transitions, understood as changes in functionality brought forth by regulatory mechanisms in the context of the continued maintenance of organizational viability at every step.

L. Bich (✉)
Department of Philosophy, IAS-Research Centre for Life, Mind, and Society,
University of the Basque Country (UPV/EHU), San Sebastian, Spain

Center for Philosophy of Science, University of Pittsburgh, Pittsburgh, PA, USA
e-mail: leonardo.bich@ehu.es

D. Skillings
Department of Philosophy, University of North Carolina at Greensboro,
Greensboro, NC, USA

© The Author(s) 2024
M. Mossio (ed.), *Organization in Biology*, History, Philosophy and Theory of
the Life Sciences 33, https://doi.org/10.1007/978-3-031-38968-9_11

Keywords Organization · Regulation · Functions · Developmental change ·
Adultocentrism

11.1   Introduction

Accounts of development are characterized by a common focus on changes taking
place during the lifetime of a biological system, usually centered on multicellular
organisms. These accounts may address how complex forms are generated, the origin of differentiation and morphological variation, growth, regeneration, metamorphosis, and other related phenomena (see, e.g., Muller & Newman, 2003). In some
cases, development is identified with any changes taking place throughout the entire
life cycle of an organism, including phenomena such as the production of new blood
cells or senescence (Gilbert & Barresi, 2018).

   Yet, while differing on the type of change involved, development is usually identified with a process that gives rise to a complex multicellular organism (Barinaga,
1994; Martinez-Arias & Stuart, 2002; Wolpert & Tickle, 2011), more specifically
with the achievement of an adult form with the capability to reproduce (Minelli,
2011; Griesemer, 2016). As explicitly stated by Griesemer (2016): "Development is
the recursive acquisition, refinement, or maintenance of a capacity to reproduce";
and "development can be seen as relatively continuous growth and differentiation of
shapes and sizes of parts along with the maturation necessary to reproduce." On
these views, development is considered part of a larger life cycle defined by reproductive events. As a consequence, explanations of change across whole life cycles
are usually focused on evolutionary considerations at the scale of lineages. This
directionality of developmental processes from the zygote to an adult form capable
of reproduction underlies these characterizations of development. By doing so,
developmental biology has become a field in which scientists and theoreticians
(implicitly or explicitly) face the issue of biological teleology and employ a teleological terminology.

   Accounts of development face several internal challenges or puzzles. We introduce some of them in Sect. 11.2. These challenges include issues deriving from
adopting notions of directionality and potentiality, especially if cases of reversible
and multidirectional development are considered. Further tensions concern the
problems of identifying the boundaries of development, i.e., the start and end points
of development. An important issue in this respect is identifying the distinctive
character of developmental change compared to other changes that living systems
undergo during their lifetime, such as individual adaptivity, growth, regeneration,
plasticity, acclimation, etc. Development is commonly defined in terms of the
organism. It is the history of a particular organism from its earliest form, e.g., egg,
seed, or clone, to its end, usually either reproductive maturity or death. Another
important question is what are the types of systems that can develop and whether
those systems should include unicellular organisms, colonies, superorganisms,
symbiotic associations, or ecosystems.

In Sect. 11.3 we argue that the organizational framework may provide a shift in perspective that may be helpful in relieving these tensions and contribute to a better understanding of development by complementing existing accounts. A deep change in strategy concerns how to pick out the relevant system. We decenter the organism to focus on living systems characterized by organizational closure and regulatory mechanisms. Focusing on a wider phenomenon realized on a longer time scale – the reproductive cycle or the lineage – misses the importance of the explanation of development from the point of view of the organization of the system undergoing processes of change. The organizational approach accounts not only for the mainte-nance of the lineage through reproduction and selection but also of the organization of the system.

Adopting an organizational approach means switching focus to how organisms – and possibly other biological systems – maintain their viability at each moment of the developmental process instead of considering them as going through intermedi-ate stages of a process directed toward a predetermined goal state. How an organiza-tion is maintained through a series of radical changes and what types of changes are involved are issues that still require clarification. We argue that from an organiza-tional perspective, development is a regulatory process that changes the number or types of functions of a regime of closure of constraints. The reasons why we recom-mend this conceptual switch is because it resolves some of the outstanding tensions or problems of generalization emerging in comparative work that tries to circum-scribe developmental phenomena across the entire tree of life. These are the chal-lenges we take up in Sect. 11.4. In Sect. 11.5 we conclude by summarizing the main conceptual points of our proposal for an organizational view of development and some remaining open questions for future work.

11.2 Puzzles and Challenges Within Theories of Development

There are a cluster of phenomena related to development that make the boundaries of this phenomenon difficult to define. Here we introduce a set of puzzles regarding development and development-like phenomena. They constitute challenges for any unitary theory of development. Few, if any, of these challenges are new, but neither are they universally accepted. The purpose of addressing them is not to provide a definition of development through a definitive set of necessary and sufficient condi-tions, which may not be desirable given the breadth and variety of work on develop-ment. Our aim is to build a conceptual account able to provide a different, theoretically coherent perspective from which it may be possible to better under-stand the implications of these puzzles and address them. We take as a starting point the work of Alessandro Minelli (especially Minelli, 2003, 2011, 2014). Minelli presents a definition of development that he thinks is often taken for granted: "devel-opment is a sequence of changes through which a multicellular adult is produced, through an increase in complexity more or less strictly programmed in its genes, starting from a single cell which in most instances is a fertilized egg" (p. 5). He uses this definition as a jumping off point for showing how inadequate it is and to further argue for "what development is not" (Minelli, 2011, p. 6). Minelli argues that the given notion of development is inadequate in five ways, because "development (1) is not restricted to the multicellular organisms, (2) does not necessarily start from an egg, (3) does not necessarily start from a single cell, (4) does not necessarily imply an increase in structural complexity, and (5) does not necessarily end with the achievement of sexual maturity" (p. 5).[1] We agree that these five points present contentious problems for a comprehensive theory of development and will use them as a starting point, taking and expanding on them one-by-one. We also introduce two more problems for the given definition of development: development (6) can proceed across complex life cycles punctuated by multiple reproductive events, and (7) does not necessarily exclude multispecies complexes (e.g., lichens) and host-microbe systems (e.g., aphids, corals, cows, or humans).

11.2.1   Challenge 1: What Is Developmental Change, and Is It Restricted to Multicellular Organisms?

Development is some sequence of changes, but what kinds of changes can count as development? Organisms, for example, are constantly going through metabolic changes in order to maintain themselves from moment to moment. However, metabolic change is not sufficient on its own to count as development if development is to retain any specialized meaning. To differentiate developmental changes from metabolic changes, development is usually restricted to changes that have to do with growth or a change in morphology or function. Acquiring new functions or changing the function of existing structures is especially central. The first challenge is to establish whether or not single-celled organisms go through the right kinds of changes to count as development. The answer might not be univocal due to the diversity of organisms under consideration. Asexually reproducing bacteria might only go through metabolic changes, some growth, and then reproduction, whereas many parasitic eukaryotes have complex life cycles transiting through very different morphologies. In these latter cases, the border between reproduction and development gets fuzzy. We will return to this in challenge 6.

---

[1] An additional source of inadequacy may be constituted by the idea that development is programmed in the genes. For criticisms of this idea, see Sonnenschein and Soto (1999) and Veloso (2017).

11.2.2   Challenge 2: Does Development Necessarily Start at Fertilization?

Pregnancy is a challenge when trying to draw clear cut lines between biological individuals (Grose, 2020; Kingma, 2020; Nuño de la Rosa et al., 2021). A fertilized egg makes for a clear starting point in the life cycle of a sexually reproducing species and is perhaps the obvious choice for where to pin the start of development. But this is clearly a byproduct of focusing on development in sexually reproducing organisms, primarily animals. It doesn't work for a general account of development because it ignores vegetative reproduction (e.g., cloning, budding) in plants, fungi, and many animals (Minelli, 2011). It doesn't even work for paradigmatic cases like humans because of monozygotic twins that develop from the splitting of a single embryo. Even more challenging is the case of the armadillo, which almost always produces four identical quadruplets, splitting at an even later developmental stage (Enders, 2002). Cases like these complicate drawing the line between developmental processes. For example, in the case of monozygotic twins, it seems like there are two equivalent interpretations of the developmental process: (1) there is one developmental process, and it is split between two separate entities, and (2) new developmental processes split off from an ongoing developmental process, but that splitting is not reproduction. Either interpretation produces a problem. In the first option, there is an apparent contradiction, as one and the same (token) developmental process is carried out within two distinct organisms. The second option undermines the initial claim that development necessarily starts at fertilization. Such claim is also undermined by phenomena such as parthenogenesis (Sonnenschein & Soto, 1999). There is a third option: the development of each twin starts at a stage later than fertilization. We will explore this option in Sect. 11.4.

11.2.3   Challenge 3: Does Development Always Start at a Unicellular Bottleneck?

This question is related to the previous one or is perhaps an expansion of it. Identifying the initiation of development requires identifying a new individual that is about to undergo development. A single cell that multiplies and transforms into a multicellular organism is a good candidate. Such a cell, especially when it is at least partially independent, is easier to delineate. Citing a multicellular clump as the start of development raises a few questions: (1) what is the origin of the clump, and what changes did it undergo to get to the multicellular stage?, (2) what kinds of changes did it undergo, if they were not developmental changes?, and (3) are there bounds on the size and complexity of an entity that both counts as a new individual and is able to undergo development?

In principle, is there a restriction on the types of mereological structures that can serve as starting points for developmental processes? The case of the armadillo, for example, seems to point to a later beginning of the development of distinct embryos. One possible restriction is that a system must be unified as an individual whole, as opposed to a colony or a collection of individuals. Furthermore, such a system must not have yet started to undergo changes in the functional/structural relations of its parts since its formation as a new and discrete individual whole.

If it were the case that the formation of a new cell through the fusion of sperm and egg marks the beginning of development, as is commonly assumed, then what does that leave out? One possibility is asexual reproduction through budding or parthenogenesis, which are both common in multicellular systems. A second is symbiotic associations such as biofilms, holobionts, lichens, or other multispecies systems that seemingly fuse to form new individuals (Skillings, 2016). The symbiotes don't fuse into a new single cell that then begins development. The creation of these entities happens when the cells of different species join together. Thus, the initial condition of the association is multicellular.

Challenges 2 and 3 can then be generalized: instead of looking for specific structural features, is there a common functional starting point for every possible developmental process?

11.2.4   Challenge 4: Does Development Imply an Increase in Complexity?

There are two different ways the answer to this question is no, or at least ambiguous. The first answer is that different kinds of complexity can appear during a life cycle, and it is unclear how to compare them. For example, in species that go through metamorphosis, organisms will often lose some functions while gaining others. Through metamorphosis an organism might lose the capacity to feed but gain the capacity to sexually reproduce, such as in mayflies (Skillings, 2019), or behavioral complexity found in the larvae might be lost while structural complexity increases in the sessile adult form, such as in tunicates (Holland, 2016). Tunicates lose complex and energetically expensive structures like a head/brain that become unnecessary once they transform into sessile adults. In cases like these, it is unclear if there has been an overall increase in complexity.

The second, more definitive, answer to the question of whether development implies increase in complexity appears to be a straightforward no. Parasitic rhizocephalan barnacles transform from a free-swimming larval stage to a larval "injection" stage, where the organism essentially acts as a giant hypodermic needle. This stage then injects a small group of poorly differentiated amoeboid cells into the hemolymph of a crab host. Those cells later metamorphose into the adult form, which does not inherit any organs from the larval stage and consists of two parts: an interna, which is a system of ramifying rootlets spanning the body of the host, and an externa, which is a structure containing the reproductive system (Høeg et al., 2012; Miroliubov et al., 2020). In a more extreme case, the immortal jellyfish *Turritopsis dohrnii* (see Matsumoto et al., 2019) is able to reverse its developmental trajectory from medusa back to polyp in response to stress without going through the whole cycle (i.e., through reproduction and the unicellular stage). It does so by going through a different intermediate stage, the cyst, constituted by a cluster of poorly differentiated cells. Is rejuvenation, with or without simplification, a kind of development? If it were the case, it would put into question the very idea of development as a unidirectional or irreversible process (of which an increase in complexity is one example).

11.2.5   Challenge 5: Does Development End at Reproductive Maturity?

According to Griesemer, development is intrinsically related to reproduction "Development is the recursive acquisition, refinement, or maintenance of a capacity to reproduce. Reproductive capacity is realized in diverse ways and modes of development in extant lifeforms on Earth" (Griesemer, 2016). He writes elsewhere that "reproduction involves the conveyance or conferral of developmental capacities. Not every mereological change achieves that. Moreover, since development is the acquisition of a capacity to reproduce, only lineage- forming (or terminating) mereological changes in development count" (Griesemer, 2016). It appears that Griesemer is using reproduction to explain development.

This focus on reproductive maturity as the endpoint of development can lead to a kind of "adultocentrism" (Minelli, 2011). This adultocentrism is an improper fixation, or essentialization, of the adult form of an organism as the true or proper form. This can create the view that the adult form is the form the "organism works toward," injecting a kind of teleology or forward-lookingness into all developmental processes. This risks overlooking the importance and distinctive features of other non-terminal forms, especially when studying organisms that are not amniotes (reptiles, birds, mammals). Let us think of organisms that have different free-living forms and may undergo metamorphoses, or that go through complex life cycles, where it is not clear if there is even an adult or terminal form.

It is not hard to find examples of adultocentrism; it is rooted in our language and maybe even our psychology. Picture a sea star, butterfly, or frog. It is probably the adults, not the larvae or juvenile forms, that come to mind. This is also the case for how organisms like mayflies are characterized, where the adult form is present for only a fraction of the life cycle. This is all fine for everyday use. It is the adult forms that we are most likely to encounter, because they are either larger or more active and consequently easier to see or come across. But this becomes misleading when thinking about the development or evolution of an organism. The organism has evolved to maintain itself at every point of the life cycle, and the maturation process (developing into an adult) constitutes an important part (and often a major one) of the whole life cycle. The question (for development) isn't only how did these features evolve in order to increase future reproductive function or output but how/why did these features evolve in order to maintain the organism at that stage of the life cycle. The proximal selection pressure is at the maintenance of the organism at that stage, not some future reproductive stage that isn't realized at that point. The tadpole is an adapted self-maintaining organization subject to selection and lives freely as an agent in its environment. In the case of the axolotl, a pedomorphic salamander, it can even undergo reproduction. So, the tadpole's tail is just as important as the frog's four-legged form.

11.2.6   Challenge 6: Are the Transitions Between Multicellular and Unicellular Forms in a Complex Life Cycle Development or Reproduction?

Complex life cycles are probably the most common type across the spectrum of life. This includes sequences of forms that are divided by metamorphosis (like between the caterpillar and the butterfly), by both asexual and sexual reproduction (e.g., corals, and parasitic flatworms, etc.) and transitions between multicellular and unicellular forms (e.g., algae, ferns). Let us think of a life cycle with multiple stages divided by reproduction, where the same type of form doesn't come back until it has gone through different stages separated by reproduction. It is hard to parse a life cycle like this on an account of development that focuses on development as a unitary process that moves solely toward reproduction. Moreover, it makes it extremely problematic to distinguish development from reproduction. Is a single life cycle – say from haploid form to sexual reproduction to diploid form to asexual reproduction with multiplication at each step – made up of multiple and vastly different, reproductive and developmental cycles attached end to end? Or is it a single developmental process measured by one turn through the entire cycle but punctuated by "minor" forms of reproduction along the way? There isn't a knockdown argument for either interpretation (Godfrey-Smith, 2016). But there needn't be. What is needed is a conception of development that can make sense of this problem and the rest of the other problems.

11.2.7   Challenge 7: Can Multispecies Assemblages Develop as One System?

The deeper question behind this challenge asks what kinds of systems can develop: Is it only organisms that develop? Do symbiotic associations like lichens develop at the level of the whole? Can ecosystems develop?

Accounts of development focused on achieving reproductive capabilities as the end point of the process fail to be satisfactory when dealing with integrated symbiotic assemblages. These are systems where developmental phenomena appear to be present in the more comprehensive system (the assemblage) but that do not repro-duce at the level of the comprehensive system. Examples include symbiotic assem-blages like lichens, biofilms, and holobionts. Lichens don't reproduce to form new lichens; the algal and fungal partners reproduce separately and then disperse and rejoin to form new lichens. Yet they undergo developmental changes at the level of the system as a whole. These systems are contradictory for developmental accounts focused on reproduction. Moreover, the boundaries might be drawn in the wrong places, because the overall developmental process of the assemblage would be over-looked. To address the question whether these associations undergo development, one needs to focus on how they change as integrated entities and abandon the idea of development as a tendency toward reproduction.

11.3   Gestalt Switch: Adopting the Organizational Perspective

The challenges discussed in the previous section put into question accounts of development as a progressive irreversible process directed toward the production of an adult organism capable of reproduction. They bring to the surface the need for a gestalt switch: development needs to be addressed also from a different point of view, one that is not directed toward some defined state in the future. A possible way of answering these challenges is through a change in perspective that centers on the maintenance of the living system from the very beginning of development. Such an account should be able to provide a general characterization of what development is, what type of changes it implies, and when development starts and ends. At the same time, it should be precise enough to address the challenges posed by phenom-ena such as rejuvenation, complex life cycles, and multispecies assemblages. Moreover, it needs to distinguish development from other types of change taking place during the life of an organism. By this, we do not claim that a different approach should replace those currently available but provide a complementary coherent theoretical perspective.

In the second part of this paper, we argue that this gestalt switch can be accom-plished by adopting an approach focused on organization. We sketch a proposal of an organizational account of development, and we discuss how it can address the challenges presented in Sect. 11.2. The central idea is to focus on how the organiza-tion of living systems is maintained during the transitions that characterize develop-ment. This is a radical shift from a view of development as a process aiming toward a final state, or as an actualization of an intrinsic potentiality. Our focus is on what is maintained and on the developmental changes occurring at each moment, rather than interpreting them as early stages work toward constructing the adult form and achieving reproductive capabilities. According to the organizational approach, there are no intermediate stages, ones defined by their relation to some future goal state.

Every stage is equally important, because the system must build and maintain itself at every point of its existence.[2]

The organizational framework was built upon pioneering work on biological autonomy carried out by Jean Piaget (1967), Robert Rosen (1972), Humberto Maturana and Francisco Varela (Varela et al., 1974), and Howard Pattee (1972), among others. More recently it was further developed by Stuart Kauffman (2000) and by Alvaro Moreno and collaborators (Ruiz-Mirazo & Moreno, 2004; Moreno & Mossio, 2015), among others. The organizational account characterizes a biological organism as an autonomous system capable of producing its own components and maintaining itself in far from equilibrium conditions while interacting with its environment. To explain this capacity, this tradition appeals to the internal organization of the organism, which is maintained despite the continuous transformations that the organism undergoes at the level of its components. The core feature of this approach is the focus on the organization of the system: the identification of topological relations between the operations of components and between processes of transformation within a system. Organization refers to the way production and transformation processes are connected so that they are able to synthesize the very components that make them up, by using energy and matter from the environment. In this view, the fundamental feature of the organization of biological self-maintaining systems is its circular topology as a network of processes of production of components that in turn realize and maintain the network itself. This distinctive type of generative circularity that characterizes biological systems is known as "organizational closure." The basic capability of a biological organization to self-produce and self-maintain has been explained in terms of closure of constraints (Moreno & Mossio, 2015; Montevil & Mossio, 2015). Constraints are characterized as material structures that harness processes and that by doing so specify part of the conditions of existence of those processes. According to this framework, living systems are capable to generate a subset of the constraints acting on their internal processes and realize a distinctive causal regime by which these constraints are organized in such a way that they are mutually dependent for their production and maintenance and collectively contribute to the maintenance of the conditions in which the whole network can persist.

The notion of closure of constraints focuses on the distinctive capability of living systems to contribute to their own conditions of existence and to the existence of their parts. This basic idea grounds two important biological notions: function and teleology. Within the organizational framework, a biological function is understood as a contribution of a part to the maintenance of a self-maintaining organization (e.g., a living cell) that, in turn, contributes to producing and maintaining the part itself (Mossio et al., 2009). Functional parts coincide with the constraints subject to closure. The *telos* of the system is understood in terms of

---

[2] It is important to make clear that considering different stages as equally important does not mean that they are all the same. On the contrary, it means that their distinctive features and how they are maintained should not be overlooked or interpreted in terms of a future state. However, some stages might be more relevant in relation to specific research agendas.

self-maintenance (Mossio & Bich, 2017). The focus of the organizational account of teleology is on how the activity of a biological system contributes to determining its own conditions of existence. The organization of a living system is characterized as an intrinsically teleological causal regime where the conditions of existence on which the organization exerts a causal influence are the goal (*telos*) of the system. Other accounts of biological functions and teleology centered on evolution differ from the organizational one in that (1) they take the lineage rather than the current system as the grounds of intrinsic teleology; (2) they characterize functions of traits etiologically, as contributions to the survival of the ancestors of those organisms that currently carry those traits; and (3) the goals of the system are characterized in terms of adaptation by natural selection (inasmuch as they contribute to maintain the lineage). An important implication of adopting the organizational account is that it entails this distinctive teleological framework that is focused on the maintenance of the system. Applying this framework to development means identifying the *telos* of the developing system in its current organization, rather than in a future state that contributes to the maintenance of the lineage (i.e., the adult form and reproduction). As such, it provides a different theoretical perspective which is not based on a future-oriented directionality and is not subject to the issues discussed in the previous section.

A further aspect of the organizational approach needs to be taken into consideration before building an organizational framework of development. The idea of closure alone is insufficient to ground a theoretical understanding of this biological phenomenon. There are two primary reasons. The first is an intrinsic limitation of the very notion of closure alone in providing an understanding of biological organization. The second is the limitation of the notion of closure in accounting for change in general and, therefore, even more so for developmental change. Both limits can be overcome by employing the notion of regulation. Regulation is carried out by mechanisms realized by sets of constraints that are sensitive to internal and external variation and are capable of changing their activity accordingly. Regulatory mechanisms operate as higher-order constraints in the sense that they act on other constraints in the system. What they do is to selectively shift between different available regimes of self-maintenance, in such a way as to contribute to the viability of the system (Bich et al., 2016).

Let us consider the first limitation. The capability to produce their own functional components (i.e., constraints) is not enough to understand how biological organizations maintain themselves and actually realize closure. The basic biological constraints involved in a regime of closure are not always functioning or functioning whenever their substrates and energy are available. Their activities are constantly controlled (inhibited, activated, modulated) by other constraints on the basis of the state of the system and the environment:

> Cells, for example, engage in division, but they are not constantly dividing (when they do, the result is a pathology such as cancer). Cells metabolize glucose to produce ATP, but they only do so when ATP levels drop and energy is needed. Otherwise, they convert glucose to glycogen. Protein synthesis is another process that is inhibited or activated on the basis of the needs of the cell. Neurons generate action potentials, but either do so only when they receive an appropriate stimulus or change the rate at which they generate action potentials in response to stimuli. (Bich & Bechtel, 2022a).

To maintain itself, an organism needs to *continuously* modulate and coordinate the activities of its basic functional constraints, which directly harness thermodynamic processes, in such a way that they can realize a viable regime of closure (Bich, 2018). Equally important, an organism must interact with a changing environment, which is the source of matter and energy for its internal processes. To do so, the internal organization of an organism must manage adaptively the dynamical variability available within it. As argued elsewhere, this is achieved by means of regulatory mechanisms (Bich et al., 2016, 2020; Bich, 2018; Bich & Bechtel, 2022a, b). They continuously exert a fine-tuned functional control over the exchanges of matter and energy of the system with its surroundings and over the activity of the internal constraints in such a way that the system is able to bring forth different viable responses to environmental perturbations and internal needs.

The second limitation of an account of closure without regulation concerns the capability to account for change in biological systems. This is particularly relevant because development is a specific type of change. As argued in the previous paragraph, living systems do not only and simply produce, repair, and maintain their components. Such activities are continuously undergoing regulation. Importantly, on this view the basic regime of closure does not operate in a regular manner: an organism needs to constantly change in order to maintain viability. However, as argued by Bich et al. (2016), closure alone would account only for a very limited type of change, one understood in terms of a dynamic stability that is realized as a passive network property. The basic regime of closure simply "absorbs", as a network, the effects of a limited set of perturbations or internal variations. The system compensates for perturbations by means of reciprocal adjustments between tightly coupled internal subsystems. The dynamics of the whole system are maintained in the initial attractor state or are pushed by the perturbation into a new stable attractor state. In living organisms, instead, regularity and stability in the activity of components are exceptions. A living system coordinates the activities of its components, modulates internal processes, and responds adaptively to environmental variation. The activity of each basic constraint is controlled according to the needs of the organization, starting from those basic constraints involved in transcription, translation, and protein synthesis. The system changes what activities its constraints perform in ways appropriate to the circumstances it faces and its internal state. On this view, active change is controlled by regulatory mechanisms (see Bich et al., 2016, 2020). Regulation is therefore a crucial notion to understand the organizational approach and its application to development.

In sum, adopting an organizational account implies focusing on how a living organization is currently maintained and on the functional contributions of its different components. Change, in this perspective, is understood as the result of the action of regulatory mechanisms which, on the basis of the state of the system and the environment, modify what functions are realized and modulate how they are performed.

11.4   Toward an Organizational Account of Development

What can an organizational approach say about development and its teleological dimension? As a starting point, this approach has a distinctive focus on developmental processes: on the system's current organization rather than on the realization of potentialities or the achievement of reproductive capability. The organizational account provides a conceptual framework which can be applied to development by accounting for how organisms, or biological self-maintaining organizations in general, maintain their viability at each moment of the developmental process. On our view it allows characterizing development as a regulatory process that changes the number and type of functions available to the system at a given moment while the system itself maintains its viability. From this perspective, development is not addressed as an adult-oriented process. Instead, the telos of the system is grounded in the maintenance of the viability of the organism at each stage of development, rather than in a future state.

To date, few contributions belonging to the organizational framework have addressed aspects of developmental processes. Arnellos et al. (2014) and Veloso (2017) focus on the role of intercellular signals and constraints in cell differentiation at early stages of development as an important factor to achieve integration. They contrast it with accounts of cell differentiation processes focused on intracellular factors such as genetics. Bich et al. (2019) point out the limits of cell differentiation alone to address multicellularity and development.[3] They focus on what is a minimal multicellular organization capable of maintaining itself as an integrated system and what are the types of mechanisms that control individual cells and ensembles of cells to realize tissues and organs. They emphasize the importance of the control of spatial organization and the role of the extracellular matrix (ECM) in development.

The only work entirely centered on development within an organizational perspective is by Nuño de la Rosa (2010). She focuses on vertebrate development. She characterizes development as the generation of a fully-fledged autonomous organization, which happens in the later stages of this process. According to this view, development is considered as the *explanans* for biological autonomy.

This pioneering and detailed work has the merit to be the first and only to fully focus on development. However, it exhibits some features that make it incompatible with current organizational accounts. Moreover, it shares some of the limits exhibited by the other accounts of development discussed in the previous sections. In the first place, it focuses on a small subset of vertebrates: mammals and more specifically humans. Therefore, it might not be representative of development as a general biological phenomenon. In the second place, it focuses on autonomy considered as a form of independence from the mother organism, rather than a general form of self-maintaining organization characterized by a regime of organizational closure.

---

[3] Moreover, cell differentiation itself, is determined by the surrounding ECM and by tissue in which cells reside (see Sonnenschein & Soto, 1999).

The problem is that biological autonomy does not imply independence. In principle, closure is not incompatible with forms of dependence, and a system can be autonomous in the sense that it realizes closure even though it is not independent from other systems. Examples are symbiotic associations, multicellular organizations, and, possibly, ecosystems (see Montevil & Mossio, 2015; Nunes-Neto et al., 2014, Bich, 2019). This is an important aspect of the organizational framework and allows it to account for forms of nested closure. In the third place, this account is in tension or even in contradiction with organizational approaches. Nuño de la Rosa's view is explicitly Aristotelian: development is characterized as a process of progressive actualization of autonomy. This account is characterized by a future-oriented teleology incompatible with the teleology that is characteristic of the organizational framework, based on current contributions to the persistence of an organization.[4] Like other accounts, here again the goal of development is producing the adult organism. Finally, an implication of this focus on the progress from potency to actuality is that this account assumes change as an *explanans* of autonomy instead of an *explanandum*: on this view developmental change is what bring forth and explains the origin of an autonomous system. It is not the object of analysis.

The developmental dimension of biological organization has yet to be given a detailed analysis within the organizational theoretical perspective. The idea that an organization is maintained through a series of radical changes or transitions, such as those that take place in developmental processes, is an issue that still requires clarification. The application of this idea faces several internal tensions, insofar as the organizational approach is mainly focused on what is currently maintained – the whole organization – rather than what changes over a long sequence of often radical transitions. The first conceptual problem is determining what kinds of organization can undergo development as opposed to mere change. The second problem is how to account for the specificity of developmental change. It consists in distinguishing developmental changes from the other types such as metabolic changes. The third problem concerns the boundaries of development. To address it requires establishing when development starts within an organizational framework by identifying what is the initial self-maintaining organization that undergoes developmental change. It also requires establishing whether and when development stops and why. Common options have it ending with the realization of the adult form versus continuing through ageing or senescence.

The organizational approach can provide a principled way to address these problems and helps refocus those tensions and puzzles discussed in the previous sections. We offer here the beginnings of a supplementary analysis of development that

---

[4] The criticism of future-oriented approaches and potentialities is shared also by early work on the organizational framework, such as the autopoietic theory, which is explicitly focused on the current system (Maturana & Varela, 1980). Change is understood in terms of "structural determinism," that is, all changes a living organization undergoes at a given moment are determined by its structure at that specific moment. It is important to mention that autopoietic theory rejects teleology insofar as in the interpretation of Maturana and Varela teleology is future or past oriented (see Mossio & Bich, 2017).

focuses on the continued maintenance of organizational viability at every step. The starting point is the idea that during the life of an organism, what is maintained through the deep and continuous changes of its components is the organization of the whole. The conservation of organization unifies the biological processes an organism undergoes, which includes development, growth, senescence, etc. This idea has been expressed by Di Frisco and Mossio (2020) through the notion of organizational continuity, that is, "the presence of a continuous causal process linking successive organizational regimes, irrespective of material and functional changes." This is the foundational assumption that it is to be adopted in order to understand biological phenomena from an organizational perspective. However, it is a very general notion. Alone, it does not provide conceptual tools to distinguish between development and other phenomena such as reproduction and aggregation. To do so it requires additional assumptions such as on the necessary variation in the number of organizations.[5] Even more importantly, it does not provide an account of change. It focuses on what is maintained. Understanding development exactly requires understanding change, a specific type of change, within a scenario of organizational continuity.

The first problem to face in order to address development is how to pick out the relevant system. Focusing on a system realized on a longer time scale than the organismic organization – the reproductive cycle or the lineage – would miss the importance of the explanation of development from the point of view of the organization of the system undergoing a process of change. The organizational approach focuses on the organization of current biological systems capable of maintaining themselves. In this context the relevant system is a functionally integrated self-maintaining organization. Functional integration on this view consists in the degree to which the different components that collectively realize a biological regime of self-maintenance depend on one another for their own production, maintenance, and activity. In principle, an integrated system can be a unicellular or a multicellular organization (see Challenge 1 in Sect. 11.2) or even a symbiotic association if it satisfies the requirements (Bich, 2019; see Challenge 7 in Sect. 11.2). Let us focus on the second type of organization, given that it is the one usually discussed in relation to development. To achieve functional integration, a multicellular system requires some internal differentiation, the basic requirement for division of labor. For example, internal differentiation depends on the presence of components that contribute in different ways to the realization of the system, such as cells and an extracellular matrix (ECM). Through functional differentiation multicellular systems become, in principle, capable of harboring components that have different functional roles. Hence, functional differentiation realizes division of labor under certain conditions. Integration between these different tasks is achieved when functions are coordinated at the system level such that the differentiated components actively contribute to the maintenance of the system while their activities are being

---

[5] Counting the variation in number of organizations might be problematic as well. For example, it does not respond to the question whether the first steps of cell division in the embryo count as reproduction or development and why.

activated, inhibited, or modulated at different moments in time depending on the state of the system. This is achieved by means of mechanisms of control and spatial organization acting at different ranges or time scales. Examples include cell-to-cell interactions, an ECM dynamically constraining groups of cells, biomechanical forces, and long-range control exerted by the vascular, nervous, or immune systems (Bich et al., 2019; see also Sonnenschein & Soto, 1999; Montevil & Soto, this volume). To undergo development, this organization should also exhibit regulatory capabilities, that is, be able to determine its own processes of change. This means that it should be able to modify itself and modulate its internal dynamics in response to variation in internal and external conditions, rather than only passively undergoing change driven by perturbation (Bich et al., 2016).

The second problem faced by the organizational approach is that not all regulated change is developmental (see Challenge 1 in Sect. 11.2). Otherwise, development would include all possible biological dynamics. Let us sketch an account of developmental change. Regulation operates by sensing the internal and external conditions of the system. Most regulatory processes act upon available mechanisms and the processes responsible for energy production, synthesis of parts, and the like. Regulation in these cases consists in bringing forth change in the basic dynamics of the living system by selecting between available mechanisms, via activation or inhibition, or by modulating mechanisms already in operation. An example is the activation and inhibition of genes responsible for the synthesis of enzymes specific to the presence of variable food sources, as in the case of the *lac operon* in bacteria.[6] In these cases, regulatory mechanisms select between available functions or functional regimes.

Development is qualitatively different from other regulatory processes because it does not operate only on available functions but also changes the set of functions available to the system. At each developmental step, some new functional traits are generated, such as in the appearance of new tissues, organs, or limbs. In unicellular systems development might include the production of new organelles or other functional supramolecular structures. Functional traits might also be shed. Think of the transition between tadpole and frog, with the appearance of legs and lungs and the disappearance of gills and tail. These changes are different from the activation or inhibition of mechanisms which are already present in the system, and they affect the way the multicellular organization maintains itself in its new regime of closure.[7]

---

[6] While the case of the lac operon is well-known and illustrative of regulatory mechanisms, self-maintenance and regulation do not only apply to metabolic processes. Many other types of processes that are not metabolic contribute to self-maintenance and are strictly regulated: for example, behavior, movement, perception, or the activity of an organ or an organelle.

[7] This notion of developmental change has interesting implications. At a first approximation, it does not seem to necessarily apply to growth, unless a change of size of the system, or of part of it, implies the realization of a new function instead of a change in the realization of function that is already available.

Developmental regulatory change on this picture is not necessarily irreversible.[8] Nor does it imply that development necessarily tends toward some future adult state. There are cases that contradict the idea of development as either an irreversible adult oriented process or the actualization of a potentiality (see Challenge 4 in Sect. 11.2). The "immortal jellyfish" *Turritopsis dohrnii* can return to the juvenile polyp stage from the adult medusa stage through the action of regulatory mechanisms that kick in as a response to adverse conditions or damage (Matsumoto et al., 2019). It does so while maintaining its multicellular organization, that is, without going through the whole life cycle and passing through a unicellular form. On our account of development, there is no conceptual problem in including cases in which changes are reversed, if these changes contribute to maintaining organizational closure. Therefore, an organizational approach to development can go beyond the biology of vertebrates and account for controversial cases such as rejuvenation or reverse development.

Adopting a regulatory framework to understand development also addresses issues such as the role of environmental factors. On some accounts, these factors are viewed as directly regulating development (Gilbert & Epel, 2015). On our view, what triggers developmental transitions is the sensing of environmental conditions by regulatory mechanisms and the consequent changes they trigger, not the direct action of the environment. We do not deny that environmental factors modify regulatory mechanisms. But even in these cases, environmental factors engage regulatory mechanisms which then bring forth developmental change.[9] Direct change from environmental causes, such as the loss of a limb to a predator or an accident, would not count as developmental.

Let us focus now on the third problem: identifying the boundaries of development. Regarding the starting point of development, some approaches, such as Minelli (2011), have questioned the idea that development begins with the egg (see Challenges 2 and 3 in Sect. 11.2). Reproduction can be also vegetative or occur via budding, that is, through a system that is already multicellular. However, Minelli's focus is centered on reproduction and so implicitly adopts the directional teleological framework underlying the adultocentrism that he criticizes. Accordingly, when he criticizes the identification of the adult as the end point of development, he does so on the grounds that some species of animals undergo reproduction before reaching the adult form (Minelli, 2011). This doesn't alleviate the problem. The adultocentrism is just a symptom; the real issue is the directional teleology that underlies it – reproduction as the goal of development.

---

[8] Many, if not most physiological and behavioral regulatory processes, are reversible, starting from the simple case of the *lac operon*. However, it is important not to confuse thermodynamic reversibility with regulatory reversibility (physiological, behavioral, developmental, etc.). In a nutshell, the reversibility of regulatory processes requires energy, so it is a thermodynamically irreversible process.

[9] The only partial exception would be molecular compounds, such as, for example, hormones, released by other organisms and capable to operate in the receiving organism as if they were some of its own regulatory mechanisms.

Our organizational account questions the idea that the egg is the starting point of development but does so on a different basis. The egg cell divides into several cells when it undergoes cleavage. But this is not the growth of a single developmental system because these cells do not realize an integrated multicellular organization but several distinct unicellular organizations. These cells do not communicate among themselves, do not realize division of labor, and therefore do not collectively realize closure. On our organizational view, development starts when cells come together to form one integrated multicellular system – one organism – that then undergoes changes regulated at the level of the whole. So one system, the egg cell, reproduces to form multiple connected unicellular systems that only later come together as a single system with organizational closure. To illustrate an alternative trajectory, sometimes the aggregate of individual cells splits into two independent developing systems, such as in the case of monozygotic (identical) twins. Development does not start with the egg but with integrated organization capable of regulation.

What are the requirements for realizing an integrated multicellular organization? One might say when cells start signaling to one another (Arnellos et al., 2014), but this is neither necessary nor sufficient. What is needed is functional and spatial differentiation and integration, that is, a differential contribution to the maintenance of the organization. For example, when the ECM (a noncellular constraint) is deposited, it contributes to the maintenance of the system by controlling cell differentiation and behavior, cell migration, and spatial differentiation and subsequently allows different *groups* of cells to emerge that perform different activities (see Bich et al., 2019). When these functional changes taking place within an integrated multicellular organization are directed by regulatory mechanisms, development starts. These organizations need not have parts that all have the same origin (see Challenge 7 in Sect. 11.2). In principle, they can be realized also by symbiotic associations such as lichens or biofilms insofar as they satisfy requirements for integration and regulation.

When does development end? Not necessarily with the achievement of the adult form (see Challenge 5 in Sect. 11.2). Adult organisms can still exhibit functional changes, and in some cases they can undergo phenomena such as rejuvenation (reverting to a previous stage) or regeneration (reestablishing a lost function). According to the organizational view that we propose in this paper, development ends when regulated changes to the number or types of functions no longer take place.[10]

In this respect, it is important to distinguish development from senescence. Senescence is a process of loss or decrease of functionality due to a change in the properties of components (such as the ECM; see Moreau et al., 2017) or of the overall organization of the system. While development might also produce a loss of functions, the difference is that in the case of senescence the loss of functions is not

---

[10] On this view, whether and when development actually ends is an empirical question and might have different answers for different species.

determined by the action of regulatory mechanisms. It is rather a question of whether and how functions are realized, and therefore it could be fruitfully addressed in relation to the organizational view of malfunctions (Saborido & Moreno, 2015).

11.5   Conclusions

In this chapter we showed that from an organizational perspective, development is a process of regulated change in number or types of functions of a regime of closure of constraints. It starts when a functionally integrated multicellular organization endowed with regulatory mechanisms is realized, and it ends when there are no further regulated changes in functions. It is a goal-oriented process, but a special one that is focused on the present, in which at each stage the goal is to maintain a viable organization of the system.[11] Development does not aim at a future goal state, and therefore there are no intermediate stages. Each stage of a life cycle is equally important from a point of view that is focused on the persistence of that life over the life cycle. The *telos* can be found in the actual developing system at any point during the entire process. This approach does not characterize living systems as the result of development, but the system undergoing development is already considered a self-maintaining organized biological system. With respect to environmental factors, the organizational approach can explain their role in relation to the internal logic of the system that undergoes a regulatory transition in the presence of these environmental conditions. There are still several challenges that an organizational approach needs to face. Some, not included among those discussed in Sect. 11.2, concern the range of developmental systems and phenomena. They include questions such as whether biofilms or ecological systems – to which some argue it may be possible to ascribe a closure of constraints (see Militello et al., 2021; Nunes-Neto et al., 2014, respectively) – can undergo development or not. A fundamental challenge among those mentioned in Sect. 11.2, which is still open, is related to the type of organisms taken into account to explain development. Most work has been focused on metazoa and specifically on vertebrates. Vertebrates along with many other animals have the advantage of being easily individuated, often exhibiting a straightforward developmental pathway. However, they constitute only a small portion of the multicellular systems which undergo development. Some steps have been taken in this direction in this chapter. However, a sufficiently general organizational account of development needs to be able to handle a wide range of multicellular systems and provide the tools needed to address problematic cases such as facultative multicellular systems with life cycles composed of alternating and

---

[11] Focusing on the present does not mean ignoring phenomena happening in the past. A system undergoing development is the result of a reproductive event and of a history of evolution. However, here we have been focusing on the developmental process itself. A direction for future work to expand this approach within the organizational framework is to integrate development with heredity (Mossio & Pontarotti, 2019).

distinct life stages (e.g., multicellular and unicellular; see Challenge 6 in Sect. 11.2). These more basic, yet no less complex, cases constitute an important sample of all multicellular systems and might play an important role for our understanding the origin of multicellularity and development. However, they have been primarily explored only from a historical and evolutionary perspective (van Gestel & Tarnita, 2017). We have addressed this challenge by expanding the range of examples analyzed to include some of these organisms, but there is much more to do. This is surely a necessary and interesting avenue to be pursued in future work on development.

Acknowledgments This study was funded by the Basque Government (Project: IT1228-19 for LB), Ministerio de Ciencia, Innovación, Spain (research project PID2019-104576GB-I00 for LB and DS and "Ramon y Cajal" Programme RYC-2016-19798 for LB), and the John Templeton Foundation (Project 62220 – Subproject "Directedness in the physiology, ecology, and evolution of holobiont systems" – for DS and LB). The revisions were done during LB's Visiting Fellowship at the Center for Philosophy of Science of the University of Pittsburgh (Spring Term 2022).

References

Arnellos, A., Moreno, A., & Ruiz-Mirazo, K. (2014). Organizational requirements for multicellular autonomy: Insights from a comparative case study. *Biology and Philosophy, 29*, 851–884.

Barinaga, M. (1994). Looking to development's future. *Science, 266*, 561–564.

Bich, L. (2018). Robustness and autonomy in biological systems: How regulatory mechanisms enable functional integration, complexity and minimal cognition through the action of second-order control constraints. In M. Bertolaso, S. Caianiello, & E. Serrelli (Eds.), *Biological robustness. Emerging perspectives from within the life sciences* (pp. 123–147). Springer.

Bich, L. (2019). The problem of functional boundaries in prebiotic and inter-biological systems. In G. Minati, E. Pessa, & M. Abram (Eds.), *Systemics of incompleteness and quasi-systems* (pp. 295–302). Springer.

Bich, L., & Bechtel, W. (2022a). Organization needs organization: Understanding integrated control in living organisms. *Studies in History and Philosophy of Science, 93*, 96–106.

Bich, L., & Bechtel, W. (2022b). Control mechanisms: Explaining the integration and versatility of biological organisms. *Adaptive Behavior.* https://doi.org/10.1177/10597123221074429

Bich, L., Mossio, M., Ruiz-Mirazo, K., & Moreno, A. (2016). Biological regulation: Controlling the system from within. *Biology and Philosophy, 31*, 237–265.

Bich, L., Pradeu, T., & Moreau, J.-F. (2019). Understanding multicellularity: The functional organization of the intercellular space. *Frontiers in Physiology, 10*, 1170.

Bich, L., Mossio, M., & Soto, A. (2020). Glycemia regulation: From feedback loops to organizational closure. *Frontiers in Physiology, 11*(69). https://doi.org/10.3389/fphys.2020.00069

Di Frisco, J., & Mossio, M. (2020). Diachronic identity in complex life cycles: An organisational perspective. In A. S. Meincke & J. Dupré (Eds.), *Biological identity: Perspectives from metaphysics and the philosophy of biology* (pp. 177–199). Routledge.

Enders, A. C. (2002). Implantation in the nine-banded Armadillo: How does a single Blastocyst form four embryos? *Placenta, 23*(1), 71–85.

Gilbert, S., & Barresi, M. (2018). *Developmental biology* (11th ed.). Sinauer Associates.

Gilbert, S., & Epel, P. (2015). *Ecological developmental biology.* Sinauer Associates.

Godfrey-Smith, P. (2016). Complex life cycles and the evolutionary process. *Philosophy of Science, 83*(5), 816–827.

Griesemer, J. (2016). Reproduction in complex life cycles: Toward a developmental reaction norms perspective. *Philosophy of Science, 83*, 803–815.

Grose, J. (2020). How many organisms during a pregnancy? *Philosophy of Science, 87*(5), 1049–1060.

Holland, L. Z. (2016). Tunicates. *Current Biology, 26*(4), R146–R152.

Hoøeg, J. T., Maruzzo, D., Okano, K., Glenner, H., & Chan, B. K. K. (2012). Metamorphosis in balanomorphan, pedunculated, and parasitic barnacles: A video-based analysis. *Integrative and Comparative Biology, 52*(3), 337–347.

Kauffman, S. A. (2000). *Investigations*. Oxford University Press.

Kingma, E. (2020). Biological individuality, pregnancy, and (Mammalian) reproduction. *Philosophy of Science, 87*(5), 1037–1048.

Martinez-Arias, A., & Stewart, A. (2002). *Molecular principles of animal development*. Oxford University Press.

Matsumoto, Y., Piraino, S., & Miglietta, M. P. (2019). Transcriptome characterization of reverse development in *Turritopsis dohrnii* (Hydrozoa, Cnidaria). *G3: Genes, Genomes, Genetics, 9*(12), 4127–4138.

Militello, G., Bich, L., & Moreno, A. (2021). Functional integration and individuality in prokaryotic collective organisations. *Acta Biotheoretica, 69*(3), 391–415.

Minelli, A. (2003). *The development of animal form*. Cambridge University Press.

Minelli, A. (2011). Animal development, an open-ended segment of life. *Biological Theory, 6*(1), 4–15.

Minelli, A. (2014). Developmental disparity. In A. Minelli & T. Pradeu (Eds.), *Towards a theory of development* (pp. 227–245). Oxford University Press.

Miroliubov, A., Borisenko, I., Nesterenko, M., et al. (2020). Specialized structures on the border between rhizocephalan parasites and their host's nervous system reveal potential sites for host-parasite interactions. *Scientific Reports, 10*, 1128.

Montévil, M., & Mossio, M. (2015). Biological organisation as closure of constraints. *Journal of Theoretical Biology, 372*, 179–191.

Montévil, M., & Soto, A. (this volume). Modeling organogenesis from biological first principles. In M. Mossio (Ed.), *Organization in biology*. Springer.

Moreau, J. F., Pradeu, T., Grignolio, A., Nardini, C., Castiglione, F., Tieri, P., et al. (2017). The emerging role of ECM crosslinking in T cell mobility as a hallmark of immunosenescence in humans. *Ageing Research Reviews, 35*, 322–335.

Moreno, A., & Mossio, M. (2015). *Biological autonomy: A philosophical and theoretical enquiry*. Springer Netherlands.

Mossio, M., & Bich, L. (2017). What makes biological organisation teleological? *Synthese, 194*, 1089–1114.

Mossio, M., & Pontarotti, G. (2019). Conserving functions across generations: Heredity in light of biological organization. *The British Journal for the Philosophy of Science*. https://doi.org/10.1093/bjps/axz031

Mossio, M., Saborido, C., & Moreno, A. (2009). An organizational account of biological functions. *The British Journal for the Philosophy of Science, 60*(4), 813–841.

Muller, G. B., & Newman, S. A. (2003). *Origination of organismal form. Beyond the gene in developmental and evolutionary biology*. MIT Press.

Nunes-Neto, N., Moreno, A., & El-Hani, C. N. (2014). Function in ecology: An organizational approach. *Biology and Philosophy, 29*(1), 123–141.

Nuño de la Rosa, L. (2010). Becoming organisms: The organisation of development and the development of organisation. *History and Philosophy of the Life Sciences, 32*, 289–315.

Nuño de la Rosa, L., Pavličev, M., & Etxeberria, A. (2021). Pregnant females as historical individuals: An insight from the philosophy of Evo-Devo. *Frontiers in Psychology, 11*, 572106.

Pattee, H. H. (1972). The nature of hierarchical controls in living matter. In R. Rosen (Ed.), *Foundations of mathematical biology volume I subcellular systems* (pp. 1–22). Academic Press.

Piaget, J. (1967). *Biologie et Connaissance*. Gallimard.

Rosen, R. (1972). Some relational cell models: The metabolism-repair systems. In R. Rosen (Ed.), *Foundations of mathematical biology. Volume II cellular systems* (pp. 217–253). Academic Press.

Saborido, C., & Moreno, A. (2015). Biological pathology from an organizational perspective. *Theoretical Medicine and Bioethics, 36*(1), 83–95.

Skillings, D. (2016). Holobionts and the Ecology of organisms – Multi-species communities or integrated individuals? *Biology and Philosophy, 31*(6), 875–892.

Skillings, D. (2019). Trojan horses and black queens: Causal core explanations in microbiome research. *Biology and Philosophy, 34*(6), 1–6.

Sonnenschein, C., & Soto, A. M. (1999). *The society of cells – Cancer and control of cell proliferation*. Bios Scientific.

van Gestel, J., & Tarnita, C. E. (2017). On the origin of biological construction, with a focus on multicellularity. *Proceedings of the National Academy of Sciences, 114*, 11018–11026.

Varela, F. G., Maturana, H. R., & Uribe, R. (1974). Autopoiesis: The organization of living systems, its characterization and a model. *Biosystems, 5*(4), 187–196.

Veloso, F. (2017). On the developmental self-regulatory dynamics and evolution of individuated multicellular organisms. *Journal of Theoretical Biology, 417*, 84–99.

Wolpert, L., & Tickle, C. (2011). *Principles of development* (4th ed.). Oxford University Press.

Chapter 12
Modeling Organogenesis from Biological First Principles

Maël Montévil and Ana M. Soto

Abstract Unlike inert objects, organisms and their cells have the ability to initiate activity by themselves and thus change their properties or states even in the absence of an external cause. This crucial difference led us to search for principles suitable for the study organisms. We propose that cells follow the default state of proliferation with variation and motility, a principle of biological inertia. This means that in the presence of sufficient nutrients, cells will express their default state. We also propose a principle of variation that addresses two central features of organisms, variation and historicity. To address interdependence between parts, we use a third principle, the principle of organization, more specifically, the notion of the closure of constraints. Within this theoretical framework, constraints are specific theoretical entities defined by their relative stability with respect to the processes they constrain. Constraints are mutually dependent in an organized system and act on the default state.

Here we discuss the application and articulation of these principles for mathematical modeling of morphogenesis in a specific case, that of mammary ductal morphogenesis, with an emphasis on the default state. Our model has both a biological component, the cells, and a physical component, the matrix that contains collagen fibers. Cells are agents that move and proliferate unless constrained; they exert mechanical forces that act (i) on collagen fibers and (ii) on other cells. As fibers are organized, they constrain the cells' ability to move and to proliferate. This model exhibits a circularity that can be interpreted in terms of the closure of constraints. Implementing our mathematical model shows that constraints to the default state are sufficient to explain the formation of mammary epithelial structures. Finally, the success of this modeling effort suggests a stepwise approach whereby additional

M. Montévil (✉)
Centre Cavaillès, République des Savoirs UAR3608, CNRS, Collège de France et École Normale Supérieure, Paris, France

A. M. Soto (✉)
Centre Cavaillès, République des Savoirs UAR3608, CNRS, Collège de France et École Normale Supérieure, Paris, France

Tufts University School of Medicine, Boston, MA, USA
e-mail: ana.soto@tufts.edu

© The Author(s) 2024
M. Mossio (ed.), *Organization in Biology*, History, Philosophy and Theory of the Life Sciences 33, https://doi.org/10.1007/978-3-031-38968-9_12

constraints imposed by the tissue and the organism can be examined in silico and rigorously tested by *in vitro* and *in vivo* experiments, in accordance with the organicist perspective we embrace.

12.1 Introduction

Throughout the twentieth century, biology underwent changes that little by little removed concepts which up until that time were considered to be the main characteristics of organisms, such as agency, normativity, and goal-directedness. Later on, even the concept "organism" was deemed superfluous and almost disappeared from biological theory as the idea of a genetic program gained acceptance (Nicholson, 2014). At the turn of the new millennium, critical appraisals of the reductionist stance of the molecular biology revolution became more numerous, both regarding the espousing of nineteenth-century physicalism and the questionable adoption of mathematical theories of information and the notions of program and signal (Longo et al., 2012). In addition to their critical analysis of the status quo, some biologists proposed alternative stances regarding organismal and evolutionary biology (Sonnenschein & Soto, 1999; Oyama, 2000; Kupiec & Sonigo, 2003; Moss, 2003; Jablonka & Lamb, 2005; Noble, 2006). It was clear to many that the promised reduction of biology to chemistry and physics was just a misplaced aspiration that did not translate into advances in experimental biology; various authors suggested alternatives. An alternative, both philosophical and theoretical, was to abandon reductionism by returning to organicism (Gilbert & Sarkar, 2000; Greenspan, 2001; Soto & Sonnenschein, 2005). Theoretical biologists inspired by an organicist stance started to reintroduce the very notions into biology that distinguished living matter from the inert, namely, agency (Kauffman, 2001). Another proposed alternative was technological, namely, the collection of data but at a larger scale (-omics). The idea was to transfer the task of making sense of phenomena to computers and data scientists by generating hypotheses from the data patterns revealed by such analysis (Bassett et al., 1999; Brown & Botstein, 1999). Another approach used the application of mathematical modeling, particularly various forms of "pragmatic systems biology" to search for molecular interactions (O'Malley & Dupre, 2005). Neither one of these technological fixes produced the expected advances in experimental biology; the theoretical work of the organicists, instead, has started to impact experimental work via mathematical modeling based on biological principles (Montévil et al., 2016b) and conceptual analysis (Bich et al., 2020).

In spite of these critical criticisms, the current practice of developmental biology is still guided by the metaphoric use of the mathematical concepts of information, program, and signal, particularly the idea of a teleonomic genetic program, shaped by natural selection. Determination of the organism follows from this program and thus is extrinsic to the developing organism as such. The developmental program is supposed to drive the developing organism toward a final state, thus defining development as an apparently goal-oriented process. This genocentric view, which endows genes with a privileged causal role, suffers from many weaknesses (Longo et al., 2012; Longo & Mossio, 2020; Soto & Sonnenschein, 2020). It falls short of providing an understanding of how a complex, fully organized biological entity will systematically be formed from this putative "program," where such a program is located, and how it is executed. One main reason behind these shortcomings is that while there is a close relationship between a DNA sequence and the corresponding protein, there is no such correspondence between genes and phenotypes because the possible properties of phenotypes are not prestatable (Moss, 2008). Consequently, the relationship between genes and forms is not straightforward (Soto & Sonnenschein, 2005). Moreover, the genetic program fails to account for the variability observed throughout embryogenesis and morphogenesis, which contradicts the invariance expected from a "program," as exemplified by developmental plasticity (West-Eberhard, 2003). Additionally, because of this reliance on the genetic program, contemporary developmental biology tends to address causality in mechanistic terms, which conflicts with the interdependence between the whole, namely, the developing organism, and its parts (Soto & Sonnenschein, 2020). All these difficulties call for a reappraisal of the philosophical and theoretical frames that guide contemporary research in development in general and morphogenesis in particular. This essay will briefly discuss the concepts and theoretical frames that we use to construct a principle-based modeling of developmental and physiological processes. This will be illustrated by recent work on mathematical modeling of mammary gland morphogenesis.

12.2 Background Concepts

While reductionism became the dominant philosophical stance in twentieth-century biology, a movement named "Organicism" developed during the period between the two world wars. Organicism is a philosophical stance committed to the following general ideas: (1) the centrality of the organism concept in biological explanation, (2) the importance of organization as a theoretical principle, and (3) the vindication of the autonomy of biology as a science (Nicholson & Gawne, 2015).

Organicism is a materialistic philosophical stance whereby new properties that could not have been predicted from the analysis of the lower levels appear at each level of biological organization. Also, implicit in this view is the idea that organisms are not just "things" but objects under relentless change. While reductionist stances are usually derived from an ontology of unchanging substances, i.e., "being," organicist stances are usually focused on an ontology of "becoming" (Dupré & Nicholson, 2018).

In the 1970s while molecular biologists aspired to reduce biology to chemistry, advances in the understanding of dissipative non-equilibrium physical systems that self-organize influenced theoretical biologists interested in biological organization. Many of these thinkers, such as S. Kauffman, H. Maturana, and F. Varela, went beyond the notion of far from equilibrium systems and were inspired by the Kantian concept of biological organization that stressed the interrelatedness of the organism and its parts and the circular causality implied by this relationship (an organism is the cause and effect of itself). Recognizing that Kantian organization does not correspond to the spontaneous self-organization of physical systems, they worked out a new regime of circular causation. In this circular organizational regime, the parts depend on the whole and vice versa; this regime not only produces and maintains the parts that contribute to the functioning of the whole integrated system, but the integrated system also interacts with its environment to promote the conditions of its own existence. This view of organization neatly leads to conceiving intrinsic teleology as a concept compatible with scientific causality (Mossio & Bich, 2017). We can understand organisms as normative agents with the main aim of keeping themselves alive; their proper understanding requiring teleological principles of explanation. In the remainder of this section, we briefly delineate the main concepts in addition to organization and teleology that guide our efforts.

Historicity While physical self-organizing systems like flames and micelles appear spontaneously, organisms are generated by the reproduction of a preexisting organism. Historicity is fundamental to phylogenesis and ontogenesis. Historicity particularity establishes a difference from the theoretical frameworks of physics and creates methodological and theoretical challenges for mathematization in biology. Moreover, the historicity of organisms encompasses two time scales, the long scale of phylogeny and the short of ontogeny. Consequently, historical analysis is central to the understanding of biological organization (Longo & Soto, 2016; Montévil, 2020).

Distinctive Materiality Organisms are made up of chemicals such as DNAs, RNAs, proteins, and membranes. Unlike computer programs (software) that are independent of the materials of the "hardware," the functions an organism accomplishes cannot be dissociated from the particular materials the organism is comprised of (Longo & Soto, 2016). This view precludes the software-hardware dualism from biological entities. The materiality of biological objects also has an epistemological dimension. This is evidenced by comparing physical objects with biological ones. In physics, objects are primarily defined by abstract mathematical constructs, as illustrated by the definition of the speed of light in a vacuum being the speed of any light ray. In contrast, biological objects are defined by referencing a particular specimen of an organism, the type, to which the scientific name of a species is formally attached. This specific materiality trickles down to all biological practices, so that biological objects are always defined in reference to concrete objects rather than to theoretical abstractions (Montévil, 2019).

Agency and Normativity Teleology is associated with the notions of autonomy and normative agency. The purposiveness of living entities is considered a consequence of the architecture of adaptive systems (Walsh, 2015). Organisms are normative agents, namely, they have the capacity to generate actions and their own rules. As extensively discussed by G. Canguilhem, normative agency is a major characteristic that differentiates living from inert objects. Organisms undergo individuation which is manifested in their ability to change their own organization, that is, change their own rules. Another remarkable characteristic of organisms is their propensity to become sick and to overcome disease; pathology is an exclusively biological discipline (Canguilhem, 1991).

Specificity Physical theories describe generic objects fitting a mathematical construct; for example, as mentioned above, when one refers to the speed of light in a vacuum, there is no need to refer to a specific ray of light, as all travel at the same speed – an invariant of Einstein's relativities. Of course, the methodological approach of physics can accommodate a variety of situations, like phase transition and crystallization, however, always under the umbrella of a generic description that goes with mathematization. In contrast, biological objects are specific, for example, organisms are individuals in the process of undergoing further individuation. In other words, they are the result of history and continue to generate historical novelties. While variation in physical objects is merely a result of quantitative changes, in biology, in addition to the latter, variation is an intrinsic characteristic of organisms which plays a major role in evolutionary biology as the substrate of natural selection and in ontogenesis as the source of functional novelty (Longo & Montévil, 2011; Longo & Soto, 2016, Montévil et al., 2016a). Reductionist attribute a form of specificity to molecules (which are assumed to be defined by their structure; thus, they are ultimately generic), consequently eluding the epistemological challenge of working with specific objects. In contrast, the organicist perspective locates specificity in biological objects endowed with autonomy, that is, organisms and their cells. Cellular specificity is the result of the particular trajectory of each cell during embryogenesis, namely, its interactions with other cells as it proliferates and migrates during histogenesis and organogenesis.

Constraints Biological specificity does not negate the idea that aspects and parts of organisms are endowed with a kind of restricted genericity, namely, limited invariance. We call these elements constraints. An example of a constraint is the structure of articulations between bones which preclude certain movements and allow others. Typically, constraints may change over a longer time scale than the process they constrain. For example, the concentration of an enzyme does not change during the time it takes to catalyze the conversion of a substrate into products. Unlike physical invariants that are postulated and stem from fundamental principles, the existence of biological constraints requires explanations (by evolution and organization).

12.3   From Organicist Ideas to Principles for a Theory of Organisms

Scientific theories provide organizing principles and construct objectivity by framing observations and experiments (Longo & Soto, 2016). Theories construct the proper observables and provide the framework for studying them. The usefulness of theories is not determined by their being "right." Even a "wrong" theory can be useful if, when proven incorrect it is modified or dismissed. The limiting factor for being useful is that a theory should not be vague, as vague theories cannot be proven to be incorrect (Feynman, 2017).

A theoretical principle of biological "inertia," the default state of cells. A method used to develop a theoretical framework consists of positing what takes place when nothing is done to a system, that is, when discussing default states. For example, the inertial state of classical mechanics corresponds to the trajectory of an isolated object. In biology, we posit that the *default state* of cells is proliferation with variation and motility. It is based on the cell theory, and it relates to the specific materiality of the alive. The *default state* is a manifestation of the agency of living objects and, thus, a cause (Longo et al., 2015). In contrast to physical objects, the presence of sufficient nutrients is required to maintain the metabolic needs, keeping the biological object alive. In these inertial conditions, cells move and proliferate generating variation (Soto et al., 2016a, b; Sonnenschein & Soto, 2021). Moreover, in the same way that the departure of inertia enables physicists to define classical forces as cause, the departure from the default state defines what causes are. It follows that there are two causal levels in the default state: the level of proliferation and motility that comes from objects understood as specific objects (i.e., causality at the level of cells as such), and the level of constraints acting on the default state (i.e., constraints acting on cells).

The principle of organization by closure of constraints. In an organism, constraints depend collectively on each other thus generating a circle of dependencies called closure (Montévil & Mossio, 2015; Mossio et al., 2016). In turn, closure provides an understanding of the relative stability of constraints and more generally of biological organizations. Moreover, the principle of organization leads to the identification of specific constraints in an organism and to assess whether a given constraint is functional, that is, it participates in closure.

The principle of variation. An implicit but overarching principle in physics is that we can understand the changes of an object by means of invariants and invariant preserving transformations (symmetries). For example, an inertial trajectory preserves momentum, energy, etc. This perspective is the basis for understanding physical objects as generic objects. By contrast, the principle of variation posits that biological objects are specific, and therefore relevant invariants and symmetries typically change over time. Modelers sometimes propose to accommodate biological objects with mathematical constructs that would change over time; these changes are somewhat similar to the phase transitions of physics. However, such a construct would again define a generic object, and assume that we can prestate the possible changes taking place. Instead, it is not possible to identify the objects of an experiment, let's say a group of mice, with a mathematical construct that would accommodate the way they are organized on theoretical grounds. In other words, alternatives are always possible. As a result, biology must reason with a different kind of object when compared to physics, namely, specific objects.

Variation relates to the historicity of biological objects and their contextuality. Historicity stems from the historical accumulation of variations that, by creating novelty, co-define present biological organization. Contextuality is related to historicity because understanding the historical changes that formed current organisms requires knowledge of the context that facilitated these changes. Contextuality is obviously also relevant at the time of observation because the definition of experimental objects depends on the context in which they are found. Different contexts may entail different organizations. For example, during embryogenesis the relationship of a cell with its environment, namely, the surrounding extracellular matrix and the neighboring cells, is a major determinant of the morphology and function of this cell within the organ in which it resides. Indeed, understanding a biological organization requires taking into account its interaction with the surrounding environment, both at a given time-point and through the successive environments that the biological object traverses (Soto & Sonnenschein, 2005; Miquel & Hwang, 2016; Montévil et al., 2016a; Sonnenschein & Soto, 2016; Montévil, 2019).

Overall, these three principles provide a framework for understanding both general aspects of biology and particular biological situations. Building on the organicist and evolutionist traditions, they represent the beginning of novel thinking about principles and their applications (Soto et al., 2016a, b).

A recent addition to this theory-building process is a symbol, χ, to accommodate specific objects as such. The crucial point is that this symbol does not play the same role as the variable of mathematics; instead it refers to a material object and the objects that are related to it, in a manner that is compatible with the phylogenetic method of classifying living beings (Montévil & Mossio, 2020). It follows that this symbol is also a way of writing about specific objects such as cells on which constraints may act. Additionally, χ is a point of entry for modifications of an organization. As such, it represents the entry of diachronicity into the synchronic closure of constraints.

12.4 The Mammary Gland as an Organ Model for the Study of Morphogenesis

Let us now show how the theoretical framework summarized above can be applied to the study of morphogenesis in general, as well as that of different organs, for example, the mammary gland. Mammary glands are an evolutionary novelty of such importance that they define the class Mammalia. The gland is made up of two main components, namely, (1) the epithelial parenchyma, represented by the epithelial cells, whose function it is to produce and secrete milk to nourish the growing newborn, and (2) the stroma which surrounds the epithelium. The epithelium is composed of two layers of cells: a continuous luminal cuboidal cell layer and a basally located discontinuous myoepithelial cell layer. The stroma surrounding the epithelium is composed of various cell types (fibroblasts, adipocytes, and immune cells), blood vessels, nerves, lymph vessels, and an extracellular fibrous matrix of which the main component is collagen (Howard & Gusterson, 2000; Masso-Welch et al.,

2000; Richert et al., 2000) (Fig. 12.1). In the resting gland, the epithelial compartment consists of a ductal system. During pregnancy alveoli grow from the ducts and these structures produce and secrete milk. Reciprocal interactions between the epithelium and the stroma mediate the development, function, and remodeling of the mammary glands. The development of the organ can be divided into the following stages: fetal, pre-pubertal, pubertal, pregnancy, lactation, and involution. Ovarian and pituitary hormones regulate the morphology and function of the gland during puberty and adult life, but the fetal and prepubertal isometric development is not hormone-dependent (Soto et al., 2013). Disruption of epithelial-stromal interactions results in various pathologies including neoplasms (Soto & Sonnenschein, 2011; Sonnenschein & Soto, 2020).

12.4.1   A 3D Culture Model for the Study of Mammary Gland Morphogenesis

3D models aim to mimic *in vivo* conditions while reducing the number of organismal constraints to those which are hypothesized to be the most relevant ones for the purpose of the study. This approach allows the researcher to obtain results from which to estimate the contribution of these components to morphogenesis and/or physiology of the gland inside the organism. Simpler models may then be compared to more complex ones by adding other components. Ultimately, these models must be compared to the behavior of the gland in situ.

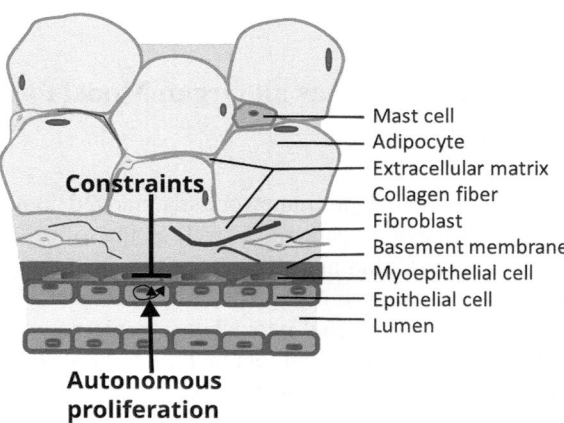

Fig. 12.1 Schematic representation of a mammary gland. In the resting mammary gland of adult females, the epithelium is organized into a branching ductal system. Epithelial cells proliferate spontaneously unless constrained; here they are constrained by the stroma containing extracellular matrix and connective tissue cells

Let's now discuss how our theoretical frame guides our strategy. Our theoretical proposition profoundly modifies both modeling and experimental practices. A main objective of this section is to discuss the theoretical determination of the object of study. It requires locating the part (i.e., the mammary gland) into a model of the whole (i.e., the organism). *Prior* to working on the isolated part (*in vitro* or *in silico*), choices must be made regarding what to extract from the whole (Bich et al., 2020). Then, we identify the process that we aim to elucidate: in this case, ductal morphogenesis, where given classes of constraints emerge, such as epithelial structures similar to ducts (which have a geometric feature and undergo cell polarization while developing a lumen). We next hypothesize that some elements are critical, and to an extent, sufficient for this process: some constraints, such as collagen type I fibers, and some specific objects, here epithelial cells from suitable cell lines. This simplification is only possible in a given context that roughly mimics the outcome of critical physiological processes: an incubator for temperature, CO2, sterility, and humidity; media for the chemical milieu, including nutrients; and an extracellular matrix that allows the growth in 3D of the cells into structures. Now, even if such conditions are sufficient for the intended constraints to emerge *in vitro*, it does not follow that these elements provide a full understanding of the actual phenomenon, and the integration in the organism (with more complex *in vitro* experiments) is critical to genuinely understand it.

Herein we use a human breast epithelial cell line, MCF10 cells embedded in 3D matrices containing only collagen I or constant concentrations of collagen I and variable concentrations of a mixture of basement membrane proteins (Matrigel); these components of the mammary stroma allow for breast epithelial cells to organize into structures that closely resemble those observed *in vivo* (Fig. 12.2) (Krause et al., 2008; Dhimolea et al., 2010; Krause et al., 2012; Barnes et al., 2014; Speroni et al., 2014).

12.5   From the 3D Culture Model to a Mathematical Model

To understand the morphogenesis taking place in 3D culture, we methodically used the principle of the default state to build a first mathematical model and then a computational one (Montévil et al., 2016b).

12.5.1   Proliferation

Breast estrogen-target epithelial cells express their default state proliferating maximally in serumless medium. Addition of hormone-free serum (or serum albumin, the inhibitor of cell proliferation present in serum) to the culture medium results in a dose-dependent inhibition of cell proliferation. This inhibitory constraint could be removed by lowering the albumin concentration or by adding estrogens

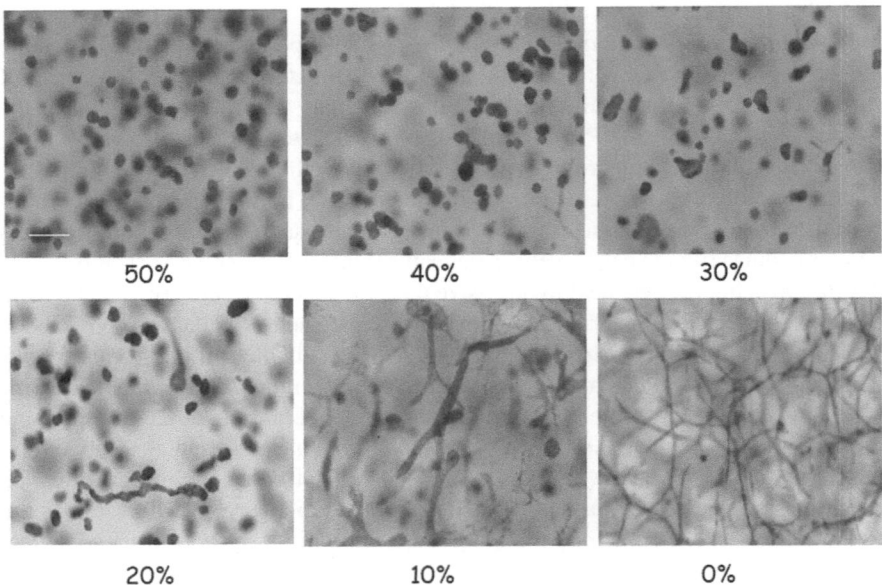

Fig. 12.2 Mammary epithelial morphogenesis in 3D culture. Mammary epithelial MCF10 cells were seeded in matrices containing a constant concentration of collagen type I (1 mg/ml) and varying concentrations (0–50%) of a basement membrane preparation (Matrigel™). High concentrations of Matrigel resulted in the formation of acini (spherical structures), while ductal elongated branching structures became increasingly prevalent as the Matrigel concentration decreased. Scale bar: 200 μm

(Sonnenschein et al., 1996). Additional constraints are those imposed by cell-cell contact and more generally the mechanical properties of the cells and the matrix in which they are embedded (Barnes et al. 2014).

12.5.2   Motility and Constraints to Motility

In biology, cells are agents, they generate forces and initiate motion. They proliferate and move unless there are constraints which prevent them from doing so. In general, classical mechanics imposes that cells exert forces on something to move, and the way they can exert forces depends on their history, both history at the evolutionary level and the history of their lineage inside the organism (and in laboratories in the case of established cell lines). Specifically, breast epithelial cells need a support to crawl on since they do not have a flagellum or a functionally analogous set of constraints. Notably, they use fibers to which they can attach and that they can pull in order to move. Moreover, cells are not simple mechanical structures that remain invariant over time; they react in a diverse manner to a mechanical force,

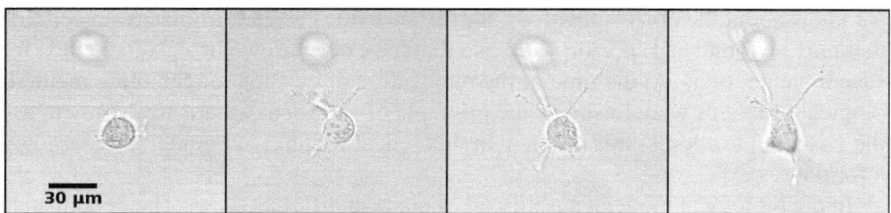

Fig. 12.3 A cell emits projections, here in a fibrillar matrix of collagen type I. [Reprinted with permission from Elsevier (Montévil et al., 2016b)]

depending on their history and normativity. For example, mechanical compression induces the expression of a set of genes (Soto et al., 2008; Longo & Montévil, 2014).

The constraints to motility that cells experience in situ can be modeled in a 3D culture system. The matrix in which the cells are seeded mimics the tissue environment. Once embedded in a matrix, breast epithelial cells emit projections, like filopodia and pseudopodia, which are used for motility; matrix composition may facilitate or hinder the ability of these projections to generate locomotion (Fig. 12.3).

In a fibrillar matrix, these projections can attach to fibers and exert forces on them. This activity leads to cell elongation and later to the appearance of structures geometrically akin to ducts (Barnes et al., 2014). Similarly, cells use these projections for locomotion. The latter is constrained notably by adhesion to other cells but also by the space occupied by the matrix. Specifically, pore size and matrix rigidity are constraints on cell migration. Pores are larger in the fibrillar matrix than in the globular matrix, while the latter is stiffer than the fibrillar matrix (Barnes et al., 2014). It follows that these properties contribute to morphological differences among epithelial structures.

Breast epithelial cells growing in a globular matrix emit short projections into the matrix that retract soon afterward and display limited motility (Montévil et al., 2016b). Cells rotate and divide resulting in the formation of an acinus, a sphere with a central lumen (Tanner et al., 2012).

Cells that touch each other, whether as a result of migration or after cell division, can attach to each other. Adhesion, and more specifically the physicochemical structures involved, constrain cell movements. Moreover, during morphogenesis, cells may detach from a structure and later reintegrate with it (Barnes et al., 2014).

12.5.3   Determination of the System

Cells are specific objects and should therefore be modeled by including the χ symbol (Montévil & Mossio, 2020). Unlike properties in physics, which are described by their causal relations and their underlying invariants, χ is defined by its past, including past contexts, for example, the common ancestor of a population of laboratory animals. This symbol enables us to transcribe with theoretical accuracy what we know about the objects involved, for instance, the cells are from a given cell line that may be found at a specific place and that have been grown in a given context for several generations. At the time of the publication of our first model, these methodological problems were raised by the principle of variation; we are now ready to use the χ symbol to address this problem in theoretical writing; our model is undergoing a formal rework.

In the biological model, causality takes place in different ways. The default state of cells frames how objects designated by χ proliferate. The departure from the default state describes how constraints act on cells, that is, objects designated by χ. Finally, constraints acting together, here mainly in the matrix, are analyzed in a more standard biophysical manner – except that they are in relation to cells. An example of such a constraint is collagen orientation with respect to force transmission.

Specifically, following the default state, cells proliferate, leading to an increase in cell number. Cell accumulation has several consequences: the redistribution of fluids, compression of matrix, and/or matrix degradation. Cells exert the other component of the default state, motility, by exerting forces on the matrix if they can do so. In Matrigel rich matrices, cells cannot attach to the matrix, and this component of the default state is constrained. That is, cells emit filopodia and exert their motility but cannot migrate. By contrast, in collagen matrices, cells grab fibers and exert forces on them, leading to changes in fiber organization [orientation notably, but also density (Dhimolea et al., 2010)]. The forces propagate in the matrix depending on its specific state (i.e., fiber orientations) and can reach over long ranges (Guo et al., 2012). As fiber organizations change, so do the constraints that they exert on cells. At the beginning of the formation of a structure, there is a symmetry breaking that leads to the emergence of a main direction in which forces are exerted (the direction of the elongated structure). In particular, forces exerted by cells on each other and on the structure's tips also constrain the default state due to the strain that follows from this force (Fig. 12.4). Collagen bundles facilitate the merging of epithelial structures initially positioned at a long distance range (Guo et al., 2012).

12.6   Mathematical Model

Mathematical modeling of biological phenomena is usually practiced using principles from one discipline (i.e., physics) and applying them to biology without evaluating the theoretical meaning these principles have when transported into the theoretical context of biology. It follows that, when models include cells as elementary components, the latter are described by ad hoc hypotheses that we reviewed elsewhere (Montévil et al., 2016b). This modus operandi is properly interpreted as imitation (Turing, 1950); stricto sensu mathematical modeling must be based on the theoretical principles of the discipline being studied. Below we describe the mathematical model both from the theoretical framework provided by the principles and the analysis briefly described above.

BIOLOGICAL COMPONENT   **PHYSICAL COMPONENT**

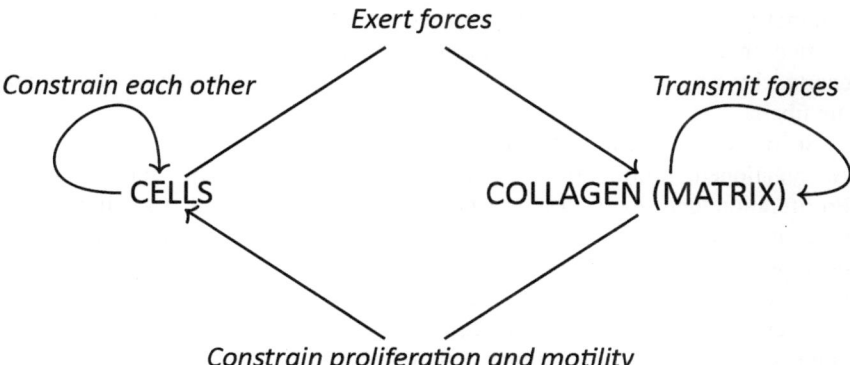

Fig. 12.4 Schema of the determination of the system. The biological component is determined by the default state, while the physics component is determined by the physics of material. The two are related since the matrix constraints the default state, and cellular activity, notably motility, affects the fibers. [Reprinted with permission from Elsevier (Montévil et al., 2016b)]

The theoretical framework restricts what is acceptable in order to model cellular behaviors. For example, the absence of proliferation requires constraints and quiescence cannot follow from ad hoc rules describing cells in agent-based modeling. More generally, it means that mathematical modeling, in this iteration, is about the interplay between the *default state* and the constraints acting on it (principle of organization); thus, it is not admissible for models of cells to follow arbitrary computational rules.

12.6.1   Description of the Model

In this initial model, we opted for a macroscopic and mesoscopic description of the 3D cultures, meaning that we described cells as elementary units and the fibers by their local orientation in a small spatial volume. We used agent-based modeling for cells and lattice modeling for fibers (limited to fiber orientation), mechanical forces, and a hypothetical chemical inhibitor of cell proliferation. The later seemed to be required to understand some aspect of the biological model, and this fact is also an illustration that theoretical principles constrain mathematical modeling and lead to the formulation of hypotheses.

The core and the originality of the model reside in our the method of understanding cell behavior. First comes the modeling of the default state, a modeling that evolves and expands in future works with the introduction of χ. Cells proliferate after a fixed time, unless constrained. One of the two cells produced by cell division occupies a random adjacent position to the mother cell, while the other occupies the position of the mother cell. Motility, instead, is more complex to model. Cells move unless constrained, according to the default state. When the cell environment is symmetric, this motion is random. Moreover, motility also encompasses the forces exerted on adjacent cells and extracellular matrix. The latter depends on the force exerted by cells, the orientation of the cytoskeleton, and that of the neighboring fibers.

Second comes the modeling of the constraints on cell proliferation and motility. As mentioned, proliferation requires that space is available for the new cell. Proliferation tends to occur along the direction of forces, so that a cell under a significant mechanical strain may not be able to proliferate even when an adjacent free position exists. Third comes the modeling of the hypothetical chemical inhibitor which slows down proliferation and lessens movements.

Overall, even in this simple iteration, the default state leads to a practice of modeling where spontaneous cellular activity, endowed with randomness, is central. Constraints limit this randomness and orient cellular behavior toward structures that are functional in the organism's life cycle. Moreover, the relationship between the default state and constraints is not just a molding of cell behavior by constraints because the constraints are transformed by cells exerting their default state in a manner that depends on their historical path (both evolutionarily and inside the organism) – the outcome of this historical path is made explicit to an extent by intracellular constraints such as the cytoskeleton.

12.6.2   Outcomes of the Mathematical Model

Here, we are discussing the outcome of the initial model as described in Montévil et al. where the details of the model and the analysis can be found (Montévil et al., 2016b).

12.6.2.1   In a Globular Matrix

In globular matrix, cells cannot attach to the matrix, and, therefore, cannot use it to move nor rearrange it. It follows that cells only exert forces on each other and crawl on each other when not attached. As a result, cells proliferate and remain tightly together, leading to a spherical structure (Fig. 12.5). Proliferation takes place at the periphery of the structure because cells inside stop proliferating due to the lack of available space. The structure stops growing after some time (due to the chemical inhibitor).

12.6.2.2   In a Fibrillar Matrix

In fibrillar matrices, things are a bit more complex because cells interact actively with the matrix and the latter constrains them. In the beginning, a single cell is surrounded by collagen, and it starts to pull on fibers, possibly moving, and the collagen tends to align with the direction of the force exerted. The structure gains additional cells by cell proliferation, and the new cells tend to remain together by cell adhesion (though some may escape the structure). By pulling on each other and on fibers, a dominant direction emerges. This direction is both influenced by the direction in which the first cells pull but also by the random initial orientation of every part of the collagen. Mathematically, it comes from an instability leading to a symmetry breaking, so that any small asymmetry in the initial condition is amplified leading to a large system-wide dominant direction (Longo & Montévil, 2018). Motility and proliferation are mostly constrained in this direction (due to the mechanical constraint imposed by this force). It follows that the structure becomes elongated. The chemical inhibitor, in combination with the mechanical forces, leads

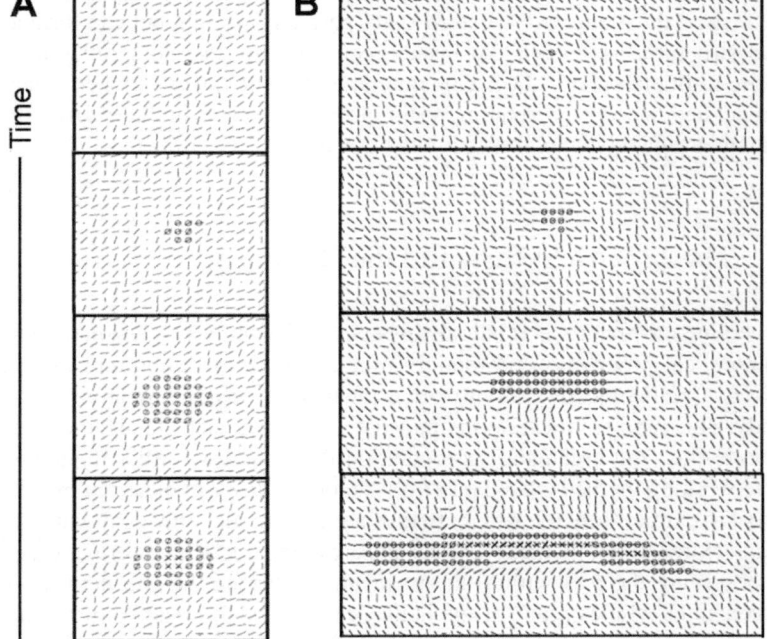

Fig. 12.5 Epithelial cells and collagen orientation in a plane of the simulation. (**a**) A case of a globular matrix: the cells cannot attach to the matrix nor reorganize it, leading to a spherical structure, an acinus. (**b**) A case of a fibrillar matrix: the cells reorganize collagen along a dominant direction, leading to the progressive formation of a duct. [Reprinted with permission from Elsevier (Montévil et al., 2016b)]

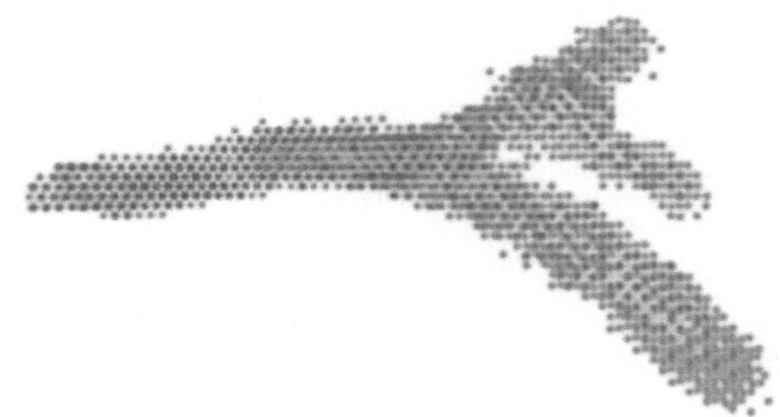

Fig. 12.6 Example of a branching duct resulting from a simulation run

to a stop of the proliferation in the middle of the structure, while the tips can continue to expand (Fig. 12.5).

Due to the randomness used to model cellular behavior under constraints and the initial matrix, the elongated structure is not perfectly straight but can form a curve-shaped structure. Moreover, the instability at the tip also sometimes allows the structure to branch (Fig. 12.6). This outcome was not expected when establishing the model and is a very interesting result of the method, as in the *in vivo* condition, the mammary gland ductal tree exhibits branching.

12.7   The *In Vitro* System and the Organism

By accepting the reciprocal relationship between the whole (organism) and its parts, our theoretical proposition profoundly modifies both modeling and experimental practices. A main objective of our work is the theoretical determination of the object of study. This requires locating the part (i.e., the mammary gland) into a model of the whole (i.e., the organism), an operation that requires further modeling work. Prior to working on the isolated part (*in vitro* or *in silico*), choices are made regarding what to extract from the whole. In this case, our model only dealt with epithelial cells and extracellular matrix. Next, results are compared with information gathered from observing the part within the organism. To bridge the gap between what is observed in the whole organism and in the *in vitro* model, we add other components of the mammary gland stepwise such as relevant cell types (i.e., mammary gland stromal fibroblasts). To grasp the organismal constraints that affect mammary gland development and function, we add hormones to the model consisting of epithelial cells, fibroblasts, and different matrices. We aim to identify primary constraints (Bich et al., 2016) which in our model are the matrix with or without stromal fibroblasts and regulatory constraints, which in our model are the mammotropic hormones (estradiol, progesterone, prolactin) (Bich et al., 2020).

Regarding the role of mammotropic hormones, at the onset of puberty, estrogen influences the formation of terminal end buds, the structure at the end of the ducts that invade the stroma and guide ductal growth. Progesterone promotes side-branching, and prolactin facilitates alveolar development in preparation for lactation. The dominant reductionist approach focuses on the hormone-receptor interactions and consequent induction of gene expression inside the cell rather than searching to explain the shape changes of the epithelial structures resulting from these hormonal influences in the epithelial cells. Instead, by applying an organicist perspective using a hormone responsive cell line, we found that exposure to hormones leads cells to modify the collagen fiber organization of the matrix in which they are embedded. This, in turn enables the cells to generate the distinct epithelial organization patterns observed in situ, namely, estrogen-mediated ductal elongation, progesterone-mediated lateral branching, and prolactin-mediated budding (Speroni et al., 2014). *In vitro* 3D models can also be used to manipulate constraints beyond the range operating *in vivo*. For example, to learn how rigidity affects shape beyond the limits imposed by the organism, Paszek et al. showed that by increasing the rigidity of the mammary gland model to mimic that of bone, lumen formation was inhibited and epithelial structures disorganized in a way reminiscent of neoplasms (Paszek et al., 2005).

12.8 Conclusions

Experimental research guided by our global theoretical approach addresses different questions from those guided by the metaphors of information, signal, and program borrowed from mathematical information theories (Longo & Montévil, 2011). The use of information metaphors drives experimenters to search for causality in discrete structures such as molecules. Additionally, ignoring the circular interdependency of the organism and its parts while embracing the idea that explanations need to uncover "molecular" mechanisms precludes the identification of physical "constraints" which causally contribute to the generation and maintenance of the organism.

Some of these shortcomings have been addressed by a view that, to account for the acquisition of form, combines the genetic program with physical determinants. This view facilitates the introduction of mathematical modeling of morphogenesis whereby matter plays an active role in the stability of local processes and the appearance of shapes. Nevertheless, it has shortcomings: (i) it addresses development, a phenomenon that results from a historical process, evolution, with tools designed to study spontaneous phenomena resulting from ahistorical laws, (ii) it conflates theories of physics with existing models in physics and with the method of modeling of physics (Arias Del Angel et al., 2020), and, finally, (iii) purposiveness is still understood as genetic teleonomy (Montévil, 2020).

Rather than applying the usual procedure of transferring mathematical structures developed for the understanding of physical phenomena into biological ones, we model biological processes from a biological theoretical framework. Here we base our approach on two principles (default state and principle of organization) of the three principles proposed as foundations for a theory of organisms. We have thus provided the proof of principle that *mathematical modeling* based on the theoretical framework of the discipline to which the modeled phenomenon pertains, namely, biology, is feasible and provides biological insight.

In fact, the two principles (default state and constraints leading to closure) were sufficient to show the formation of ducts and acini. Cells generated forces that were transmitted to neighboring cells and collagen fibers, which in turn created constraints to movement and proliferation. Additionally, the model pointed to a target of future research, namely, the inhibitors of cell proliferation and motility which in this mathematical model are generated by the epithelial cells. For a better integration with the principle of variation and the historicity of cells, we are introducing the use of the new symbol χ. Finally, the success of this modeling effort performed as a "proof of principle" opens the possibility for a stepwise approach whereby additional constraints imposed by the tissue (additional cell types) and the organism (hormones) could be assessed in silico and rigorously tested by *in vitro* and *in vivo* experiments.

Acknowledgments This work was conducted as part of the research project entitled "Building bridges between natural and social sciences through the prism of a theory of organisms" during AMS tenure as a Fellow of the Institute for Advanced Studies of Nantes, France. Additional support to AMS was provided by Grant ES030045 from the US National Institute of Environmental Health Sciences and the research project entitled "Toward a science of intrinsic purposiveness: shaping development," supported by the Templeton Foundation (PI, AMS). The funders had no role in the study design, data collection and analysis, decision to publish, or preparation of the manuscript. The authors are grateful to Cheryl Schaeberle and Victoria Bouffard for their critical input and to the reviewers for their helpful suggestions. The authors have no competing financial interests to declare.

References

Arias Del Angel, J. A., Nanjundiah, V., Benítez, M., & Newman, S. A. (2020). Interplay of mesoscale physics and agent-like behaviors in the parallel evolution of aggregative multicellularity. *EvoDevo, 11*, 21.

Barnes, C., Speroni, L., Quinn, K., Montévil, M., Saetzler, K., Bode-Animashaun, G., McKerr, G., Georgakoudi, I., Downes, S., Sonnenschein, C., Howard, C. V., & Soto, A. M. (2014). From single cells to tissues: interactions between the matrix and human breast cells in real time. *PLoS ONE, 9*, e93325.

Bassett, D. E., Jr., Eisen, M. B., & Boguski, M. S. (1999). Gene expression informatics – It's all in your mine. *Nature Genetics, 21*, 51–55.

Bich, L., Mossio, M., Ruiz-Mirazo, K., & Moreno, A. (2016). Biological regulation: Controlling the system from within. *Biology and Philosophy, 31*, 237–265.

Bich, L., Mossio, M., & Soto, A. M. (2020). Glycemia regulation: From feedback loops to organizational closure. *Frontiers in Physiology, 11*, 69.

Brown, P. O., & Botstein, D. (1999). Exploring the new world of the genome with DNA microarrays. *Nature Genetics, 21*, 33–37.

Canguilhem, G. (1991). *The normal and the pathological*. Zone Books.

Dhimolea, E., Maffini, M. V., Soto, A. M., & Sonnenschein, C. (2010). The role of collagen reorganization on mammary epithelial morphogenesis in a 3D culture model. *Biomaterials, 31*, 3622–3630.

Dupré, J., & Nicholson, D. J. (2018). A manifesto for a processual philosophy of biology. In D. J. Nicholson & J. Dupre (Eds.), *Everything flows: Towards a processual philosophy of biology*. Oxford University Press.

Feynman, R. (2017). *The character of physical law*. MIT Press.

Gilbert, S. F., & Sarkar, S. (2000). Embracing complexity: Organicism for the 21st century. *Developmental Dynamics, 219*, 1–9.

Greenspan, R. J. (2001). The flexible genome. *Nature Reviews. Genetics, 2*, 383–387.

Guo, C. L., Ouyang, M., Yu, J. Y., Maslov, J., Price, A., & Shen, C. Y. (2012). Long-range mechanical force enables self-assembly of epithelial tubular patterns. *Proceedings of the National Academy of Sciences of the United States of America, 109*, 5576–5582.

Howard, B. A., & Gusterson, B. A. (2000). Human breast development. *Journal of Mammary Gland Biology and Neoplasia, 5*, 119–137.

Jablonka, E., & Lamb, M. J. (2005). *Evolution in four dimensions*. MIT Press.

Kauffman, S. (2001). Molecular autonomous agents. *Philosophical Transactions. Series A, Mathematical, Physical, and Engineering Sciences, 361*(1807), 1089–1099.

Krause, S., Maffini, M. V., Soto, A. M., & Sonnenschein, C. (2008). A novel 3D in vitro culture model to study stromal-epithelial interactions in the mammary gland. *Tissue Engineering, 14*, 261–271.

Krause, S., Jondeau-Cabaton, A., Dhimolea, E., Soto, A. M., Sonnenschein, C., & Maffini, M. V. (2012). Dual regulation of breast tubulogenesis using extracellular matrix composition and stromal cells. *Tissue Engineering. Part A, 18*, 520–532.

Kupiec, J. J., & Sonigo, P. (2003). *Ni Dieu ni gène. Pour une autre théorie de l'hérédité*. Seuil.

Longo, G., & Montévil, M. (2011). From physics to biology by extending criticality and symmetry breakings. *Progress in Biophysics and Molecular Biology, 106*, 340–347.

Longo, G., & Montévil, M. (2014). *Perspectives on organisms: Biological time, symmetries and singularities*. Springer.

Longo, G., & Montévil, M. (2018). Comparing symmetries in models and simulations. In M. Dorato, L. Magnani, & T. Bertolotti (Eds.), *Handbook of model-based science* (pp. 843–856).

Longo, G., & Mossio, M. (2020). Geocentrism vs genocentrism: Theories without metaphors, metaphors without theories. *Interdisciplinary Science Reviews, 45*(3), 380–405.

Longo, G., & Soto, A. M. (2016). Why do we need theories? *Progress in Biophysics and Molecular Biology, 122*(1), 4–10.

Longo, G., Miquel, P. A., Sonnenschein, C., & Soto, A. M. (2012). Is information a proper observable for biological organization? *Progress in Biophysics and Molecular Biology, 109*, 108–114.

Longo, G., Montévil, M., Sonnenschein, C., & Soto, A. M. (2015). In search of principles for a theory of organisms. *Journal of Biosciences, 40*(5), 955–968.

Masso-Welch, P. A., Darcy, K. M., Stangle-Castor, N. C., & Ip, M. M. (2000). A developmental atlas of rat mammary gland histology. *Journal of Mammary Gland Biology and Neoplasia, 5*, 165–185.

Miquel, P. A., & Hwang, S. Y. (2016). From physical to biological individuation. *Progress in Biophysics and Molecular Biology, 122*(1), 51–57.

Montévil, M. (2019). Measurement in biology is methodized by theory. *Biology and Philosophy, 34*(3), 35.

Montévil, M. (2020). Historicity at the heart of biology. *Theory in Biosciences, 141*, 165.

Montévil, M., & Mossio, M. (2015). Biological organisation as closure of constraints. *Journal of Theoretical Biology, 372*, 179–191.

Montévil, M., & Mossio, M. (2020). The identity of organisms in scientific practice: Integrating historical and relational conceptions. *Frontiers in Physiology, 11*, 611.

Montévil, M., Mossio, M., Pocheville, A., & Longo, G. (2016a). Theoretical principles for biology: Variation. *Progress in Biophysics and Molecular Biology, 122*(1), 36–50.

Montévil, M., Speroni, L., Sonnenschein, C., & Soto, A. M. (2016b). Modeling mammary organogenesis from biological first principles: Cells and their physical constraints. *Progress in Biophysics and Molecular Biology, 122*(1), 58–69.

Moss, L. (2003). *What genes can't do*. MIT Press.

Moss, L. (2008). The meanings of the gene and the future of the phenotype. *Life Sciences, Society and Policy, 4*, 38.

Mossio, M., & Bich, L. (2017). What makes biological organisation teleological? *Synthese, 194*, 1089–1114.

Mossio, M., Montévil, M., & Longo, G. (2016). Theoretical principles for biology: Organization. *Progress in Biophysics and Molecular Biology, 122*(1), 24–35.

Nicholson, D. J. (2014). The return of the organism as a fundamental explanatory concept in biology. *Philosophy Compass, 9*(5), 347–359.

Nicholson, D. J., & Gawne, R. (2015). Neither logical empiricism nor vitalism, but organicism: What the philosophy of biology was. *History and Philosophy of Life Sciences, 37*(4), 345–381.

Noble, D. (2006). *The music of life: Biology beyond the genome*. Oxford University Press.

O'Malley, M. A., & Dupre, J. (2005). Fundamental issues in systems biology. *BioEssays, 27*, 1270–1276.

Oyama, S. (2000). *The ontogeny of information: Developmental systems and evolution*. Duke University Press.

Paszek, M. J., Zahir, N., Johnson, K. R., Lakins, J. N., Rozenberg, G. I., Gefen, A., Reinhart-King, C. A., Margulies, S. S., Dembo, M., Boettiger, D., Hammer, D. A., & Weaver, V. M. (2005). Tensional homeostasis and the malignant phenotype. *Cancer Cell, 8*, 241–254.

Richert, M. M., Schwertfeger, K. L., Ryder, J. W., & Anderson, S. M. (2000). An atlas of mouse mammary gland development. *Journal of Mammary Gland Biology and Neoplasia, 5*, 227–241.

Sonnenschein, C., & Soto, A. M. (1999). *The society of cells: Cancer and control of cell proliferation*. Springer.

Sonnenschein, C., & Soto, A. M. (2016). Carcinogenesis explained within the context of a theory of organisms. *Progress in Biophysics and Molecular Biology, 122*(1), 70–76.

Sonnenschein, C., & Soto, A. M. (2020). Over a century of cancer research: Inconvenient truths and promising leads. *PLoS Biology, 18*(4), e3000670.

Sonnenschein, C., & Soto, A. M. (2021). Control of cell proliferation: Is the default state of cells quiescence or proliferation? *Organisms, 5*(1), 33–42.

Sonnenschein, C., Soto, A. M., & Michaelson, C. L. (1996). Human serum albumin shares the properties of estrocolyone-I, the inhibitor of the proliferation of estrogen-target cells. *Journal of Steroid Biochemistry and Molecular Biology, 59*, 147–154.

Soto, A. M., & Sonnenschein, C. (2005). Emergentism as a default: Cancer as a problem of tissue organization. *Journal of Biosciences, 30*, 103–118.

Soto, A. M., & Sonnenschein, C. (2011). The tissue organization field theory of cancer: A testable replacement for the somatic mutation theory. *BioEssays, 33*, 332–340.

Soto, A. M., & Sonnenschein, C. (2020). Information, programme, signal: Dead metaphors that negate the agency of organisms. *Interdisciplinary Science Reviews, 45*, 331–343.

Soto, A. M., Sonnenschein, C., & Miquel, P. A. (2008). On physicalism and downward causation in developmental and cancer biology. *Acta Biotheoretica, 56*, 257–274.

Soto, A. M., Brisken, C., Schaeberle, C. M., & Sonnenschein, C. (2013). Does cancer start in the womb? Altered mammary gland development and predisposition to breast cancer due to in utero exposure to endocrine disruptors. *Journal of Mammary Gland Biology and Neoplasia, 18*, 199–208.

Soto, A. M., Longo, G., Miquel, P. A., Montévil, M., Mossio, M., Perret, N., Pocheville, A., & Sonnenschein, C. (2016a). Toward a theory of organisms: Three founding principles in search of a useful integration. *Progress in Biophysics and Molecular Biology, 122*(1), 77–82.

Soto, A. M., Longo, G., Montévil, M., & Sonnenschein, C. (2016b). The biological default state of cell proliferation with variation and motility, a fundamental principle for a theory of organisms. *Progress in Biophysics and Molecular Biology, 122*(1), 16–23.

Speroni, L., Whitt, G. S., Xylas, J., Quinn, K. P., Jondeau-Cabaton, A., Georgakoudi, I., Sonnenschein, C., & Soto, A. M. (2014). Hormonal regulation of epithelial organization in a 3D breast tissue culture model. *Tissue Engineering Part C Methods, 20*, 42–51.

Tanner, K., Mori, H., Mroue, R., Bruni-Cardoso, A., & Bissell, M. J. (2012). Coherent angular motion in the establishment of multicellular architecture of glandular tissues. *Proceedings of the National Academy of Sciences of the United States of America, 109*, 1973–1978.

Turing, A. M. (1950). I. Computing machinery and intelligence. *Mind, LIX*(236), 433–460.

Walsh, D. (2015). *Organisms, agency, and evolution.* Cambridge University Press.

West-Eberhard, M. J. (2003). *Developmental plasticity and evolution.* Oxford University Press.

Chapter 13
From the Organizational Theory of Ecological Functions to a New Notion of Sustainability

Charbel N. El-Hani, Felipe Rebelo Gomes de Lima, and Nei de Freitas Nunes-Neto

13.1 Introduction

In this chapter, we will address criticisms to the theory of ecological functions introduced by Nunes-Neto et al. (2014). In doing so, we intend to further develop the theory, as a possible basis for naturalizing the teleological and normative dimensions of ecological functions. We will also take the first steps in the construction of an integrated scientific and ethical approach to sustainability that is intended to avoid an anthropocentric perspective.

The problems of teleology and normativity are two classical problems related to the ascription of functions to biological items (Cooper et al., 2016). In a causal

C. N. El-Hani (✉)
Institute of Biology, Federal University of Bahia, Salvador, Bahia, Brazil

National Institute of Science and Technology in Interdisciplinary and Transdisciplinary Studies in Ecology and Evolution/INCT IN-TREE, Salvador, Bahia, Brazil

Graduate Studies Program in History, Philosophy, and Science Teaching, Federal University of Bahia/State University of Feira de Santana, Salvador and Feira de Santana, Brazil

F. R. G. de Lima
National Institute of Science and Technology in Interdisciplinary and Transdisciplinary Studies in Ecology and Evolution/INCT IN-TREE, Salvador, Bahia, Brazil

Graduate Studies Program in History, Philosophy, and Science Teaching, Federal University of Bahia/State University of Feira de Santana, Salvador and Feira de Santana, Brazil

N. d. F. Nunes-Neto
National Institute of Science and Technology in Interdisciplinary and Transdisciplinary Studies in Ecology and Evolution/INCT IN-TREE, Salvador, Bahia, Brazil

Graduate Studies Program in History, Philosophy, and Science Teaching, Federal University of Bahia/State University of Feira de Santana, Salvador and Feira de Santana, Brazil

School of Biological and Environmental Sciences, Federal University of Grande Dourados, Dourados, Brazil

© The Author(s) 2024
M. Mossio (ed.), *Organization in Biology*, History, Philosophy and Theory of the Life Sciences 33, https://doi.org/10.1007/978-3-031-38968-9_13

explanation, causes are presented in order to explain effects that are assumed to follow from them. Functional explanations are suspected to invert the temporal order of causes and effects. When one speaks of functions in biology, one often assumes that the explanation for the presence or existence of a given trait lies in its future utility. This is suggested, for instance, when one says that the function of sea turtles' paddle-shaped limbs is to increase swimming efficiency, which implicitly amounts to saying that they are born with such limbs for swimming efficiently in the future. This is the problem of *teleology*. After all, teleological explanations point to the fulfilment of a given goal. In a teleological explanation, it is claimed that an event takes place for a given purpose, i.e., that it occurs because it is the *kind of event that brings about that goal*. The fact that this is the necessary event for a given goal to be obtained in a certain state of affairs is regarded, in this mode of explanation, as a *sufficient condition for* the occurrence of the event (Taylor, 1964). Bearing in mind this temporal reversion problem, one of the philosophical challenges of ascribing functions to a trait or other biological item is to do so from a scientifically acceptable, naturalized perspective that implies a legitimate and admissible conception of causality from the standpoint of the natural sciences, not appealing to ontological conceptions inconsistent with scientific knowledge and practices (Mossio et al., 2009; Moreno & Mossio, 2015).

The second problem is that of *normativity*. When one ascribes a function to a trait, one refers not merely to what the trait does but to what it arguably *should do* (Cooper et al., 2016). Increasing swimming efficiency, for instance, is not simply something that the sea turtles' paddle-shaped limbs do but what they should do, as their function. That is, "attributing functions to traits implies a reference to some specific effect, which constitutes a criterion against which the activity of the trait can be normatively evaluated" (Mossio et al., 2009, p. 814). This normative evaluation, in turn, seems to depend on the teleological relationship expected to be fulfilled. If the expected specific effect does not take place, this entails *mal*functioning (Davies, 2000; Cooper et al., 2016; Saborido et al., 2016), which is not an all-or-nothing feature but rather a matter of degree (Krohs, 2010, p. 342). A particular sea turtle limb, say, can be said to be malfunctioning (to some specific degree) when its activity fails in fulfilling the expected norms for efficient swimming. When one accounts for the normative dimension of functional ascriptions, it will be necessary, thus, to theoretically justify why a specific means-end relationship is the norm in that ascription. It is important to bear in mind, however, that it is not a moral sense of normativity that is at stake, but just an expectation about a given acceptable relationship of causality.

A scientifically compatible theory of functions that intends to preserve their teleology and normativity should do so in the context of a naturalized approach to purposefulness. This can be done, for instance, by appealing to the notion of "intrinsic purpose." This notion entails the idea that the organization of living beings is inherently teleological, i.e., that their own activity is, in a fundamental sense, first and foremost oriented toward an end, which is to determine and maintain themselves. The concept of self-determination connects biological organization to intrinsic teleology: biological organization determines itself in the sense that the effects of its activity contribute to establish and maintain its own conditions of existence (Moreno & Mossio, 2015; Mossio & Bich, 2017). This framework establishes a biologically distinctive notion of purposiveness: teleology is intrinsic in the case of biological systems, while it is extrinsic in the case of artifacts (Jonas, 1966; Aristotle, 1984).[1]

The question is how to build a theory of biological functions that can take in due account the intrinsic teleology of living systems and properly justify the teleological and normative dimensions of functional ascriptions and explanations. To address this question, we will begin by considering three approaches to function, namely, dispositional, etiological, and organizational theories, and how they deal with the teleology and normativity of functions. Then, we will introduce the organizational theory of ecological functions proposed by Nunes-Neto et al. (2014). We will then tackle the main criticisms raised against this theory, related to difficulties in individuating ecosystems such that they can be treated as organizationally closed, the importance of integrating evolutionary considerations into an organizational understanding of ecological functions in order to support the conceptual role of functional explanations in contemporary ecological research, and the ascription of functions to abiotic items. Finally, we will explore the implications of the organizational theory to environmental ethics, taking the first steps toward an integrated scientific and ethical approach to sustainability. This is intended to lead to a new notion of sustainability that offers an alternative to its common interpretation in anthropocentric and economically based terms.

13.2   Philosophical Theories of Function and Their Approach to Teleology and Normativity

Two philosophical approaches have been typically used for understanding functional explanation in biology (Cooper et al., 2016). On the one hand, dispositional theories explain functions in terms of the contribution(s) or causal role(s) of a system's part to an emergent capacity at the level of the whole (e.g., Cummins, 1975; Adams, 1979; Bigelow & Pargetter, 1987; Craver, 2001). To use a classical example, from this perspective, the function of the human heart is to pump blood because pumping blood is what the heart does that contributes to a specific human systemic capacity, namely, the circulation of gases and nutrients. This is an approach that relies on current means-ends relationships to conceive of functional explanation, intending to ground normativity without appealing to teleology – an approach that has been argued to be able to support only an epistemic normativity, dependent on the researchers' choices about which systems and which systemic capacities to

---

[1] Babcock and McShea (2021) argue that this distinction between externalist and internalist teleological explanations, which comes back to Aristotle's *Physics*, has been a misstep in the debates on teleology. They argue, in contrast, for a single type of legitimate teleological explanation, an artifact model of teleology in which goal-directed entities are guided by a nested series of upper-level fields (McShea, 2012, 2016). In this chapter, we will keep reference to the usual distinction between intrinsic and extrinsic teleology, leaving Babcock and McShea's proposal to be discussed elsewhere.

study (see below). It intends to dissolve the problem of the teleology of functions by reducing them to any causal contribution to a higher-level capacity that a trait/part may give, such that the normative dimension of functions is reduced to the claim that the causal effect must contribute to a higher-level capacity, with no reference to a "benefit" for the system. For those committed to this approach, teleological reasoning is merely an element of a superseded worldview, which should have no application in the way modern science explains natural phenomena. However, a common criticism of this way of understanding biological functions follows from the fact that it does not include a teleological element, namely, that it underdetermines the normative dimension of functional ascriptions, being unable to distinguish proper functions from accidental effects and to account for malfunctionality, because in the end functional ascription depends on the observer's choice of the phenomenon to be accounted for in functional terms (Millikan, 1989; Kitcher, 1993; Mitchell, 1993; Godfrey-Smith, 1994; Mossio et al., 2009). Novel versions of the dispositional theories of function have been proposed in order to include additional requirements in an effort to avoid the drawbacks pointed out by critics (e.g., Weber, 2005).

On the other hand, etiological theories seek to naturalistically ground both the teleological and normative dimensions of functions by appealing to an evolutionary perspective, i.e., turning to the selective causal history (or etiology) of organisms' traits/parts (Wright, 1973, 1976; Millikan, 1984, 1989; Neander, 1991; Godfrey-Smith, 1994). From this perspective, organisms have functional traits because those traits have increased the fitness of past organisms in their respective lineages. Accordingly, function is not a mere effect of a trait but a selected effect that explains its current presence or prevalence. From this perspective, the function of the human heart is to pump blood because pumping blood is the selected effect that explains the current presence of hearts in humans. A causal loop between the functional effect of a given trait and its persistence through time grounds the teleology and normativity of functions: Fitness-enhancing effects of past tokens explain the presence of the contemporary trait type and provide a normative standard for evaluating present tokens. This approach has been criticized, however, for being too narrow to accommodate all functional talk in biology, particularly because it makes the current contribution of a trait irrelevant to determine its function and, thus, does not account for functional ascriptions that are often made in several areas of biology in relation to current rather than past effects. This is at odds, in short, with the fact that functional attributions to biological items do seem to bear some relation to what they currently do that increases an organism's survival and reproduction chances, and not only to what explains their current existence. It is in the sense that it has been argued that etiological theories seem to offer a problematically epiphenomenal account (Christensen & Bickhard, 2002). Another criticism concerns the fact that this sort of explanation appeals only to natural selection, while this is neither the single evolutionary process important to explain how organisms came to be how they are nor the single explanation for the presence of all traits to which we ascribe functions (Cooper et al., 2016). These criticisms have been discussed and addressed by advocates of the etiological theories, which have developed them in different versions in an attempt to overcome the pinpointed problems (see, *e.g.*, Garson, 2015, 2016).

Organizational theories offer a third way for building an understanding of functional ascriptions and explanations (e.g., Schlosser, 1998; Bickhard, 2000, 2004; Collier, 2000; McLaughlin, 2001; Christensen & Bickhard, 2002; Delancey, 2006; Edin, 2008). In particular, we rely here on the theory developed by Mossio et al. (2009), which aims at explaining at the same time the persistence of a trait through time and its current contribution to the maintenance of a system.

As formulated by Saborido et al. (2011), a trait T has a function if, and only if, it exerts a constraint subject to closure in an organization O of a system S, which entails the fulfilment of three conditions:

C_1: T exerts a constraint that contributes to the maintenance of the organization O.
C_2: T is maintained under some constraints of O.
C_3: O realizes closure.

These conditions naturalize teleology as they state how the system realizes a circular causal regime that can be grasped through the concept of "closure" (Varela, 1979; Moreno & Mossio, 2015).[2] If the heart pumping blood makes it possible that the organization of a living system and, consequently, the heart itself be maintained, then that activity of the heart is a cause of its very existence and can be identified as its function. Normativity is also naturalized by these conditions, since the expected behavior of an organism's trait is related to the production of the specific effect that contributes to the systemic organization in which the trait is included and that is responsible for its very maintenance. The specificity of this effect allows for a distinction between function and nonfunctions, as well as between proper and accidental functions.

The causal loop involved in the intrinsic teleology of living systems shows the distinctive property of being a closure of constraints, rather than merely a closure of processes, as we observe in a number of physicochemical systems showing mutual dependence of entities and processes. Constraints are local and contingent causes that reduce the degrees of freedom of the dynamics on which they act (Pattee, 1972) but remain conserved at the time scale relevant to describe their causal action with respect to those dynamics (Mossio & Bich, 2017). Thus, the kind of closure expressed in conditions C_1 and C_2 is a closure of constraints, i.e., an organization in which each constraint is involved in at least two different dependence relationships, playing the role of enabling and dependent constraint, respectively (Moreno & Mossio, 2015). Therefore, as developed in detail by Mossio and Bich (2017), it is not any form of causal circularity that will show intrinsic teleology. Rather, it should be a circular causal regime of constraints that are collectively able to self-determine

---

[2] In very general terms, by "closure" one means a feature of systems by virtue of which their constitutive components and operations depend on each other for their production and maintenance and, also, collectively contribute to determining the conditions under which the system itself can exist (Mossio, 2013).

(or, more specifically, self-maintain) through self-constraint. Or, to put it differently, circularity is a necessary but not sufficient condition for intrinsic teleology, and biological organization shows this distinctive property because it realizes self-constraint. In these terms, the idea of biological function does not rely only on teleological and normative dimensions but also on the idea of organization. Or, to put it differently, the idea of organization as closure necessarily includes teleological and normative dimensions.

13.3   The Organizational Theory of Ecological Functions

Functional language is ubiquitous in ecology. Ecologists commonly talk about the function of a given tree species in a forest, or the function of decomposers in relation to soil properties, or the functional role of organisms' traits in a given ecological process, among many other possible examples (for detailed analyses of the uses of function by ecologists, see Jax, 2005; Nunes-Neto et al., 2016a). However, in spite of this extensive reliance on functional language in both descriptions and explanations in ecological research, it is not clear yet how to properly justify the use of functional language in ecology in scientifically compatible terms. However, several steps have been taken in this direction in a number of recent works (e.g., Jax, 2005; Nunes-Neto et al., 2014; Dussault & Bouchard, 2017; Odenbaugh, 2019; Millstein, 2020; Lean, 2021).

We have proposed a theoretical perspective to justify functional ascriptions and explanations in ecology from an organizational point of view (Nunes-Neto et al., 2014; El-Hani & Nunes-Neto, 2020). In order to explain it, let us begin by considering the ways in which the concept of function is used in ecology (see Jax, 2005; Cooper et al., 2016). Jax, for instance, differentiates between four different and complementary ways this concept is employed by ecologists: (1) as a purely descriptive meaning that refers to some change of state or to what happens in the relationship between biotic or abiotic objects; (2) to refer to the functioning of a whole ecosystem; (3) to refer to the role functions of biotic and abiotic components of an ecosystem in relation to its functioning as a whole; and (4) to refer to ecosystem services to some human need or purpose. Here we are specifically interested in use (3), related to the role functions of ecosystems' parts in relation to ecosystem processes (e.g., the role of plants as primary producers within an ecosystem).[3] These

---

[3] When we refer to ecological role functions, this is not in opposition to thinking on individuals or groups/types from an organizational perspective. Rather, as we make explicit in the organizational account, functions are specific roles ascribed to items of biodiversity or abiotic items (under the influence of the biotic community) that constrain the thermodynamic flows in an ecological system. These parts are identified through decomposition/localization analyses (Bechtel & Richardson, 2010), but these are not arbitrary, or under the mere discretion of the researcher, since they should be guided by hypotheses or models on the contributions of the components to the norms of the ecological system's behavior, i.e., to the maintenance of its conditions of existence. Therefore, the components fulfil the causal roles defined in a given decomposition/localization model or hypothesis when they do what they are supposed to do in relation to those norms.

role functions are connected, in turn, with the use of functional reasoning to classify organisms or species according to their effects on ecosystem processes, as we see in the common reference to functional traits and functional groups in ecological research (see, e.g., Hooper et al., 2005; Petchey & Gaston, 2006).

Based on the organizational theory of functions developed by Mossio, Saborido, Moreno, and colleagues, we have defined an ecological function as "a precise (differentiated) effect of a given constraining action on the flow of matter and energy (process) performed by a given item of biodiversity, in an ecosystem closure of constraints" (Nunes-Neto et al., 2014, p. 131). At the same time, assuming this definition as a starting point for an organizational theory of ecological systems under construction, we have recently proposed to broaden the range of organizational functional items in the ecological domain in order to include abiotic items, if one shows how they can play the role of constraints (El-Hani & Nunes-Neto, 2020).[4] In other words, an adequate set of functional items should include not only items of biodiversity (i.e., organisms, populations, functional groups, guilds, etc.) but be more encompassing, including abiotic items. Looking at individual organisms helps making this clear: a honey bee nest is an abiotic, non-biological structure (in the sense that it is not made of living cells) but at the same time is clearly functional (or at least it is typically assumed to be so by biologists). The same seems to be true of ecological systems: abiotic parts of ecosystems (for instance, fire) may play relevant functions in the whole system of which they are parts. The key point when ascribing functions to abiotic items in either organismic or ecological systems is to show how they can act as constraints internal to the organization of the systems, involved in the maintenance of their conditions of existence.

To consider an example of how the organizational approach works, let us look at an ecological system from the point of view of its main activities, decomposing it in three functional groups: producers, consumers, and decomposers. Consider, also, an abiotic factor that producers subject to their closure, namely, carbon dioxide. The functional groups form a hierarchical organization comprising two levels (i.e., a hierarchy of control, *cf.* Ahl & Allen, 1996): the level of the functional items – in this case composed by items of biodiversity – which act as constraints, and the level of the material, thermodynamic flow of carbon atoms, which is the constrained process. Considering the functional items, the producers of organic matter (plants) constrain, through photosynthesis, the flow of carbon atoms, reducing its degrees of freedom, which is something that can be clearly noticed in the building of complex biomolecules from carbon atoms as basic ingredients. The flow of carbon atoms

---

[4] The individuation of abiotic items, as components of ecological systems, poses in itself important challenges. Here we will not focus on these challenges, which will be faced, in fact, by any theory that intends to ascribe functions to abiotic items, such as Dussault and Bouchard's (2017) persistence enhancing propensity (PEP) or Odenbaugh's (2019) systemic capacity accounts. Rather, our main concern in the present work is the individuation of ecological systems. In passing, we can remark, however, that the fact that abiotic items can only be ascribed role functions according to the organizational theory if they act as constraints in relation to the organization and conditions of existence of ecological systems means that we may be able to individuate at least their role as constraints, even if it may be difficult to individuate them as entities.

becomes more determinate, more harnessed, as these atoms, initially contained in atmospheric carbon dioxide molecules, become part of plant biomass. Parts of plant biomass (leaves, fruits, sprouts, etc.) are eaten by consumers (herbivorous animals), which realize a second channeling of the flow of carbon atoms, when these atoms in the plant biomolecules, after digestion and absorption of nutrients, become part of their bodies. And the same is true of a whole network of consumers. In turn, when the consumers and producers die, the animal carcasses and plant leaves, fruits, twigs, and roots become part of the organic matter that is further processed by decomposers, which transform it into available nutrients for plants, thus closing the cycle by reducing once again the degrees of freedom of the flow of carbon atoms. Moreover, due to respiration, along the whole chain of processes, carbon dioxide molecules are sent back to the atmosphere, from where they can be cycled back to the system through photosynthesis (Fig. 13.1).

There is a clear mutual dependence between these constraints. By constraining the flow of matter (carbon atoms), the consumers, for example, create conditions of possibility (or enabling conditions) to the existence of the decomposers and, in this manner, exert an effect on the ecological system as a whole. And while, on the one

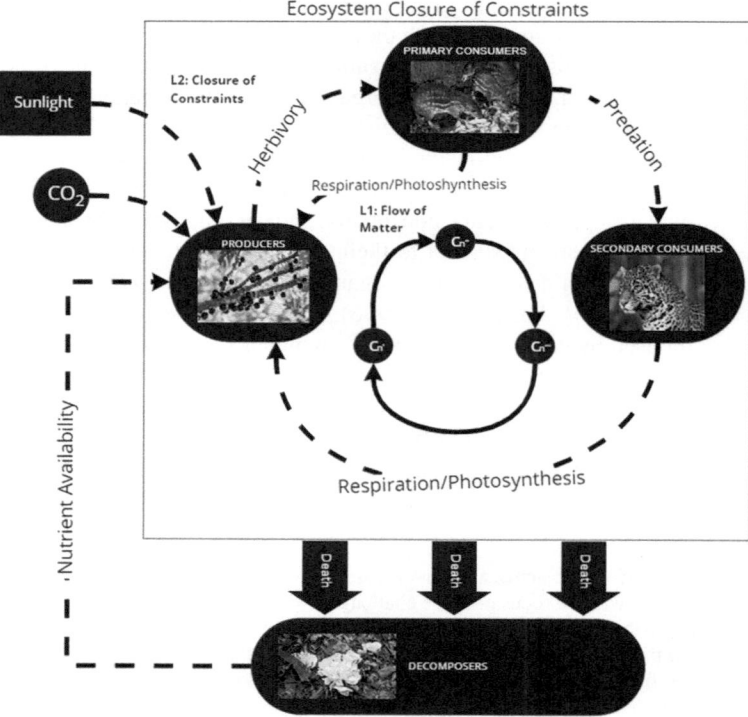

Fig. 13.1 Organizational functions in an ecosystem closure of constraints – a schematic view. (Figure elaborated by Felipe Rebelo Gomes de Lima)

hand, the consumers are enabling conditions to the existence of decomposers, they depend, on the other hand, on the producers from which they derive the matter and energy needed for their self-maintenance. Therefore, we can say that they are dependent on both the producers of organic matter and the very decomposers that mobilize nutrients to the producers. In sum, producers, consumers, and decomposers – as functional items – exert specific constraining actions that amount to the role functions they play within the ecological system of which they are parts, contributing to the self-maintenance of its organization.

13.4 Organizational Functions and the Individuation of Ecological Systems

Ecosystem individuation raises questions for the organizational theory of ecological functions (Cooper et al., 2016). As functions are ascribed in this theory to constraints subject to closure, it is a requisite to functional ascription to establish where the ecosystem closure of constraints lies. In more detail, the problem consists in that, as ecological systems interpenetrate one another at their fringes, this fuzziness of physical boundaries typically also entails a fuzziness of functional relationships, making it harder to decide which constraints are part of one or another ecosystem (or, perhaps, both) and, thus, which functions are to be ascribed to them as subject to the closure of the distinct systems. In short, to ascribe functions to ecosystem components and, accordingly, to naturalistically justify the teleology and normativity of ecosystem functions depend on the ecosystem closure of constraints and mutual dependences between items of biodiversity and abiotic items.

This does not seem to be a particular problem challenging the application of the organizational theory to ecological cases. As Bich (2019) argues, to account for limit cases in which functional closure cannot be realized from within is a more general challenge faced by this theoretical framework. This follows in fact from the thermodynamic openness of living systems and, thus, goes all the way back to Piaget's (1967) crucial conception of the complementarity in such systems between organizational closure and thermodynamic openness. In order to self-maintain themselves, biological systems often need to recruit external constraints or expand their network of control interactions to include previously external constraints, which belong to other systems, subjecting them to their own organizational closure. As we will elaborate below, the control of flammability by fire-adapted plant species in ecosystems is an example of how a boundary condition[5] external to the

---

[5] Boundary conditions are typically conceived as conditions defined externally to a system that contribute to determine its behavior and dynamics, but typically do not depend on the dynamics on which they act. When the behavior of a system is sub-specified, as it happens when it depends on variable, contingent local circumstances, boundary conditions related to these circumstances are added to its description for providing the lacking specifications. Boundary conditions are imposed on the laws of physics and chemistry and provide additional specifications by decreasing the

system has become part of its internal dynamics, once it has been subject to their closure, turning from a destructive force into a constraint that is both enabling (e.g., of regrowth processes) and dependent (on fire-adapted plant species) (e.g., Mutch, 1970; Schwilk & Ackerly, 2001; McLauchlan et al., 2020). This kind of process can blur, however, the functional boundaries of the system, jeopardizing the very idea of closure, which depends on the capability of living systems of specifying their functional boundaries from within. This threat to closure may be solved by recognizing the fact that, once an external constraint is *recruited* by a system A, it simply becomes a part of that system. Yet, as this was an external constraint, say, initially belonging to some system B, we need clear criteria to state whether the constraint is part of A, B, or both, which will affect which functions may be ascribed to it. That is, as functions are ascribed in the organizational theory to constraints subject to closure, the problem of specifying where the organizational closure of the living system lies will affect functional ascription, i.e., the identification of what can be considered a functional component of the system and what cannot. To trace the precise functional boundaries of a system can be regarded, thus, as a requisite to build functional explanations according to this theory, which is under challenge in other cases than just ecological systems. Yet, as we shall see, the theory has the resources to face this challenge.

Based on the assumption that one could rarely individuate ecosystems as organizationally closed systems, a number of criticisms of the organizational theory of ecological functions suggested that the range of ecological systems to which it applies is rather limited. Dussault and Bouchard (2017), for instance, argue that the organizational theory is too restrictive to accommodate key aspects of contemporary ecology, for instance, related to the biodiversity and ecosystem function (BEF) research program. They go on to discuss cases related to the ascription of functions to biodiversity, abiotic factors, and source-sink populations. We will engage with these cases below. Let us begin, however, by considering a critical appraisal of the organizational theory claiming that the domain in which this theory can be applied is very narrow.

Focusing on the bromeliad example chosen by Nunes-Neto et al. (2014) to illustrate the organizational theory of ecological functions, Lean (2021) argues that organizational closure is an exceptional case in ecological systems, and organizational functions will be less likely as ecological systems scale up in size, complexity, and openness. In this manner, the individuation of ecosystems as organizationally closed systems and the scope of the organizational theory can be seen as

---

degrees of freedom of the system's dynamics. In this manner, they harness the physico-chemical processes involved in such dynamics (Polanyi, 1968). A constraint is a particular kind of boundary condition, characterized by both its causal role in relation to a particular process P under its influence, such that P takes place differently under and free from the influence of the constraint, and its conservation or symmetry at the time scale characteristic of P, which follows from being locally unaffected by P (Mossio et al., 2013). A central difference between living and non-living systems is that in the former part of the constraints/boundary conditions acting over the system are produced by the system itself, and, moreover, the set of internal constraints show mutual dependence, while in the latter constraints/boundary conditions are externally produced.

interconnected issues. Lean argues that one can only take ecological systems to be closed self-maintaining units based on a strong commitment to equilibrium ecology, according to which population interactions would stabilize the composition of communities. Then, if such stabilizing interactions are coupled with stable populations, we would obtain bound self-maintaining ecological systems. However, the problem is – Lean argues – that equilibrium ecology has been extensively criticized in the history of this discipline and ultimately replaced by non-equilibrium ecology, which describes ecological communities as causally open collections of species, and local community composition as a result of path-dependent historical processes and random dispersal of populations from other local communities. The outcome of this picture would be that ecological communities are not closed systems but rather the product of many populations moving around larger biogeographic regions. With the large turnover of species within a local area, there would be changes not just in the populations playing a functional role but also in the overall causal structure of the system. Lean's conclusion is, thus, that the domain in which the organizational theory of ecological function may apply is very limited: organizational functions would only occur in some ecological systems, generally rather small ones, with just a couple of populations in close physical proximity.

Besides organizational functions, Lean (2021) also discusses selected effects, persistence, and causal role functions, concluding that all those that include a normative dimension (i.e., all of them except causal role functions) are sporadic and rare, such that ecological functions would be nearly always dispositional rather than normative. That is, they should be conceived, according to him, as descriptions of causal structure that can be used to identify features that we should preserve. As teleological arrangements of ecological systems would be extremely spotty, with just some "blips of teleological arrangement" (Lean, 2021, p. 9327), founding conservation on teleology would be a misstep. Teleological organization would not include much of what conservation biologists intend to protect and, accordingly, would not provide a strong enough scaffold to support conservation ethics. To deny intrinsic teleological grounding for conservation ethics may limit conservation decisions, however, to anthropocentric reasons, leaving instrumental values in relation to human activities as the major if not exclusive reason that would be relevant for such decisions.[6]

This is related to a key difficulty for causal role functions, which several critics have pointed out: they lack a normative component, just describing the presence or absence of a function, not whether a system's trait is malfunctional or accidental. Even though Lean's position can be described as pluralistic (see Dussault, 2022), we think his conception of ecological functions is more closely related to causal role functions. But, be that as it may, his position shares with the latter the lack of a

---

[6] This is recognized by Lean (2021, p. 9328) himself: "By deploying [Causal role] functional analysis, we can identify what supports the ecological features that we do, or should want to, protect. These could be features of the environment which have moral utility or preference. While this does not offer us a non-anthropocentric justification for intervening on ecological communities, it does offer a way of identifying which populations make a disproportionate impact on the community."

normative component. Now, this has been generally regarded as a shortcoming of dispositional theories of function, as shown by the well-known argument that these theories are too liberal for proper functional ascription (e.g., Millikan, 1989; Kitcher, 1993). However, Lean (2021) thinks differently, claiming that this is a positive characteristic of the theory, a flexibility that allows explaining any ecological system's capacity, provided there are constitutive and causal relations in the community at stake. He does not see as a weakness of dispositional theories that the system and capacity of interest are defined by the researcher. This is a defensible view, but not easily so, given the common criticisms of dispositional theories for being too broad and under-specified, incapable of capturing the explanatory force of functional ascription, or making sense of malfunction and differentiating between functioning and mere usefulness, or, else, the criticism that they allow arbitrary, subjective attribution of functions depending on which capacities of a system interest us, and on criteria to identify relevant systems that are entirely dependent on the observer (e.g., Neander, 1991; Mitchell, 1993; Godfrey-Smith, 1994; Moosavi, 2019).[7]

To our understanding, Lean's arguments about organizational functions show two major problems: first, they do not take in due account that non-equilibrium ecology does not exclude ecological interactions and ecological interactions can generate community-level functional organization; second, they do not consider that closure of constraints is not an all-or-nothing property that would necessarily require a strongly cohesive unit to obtain.[8] As we will argue later, it is sufficient that just part of the constraints exerting influence or control over the system be included in the closed organization, and, accordingly, an organizationally closed system can show different degrees of cohesion or functional integration.[9] But let us focus, first, on the idea that one might appeal to closure of constraints to individuate ecosystems only if committed to equilibrium ecology.

---

[7] Lean's approach is to justify conservation decisions based on analyses of the causal structure of ecological systems, in order to identify populations that we are interested to preserve, given their role in supporting ecological features that interest us, and populations we may want to control, due to their role in reducing biodiversity. He does not appeal to any normative reason that could justify treating systems with particular populations (say, invasive species) as malfunctioning. It is clear that these conservation decisions would be mostly based on what we are interested into, and may suffer from the same sort of arbitrariness criticized in dispositional theories. At most, Lean can introduce a justification for conservation decisions based on what "... all prudent agents should want to preserve."

[8] Here we should admit that in our original 2014 paper, we were not explicit about the idea that organizationally closed systems can show different degrees of cohesion or functional integration, such that Lean cannot really be blamed for overlooking it.

[9] This can be related to the view that biological individuality and functional integration come in degrees even in the case of organisms (see, e.g., Queller & Strassmann, 2009; West & Kiers, 2009; Clarke, 2010; Strassmann & Queller, 2010; Godfrey-Smith, 2013; Huneman, 2014a, b; Sterner, 2015; Skillings, 2016; Bich, 2019; Wilson & Barker, 2019).

13.4.1   Ascribing Organizational Functions in Non-equilibrium Ecology

Non-equilibrium models resulted from the work of neo-Gleasonian ecologists (e.g., Whittaker, 1951, 1975; Curtis & McIntosh, 1951) who proposed that typical ecological communities are composed of species which have evolved independently and were combined through chance immigration and individual suitability to ecological contexts. As Whittaker and Woodwell (1972, p. 141) argue, "communities are related by a blurred reticulateness of many intersecting strands (*i.e.*, species) relating a present community to many past communities." But, as Dussault and Bouchard (2017) emphasize, Gleasonian or neo-Gleasonian ecology does not deny community-level functional organization as depicted, say, by Elton's (1927, 1930) trophic model of ecological communities or as studied by ecosystem ecologists (Hagen, 1989, 1992). Even though Gleason and his followers adopt a population-reductionist stance concerning the migration and establishment of species in a location, they do not deny that these species interact once they are established, and their interactions can give rise to community-level functional organization (Eliot, 2011; Nicolson & McIntosh, 2002).[10] Symptomatically, we find that contemporary ecologists often do not shy away from conceiving communities and/or ecosystems as functionally organized systems, in which organisms, species, or abiotic items fulfil functional roles (e.g., Naeem, 2002a; Schulze & Mooney, 1993).

If we take into account these aspects of Gleasonian or neo-Gleasonian ecology, Lean's (2021) interpretation that one can only consider ecological systems as organizationally-closed if strongly committed to equilibrium ecology is not well supported. And, if we consider the issue more generally, the current state of knowledge in ecology does not support this interpretation either. It is truly an empirical issue which ecological systems are subject to non-equilibrium dynamics and which are in equilibrium, but it is not the case that closure of constraints would only obtain if community composition was established as depicted in equilibrium models. As in non-equilibrium models community-level functional organization can emerge once the biotic community is formed in the intersection between the distribution of several to many species, closure of constraints can also obtain even if community

---

[10] Here it is important to notice that generic interactions among species are not sufficient for functional ascription. If interactions are generic, it will be difficult to maintain that ecological function bearers contribute to their own maintenance by contributing to the maintenance of the system. Strictly speaking, a species that generically contributes to the maintenance of the ecological system as a whole cannot be convincingly described as contributing to its maintenance *per se*. Rather, it can be said to contribute to the provision of conditions that fulfil the needs of any species with sufficiently similar niches – be it itself or another species. On this issue, see Dussault (2019). For an item of biodiversity or abiotic factor to be functional within a community-level functional organization, it should have specific rather than generic effects. That's why the proviso that an ecological function amounts to a precise (differentiated) effect of a constraining action on the flow of matter and energy in an ecosystem is very important in the account proposed by Nunes-Neto et al. (2014, p. 131).

composition results from path-dependent historical processes and random dispersal of populations, provided that their interactions once together in the same community give rise to a functional organization.

Vellend's (2010, 2016) proposal of a conceptual synthesis in community ecology offers a case in point about how the role of local species interactions, which can give rise to community-level functional organization, is recognized in ecological models not committed to equilibrium assumptions. He claims that, despite the large number of mechanisms underpinning patterns in ecological communities, four distinct kinds of processes are combined in them, namely, selection, drift, speciation, and dispersal. The focus on these processes resulted from conceptual developments along the history of ecology. In the 1950s and 1960s, equilibrium ecology was consolidated, based on the idea that patterns in the composition and diversity of species in communities were the deterministic outcome of local interactions between functionally distinct species and their environments (importantly including other species). Thus, those patterns could be explained mostly by fitness differences among species, or, in other words, by selection.

Non-equilibrium ecology emerged in the 1980s and 1990s, bringing a more inclusive approach to community ecology, which recognized the importance of processes at broader spatial and temporal scales for understanding local-scale patterns. This does not mean, however, that the latter patterns could be simply dismissed, and, accordingly, it does not entail that local species interactions and their fitness consequences, or community-level functional organization, would have to be simply dropped from the picture. Rather, what was at stake was the need to take into account that the composition and diversity of species at a local scale fundamentally depend on the composition and diversity of regional pools of species, such that speciation is also a process to consider when explaining community-level features.

The next step was the incorporation of drift, with the neutral theory of biodiversity. By "ecological drift" one means random fluctuations in population size resulting from ecological equivalence in the probabilistic sense, i.e., in the sense that individuals have equal chances of reproduction or death regardless of species identity (Rosindell et al., 2012). Pure ecological drift would happen if individuals of different species were demographically identical, a very unlikely situation, but drift will be equally important when it is not the only active process at stake. In any case, there will be ecological drift, and its importance will be greater the more modest is the functional differentiation between individuals or species in a given set, or, to put it differently, the more ecologically equivalent they are. Again, the recognition of drift does not entail a denial of a role for local-scale species interactions and patterns. Finally, dispersal was incorporated into community ecology models in the form of the metacommunity concept, which concerns the influence of dispersal among local communities over community patterns at multiple scales.

As Vellend (2010, p. 185) sums up, "selection, in the form of deterministic interactions among species and between species and their environments, was always recognized as important." Accordingly, in non-equilibrium models, incorporating drift, speciation, and dispersal, local-scale species interaction is recognized, as well as the possibility that community-level functional organization emerges. The upshot is that we do not need to be committed to equilibrium ecological models to explain ecological functions from an organizational perspective. In non-equilibrium models, organized systems can be also identified, and, accordingly, the domain of the organizational theory is much larger than Lean recognizes. Moreover, it has been argued that the processes identified by Vellend (2010, 2016), which structure species dynamics within a community, can be interpreted as constraints, even though a convincing demonstration that this is really the case is yet to be done (Peck & Heiss, 2021).

Interestingly, Lean (2021) situates his own position between two extremes he identifies in ecological science: either ecological systems would be mere collections of populations, largely independent of each other (Gleason, 1926), or analogous to organisms, possessing functional organization that maintains mature organism-like individuals (Clements, 1916).[11] He associates, then, the idea that ecological systems may have functions from which conclusive statements about what is normatively functional or malfunctional can be made with the claim that they are organism-like. However, ecology is not trapped between those two extremes but also formulates an understanding of ecological systems that sits between them. Similarly, we do not need to treat ecosystems as organisms to ascribe organizational functions to their components. On the contrary, there are important differences between ecosystems and organisms, as the former typically lacks the sort of agency and regulation[12] that characterize the latter. Moreover, ecological systems do not show the same degree of stability and cohesion observed in organisms, or at least in many cases of organismality. Accordingly, there is no requirement that one is committed to an interpretation of ecosystems as superorganisms to apply the organizational theory. What is necessary to ascribe function to components of a system based on this theory is just organizational closure, conceived as closure of constraints. Or, to put it differently, what we assume is just that organisms and ecosystems can share the property of organizational closure, despite their several differences. It is sufficient, also, that a self-maintaining, organizationally closed system shows a tendency to closure,[13] and

---

[11] See Eliot (2007, 2011) for a critical appraisal of the sheer opposition between Clements' and Gleason's approaches to explaining vegetation.

[12] As defined in the organizational theory we take as a starting point. See, e.g., Moreno and Mossio (2015) and Bich et al. (2016).

[13] We use the expression "tendency to closure" following its usage by Montévil and Mossio (2015). However, this expression may be interpreted in a manner that raises unnecessary difficulties to the theory, since it may suggest that we would be referring to a process showing a tendency that closure emerges as its outcome. This interpretation may lead to a counter-argument that a tendency toward closure would be no guarantee (or support no prediction) of achieving it. Nevertheless, what is meant by "tendency to closure" is that interdependent subsystems (or modules) within a containing system (which is itself organizationally closed) showing a relatively large degree of internal cohesion (i.e., interdependent modules) can be said to tend to be closed, despite the fact that they functionally depend on one another. In this precise sense, organizationally closed systems may come with different degrees of cohesion and functional integration (either diachronically or synchronically), to the extent that it is sufficient for closure that part of the constraints affecting the system's dynamics are mutually dependent such that each of them is involved in at least two different dependence relationships in which it plays the role of enabling and dependent constraint.

even if regarded as closed, this does not mean all constraints or boundary conditions affecting its dynamics should be included within the closed organization.

13.4.2   On the Domain of the Organizational Theory of Ecological Functions

In this section, we intend to reinforce the claim that the organizational theory of ecological functions does not apply only to small ecological systems, with just some limited number of populations in close physical proximity. If the theory applied only to such exceptional cases, its utility would be surely quite limited. We do not think, however, that this is a correct assessment.

In our original paper (Nunes-Neto et al., 2014), we did not present key concepts of the original organizational theory that provide ways to respond to criticisms about the scope of the organizational account of ecological functions (Dussault and Bouchard, 2017; Odenbaugh, 2019; Lean, 2021). Valuable as these criticisms are for sharpening our ideas, there are central aspects of the theory that need to be made explicit to tackle them.

For instance, the concept of constraint has not been given by both Dussault and Bouchard (2017) and Lean (2021) the central role it has in the organizational theory. Organizational closure is explained as follows by the former authors:

> … traits have functions relative to what its proponents call the *organizational closure* of a system, which is a causal loop that occurs when the parts of a far-from-equilibrium system contribute to its self-maintenance, and the system, in turn, maintains those parts. (Dussault & Bouchard, 2017, p. 1133)

Lean (2021) also describes organizational closure without considering the concept of constraint in any detail but rather just mentioning it once in the entire explanation of the organizational theory and that as part of a quote from Mossio et al. (2009). In this manner, closure of processes and closure of constraints are not properly differentiated. This differentiation is, however, a key aspect of the organizational theory of biological functions.

Since these authors do not properly consider the concept of constraint in their arguments, they neglect aspects showing how the organizational theory is less restrictive than it might seem at first sight. Lean (2021), for instance, argues for the rarity of organizational functions based on the difficulty of satisfying the requisite of causal closure in ecological systems since they are rarely, if ever, closed systems. But for properly understanding the organizational theory, it is important to consider that a closure of constraints does not correspond to any set of causal relationships but to a rather specific state of affairs. There is indeed a clear criterion postulated in the organizational theory for constraints to be regarded as part of a closed organization, which is enunciated by Moreno and Mossio (2015, p. 20) as follows:

In formal terms, a set of constraints **C** realizes closure if, for each constraint C_i belonging to **C**:

1. C_i depends directly on at least one other constraint of **C** (C_i is dependent).
2. There is at least one other constraint C_j belonging to **C** which depends on C_i (C_i is enabling).

If one takes into account the meaning of closure, not in isolation but within the overall framework of the theory, it will not be difficult to conclude that a closed organization of constraints requires that just some but not all constraints relevant to the system's self-maintenance be included within its organizational closure. Precisely, only constraints that are both enabling and dependent are considered part of the organizational closure. If this is lost from sight, the demand that the system be organizationally closed will seem more restrictive than it is in fact.

As Bich (2019) argues, biological systems should be capable of generating within themselves *some* of the internal constraints that control their dynamics, such that they remain in far from equilibrium conditions by harnessing the thermodynamic flow. Closure is a regime of mutually dependent constraints that determines a subset of its own conditions of existence, not all of them. In these terms, we can tackle the problem posed by the expansion of the functional boundaries of an organizationally closed system that recruits external boundary conditions or extends its network of control interactions. In short, we can do so by considering how this problem follows, in fact, from an incorrect interpretation of the notion of closure of constraints, which conflates the self-specification of the functional boundaries of a system with functional self-sufficiency. Based on how the functional components of a biological organization are wired together to collectively achieve self-maintenance, one can propose criteria to characterize the *degree of functional integration* and, accordingly, the *degree of internal cohesion* of a system, i.e., the different ways and extents in which constraints are mutually dependent and realize closure (Bich, 2016, 2019). When we take into account different degrees of functional integration in organizationally closed systems, we can realize that there is a variety of ecological systems that can be described as such.

A minimal theoretical example of functional integration by means of cross-control (Bich, 2019) is found in Kauffmann's (2000) autocatalytic sets, in which a catalyst *A* is produced thanks to the action of another catalyst *B* that controls kinetically its synthesis, while *A* itself contributes, in turn, to *B*'s existence by controlling directly its production or some intermediate steps in the production of *B*. An autocatalytic set realizes a basic form of closure, given that each constraint depends for its production and maintenance on the direct action of (at least) another constraint, and together the components of the autocatalytic set (in the example, *A* and *B*) collectively realize self-production and self-maintenance. Autocatalytic sets exhibit closure because each constraint plays a function in collective self-production and self-maintenance, and we can consider the same to be true of ecological systems. Indeed, we find in the literature theoretical treatments of ecological systems as autocatalytic sets (see, e.g., Cazzolla Gatti et al., 2017, 2018). The issues related to individuation follow from the fact that, like autocatalytic systems, ecological systems are more directly determined by external boundary conditions and material constraints than more complex, autonomous systems, such as organisms. Nonetheless, ecological systems can realize a basic regime of closure, just as autocatalytic sets.

An idea that is quite helpful when discussing the individuation of ecological systems is Montévil and Mossio's (2015) "tendency to closure." Closure offers a clear-cut criterion for drawing the boundary between a biological entity and its environment, providing a fitting solution to the problem of individuation, as the set of constraints subject to closure defines the system, based on the topological property of circularity in the network of constitutive interactions, whereas all other constraints acting on the system belong to its environment. Montévil and Mossio claim that we should ascribe closure to "maximally closed systems," i.e., systems including all mutually dependent constraints in the currently available descriptions (which are, by necessity, incomplete). Thus, in the case of mutually dependent organisms, there still seems to be a fundamental organizational continuity between the interacting organisms. In this case, it seems justifiable to ascribe ecological functions to the organisms constituting the system, even if the system does not show fully-fledged functional integration or constraints closure. Montévil and Mossio were discussing cases in which an encompassing system (say, a symbiotic one) is maximally closed, such that one might say that the symbionts within that system display a tendency to closure. As they depend on each other, they are not closed strictly speaking, but one can say they "tend" to be closed. We think we can extend this notion, however, to conceive of subsystems or modules, generally speaking, which show a relatively large degree of internal cohesion but yet depend on other modules in a given network. Closure ascription can extend in this case beyond each module, insofar as a maximally closed system should include all known constraints showing the topological circular property. Yet, we can claim that the modules containing – in the case of ecological systems – functionally coupled organisms or other items of biodiversity show a tendency to closure, as elements within a hierarchical set or network of modules. That is, in this case we can introduce a somewhat more relaxed notion of internal cohesion that makes it clear how the scope of the organizational theory of ecological functions is substantially broader than just a limited number of cases showing fully-fledged closure. To make this notion more precise, we can introduce a measure of the degree of closure in a system, based on the number of constraints that are both enabling and dependent, and, accordingly, are subject to closure. A tendency to closure points, then, to a specific degree of closure measured by the number of mutually dependent constraints in a system.

Ecological systems realizing closure of constraints can indeed exhibit different degrees of functional integration. They can be rather closed systems like the phytotelmata of bromeliads, chosen by Nunes-Neto et al. (2014) as a case to develop the organizational theory of ecological functions, but not as exhausting all the possible cases to which the theory applies. They can be symbiotic systems or other functionally integrated consortia of organisms, in which control is exerted not only within but also across biological systems (Bich, 2019), as, for instance, bacterial biofilms in which bacteria exchange enzymes (or DNA sequences coding for enzymes) responsible for the control of the internal metabolic processes in response to nutritional and other kinds of stress (Davey & O'Toole, 2000), or plants integrated by mycorrhizal networks that not only exchange metabolites but mutually affect their physiology and ecology (Selosse et al., 2006). These cases are different from the bromeliad one because a new order of functional integration is realized through the control exerted by organisms upon one another's processes.[14]

Ecological systems may show, however, much less integrated and bounded configurations and, yet, realize closure of constraints, as it happens when organisms exert control upon the conditions of existence of one another, either by directly harnessing the external flow of matter and energy or indirectly generating external control constraints in the environment, such as bird nests, spider webs, beaver dams, ant nests, etc. It seems clear, thus, that the domain of the organizational theory is much larger than some critics have supposed. There are plenty of systems in which ecological functions can be naturalistically grounded, in their teleology and normativity, using the organizational theory.

13.4.3   Modularity Analysis and the Identification of Ecological Systems Showing Tendency to Closure

Surely, it is rather challenging to individuate ecological systems not as bounded as phytotelmata or beaver dams. Plant stands integrated by mycorrhizal networks, for instance, are difficult cases. However, we see this not as a fatal conceptual pitfall that the theory cannot deal with. It is rather a methodological challenge that can be tackled with its resources. Even though this is not the space to fully develop a methodological solution to the problem, we can advance some basic ideas on how to pursue it.

An analysis of modularity in ecological networks can provide at least an initial approach to identify ecological systems showing tendency to closure. Modularity – which describes the existence of subcommunities within networks – is currently regarded as a recurrent structure of many types of ecological networks (Thébault, 2013). A network shows modular structure when it consists of interconnected modules, while the extent to which species interactions are organized into modules amounts to the modularity of the network. In turn, a module in an ecological network is defined as a group of species more closely connected to each other than to species in other modules.

---

[14] Importantly, as Bich (2019) argues, the realization of a new order of functional integration does not imply that the organisms involved are not able to realize organizational closure and achieve functional integration by themselves. It just means that, while maintaining closure as functionally cohesive entities, they extend their functional networks of control constraints by realizing nested forms of functional integration that include more than one system and, we add, can also realize closure at a higher order.

In an influential paper, Olesen et al. (2007) provide a good example of the relevance of modularity analysis for understanding the structure and functioning of ecological networks, given that modularity is both a key ingredient of network complexity and plays a critical role in their functioning, e.g., in relation to species coexistence and community stability. Indeed, the modular structure of species interactions in mutualistic networks was shown, for instance, to hinder species loss and promote long-term persistence of ecological communities (Krause et al., 2003; Kashtan & Alon, 2005; Olesen et al., 2007; Guimerà et al., 2010; Stouffer & Bascompte, 2011; Gilarranz et al., 2017; Sheykhali et al., 2020).

There are several underlying processes that can explain why ecological networks show modularity, all of which can be included in non-equilibrium models: modularity may reflect habitat heterogeneity, divergent selection regimes, and phylogenetic clustering of closely related species (Lewinsohn et al., 2006). It can also result from the convergence of species on correlated suites of traits shaped by similar interaction patterns, as captured by a concept commonly used in studies on plant-animal interactions, namely, that of syndromes (Fenster et al., 2004; Olesen et al., 2007; Dellinger, 2020).[15]

Modularity is no exceptional feature of ecological networks but rather a manifestation of a common property in biological networks, which, as Kashtan and Alon (2005, p. 13773) argue, "are modular with a design that can be separated into units that perform almost independently." We can advance, thus, that modularity analysis can provide a first step to identify highly connected groups of species that may satisfy the requirements for showing tendency to closure.[16] That is, organizationally closed (sub)systems[17] in an ecological network can be searched for through the identification of modules, and the search space for those (sub)systems will be significantly reduced if we focus on modules of ecosystem parts that are more closely connected to one another than to parts included in other modules. After all, within a module, it is more likely that biological organisms/populations/functional groups will show mutual dependence due to their interactions, which are stronger than the interactions with other network components, i.e., it is more likely that they rely on one another for their own maintenance, with at least part of them being possibly both enabling and dependent constraints and, thus, being subject to closure.

The identification of modules in an ecological network can provide, thus, a first step to model organizationally closed ecological systems but needs to be complemented by an approach to investigate the within-module connections in order to

---

[15] In fact, many pollination studies implicitly assume modularity when they focus on groups of interacting species sharing a syndrome.

[16] Although we cannot develop the argument in the confines of this chapter, we advance that the approach described in the body of the text may provide a way of implementing the procedure to delimit organizationally closed systems through the drawing of their spatial boundaries derived by Montévil and Mossio (2015) from the quantitative assessment of the tendency of constraints to be "packed together" in space.

[17] We write "(sub)systems" to accommodate the fact that the whole ecological network or more inclusive parts of it may be also described as systems in a number of cases.

establish whether they take place between constraints and, moreover, between constraints that are both enabling and dependent. A possible way to model modules in ecological networks as organizationally closed systems is to ascertain whether they can be treated as ecological autocatalytic sets, as proposed by Cazzolla Gatti et al. (2017, 2018). Another way, which we are currently investigating, is to show that systems of differential equations used to describe coupled dynamics (e.g., consumer-resources, predator-prey) can provide a mathematical framework to model ecosystem closure of constraints. Surely, these two approaches can be integrated, as they offer descriptions of the same dynamical system, with networks describing the topology of the interactions, and differential equations, the dynamics of the interactions.

Olesen et al. (2007) analyzed a total of 51 pollination networks, encompassing almost 10,000 species of plants and flower-visiting animals and 20,000 links, and found that 29 of them (57%) were significantly modular.[18] In particular, all networks containing more than 150 species were modular, while all those with less than 50 species were nonmodular. The modular networks had, on average, 8.8 ± 3.7 modules, ranging from a maximum of 19 to a minimum of 5 modules. Most links in such networks were among species within the same module (on average 60% of all links), reinforcing how modularity analysis may allow us to identify organizationally closed (sub)systems in an ecological network, despite the intricacy of ecological relationships and the relative openness of such systems. Individual modules in the networks differed in size and shape because of both the variation in species number and the ratio between pollinator and plant species. A module contained on average 32 ± 34 species (on average, 26 pollinator species and 6 plant species). This suggests that the set of organizationally closed modules or (sub)systems in ecological systems may not be as small as some critics think. It was even the case that 36 (14%) of all 254 modules identified in the networks were isolated species groups without any links to the remaining network. However, this finding concerns the ecological interactions between plants and animals modeled in the networks, and there is no reason to assume that if other kinds of ecological relationships were at stake, those same species groups would be equally isolated. Only 21 of these isolates, i.e., 4% of all identified modules, were small 1:1 modules, consisting of only one pollinator species interacting with one plant species. That is, just a minority of the modules were the sort of small ecological systems, with just a couple of populations, that Lean (2021) argued would exhaust most of the domain to which the organizational theory could apply. Twenty-nine (11%) of all modules were star-shaped, consisting of one generalist hub species, most often a plant species, showing no links to other

---

[18] Olesen et al. (2007) treated all flower-visiting animals as pollinators, which, of course, is not necessarily true as several species may visit flowers without being involved in pollination but in other processes, such as nectar robbing. As the role of a species in an ecological network is defined by its topological position compared to other species, it is not central to functional ascription based on modularity analysis if the species at stake is a pollinator or not, since it may constrain the flow of energy and matter in a variety of ways and, thus, play different ecological functions according to the organizational theory.

modules, while it was linked to a range of 3–51 peripheral pollinator species connected only to the hub. Most of the hubs (189, i.e., 74%), however, varied a lot in size and shape, showing the diversity of arrangements possible in plant-pollinator networks. Some modules contained a set of species with convergent traits related to their pollination biology, i.e., to pollination syndromes, or which were closely related taxonomically.

Considering functional analysis, a rather interesting aspect of the study carried out by Olesen et al. (2007) lies in the topological analysis of the role played by each species in the networks. This role is defined by its position compared with other species in its own module and how well it connects to species in other modules. Accordingly, the analysis considers the relation between each species' within-module degree z, i.e., its standardized number of links to other species in the same module and its among-module connectivity c, i.e., the level to which the species was linked to other modules. Eighty-five percent of all species showed low z and c and were peripheral species or specialists, showing only a few links and almost always only to species within their module (72% of them had $c = 0$, with no links outside their own module). Species with either a high z or a high c value were generalists (15%), including module hubs (3%), i.e., highly connected species linked to many species within their own module (high z, low c), and connectors linking several modules (low z, high c) (11%). Species with high z and high c were network hubs or super generalists (1%), acting as both connectors and module hubs. Plants were the strongest module hubs. Connectors were mainly beetles, flies, and small-to-medium-sized bees, and most network hub pollinators were social bees, especially *Apis* spp. and *Bombus* spp., or large solitary bees, e.g., *Xylocopa* sp. and a few *Diptera* species. Even though generalists not only contribute to pack peripheral species together into modules but also connect modules together into networks, blurring in this way module boundaries, it is possible to extract modules from networks using the appropriate analytic approaches, as shown by several studies (e.g., Olesen et al., 2007; Fortuna et al., 2010; Thébault & Fontaine, 2010; Schleuning et al., 2014; Grilli et al., 2016; Sheykhali et al., 2020). This is instructive when one seeks to consider how system openness does not entail that organizationally closed systems cannot be identified.

Some ecological networks may show a greater tendency toward modularity than others, since this property is expected to increase with link specificity (Lewinsohn et al., 2006).[19] One may expect, for instance, that modularity is stronger in insect herbivory or host-parasitoid networks, which show high link specificity, than in

---

[19] Link specificity concerns the degree of specificity of the ecological interaction represented by a certain edge in a network. For instance, in a food network, the more specialized the trophic relationship considered, the higher the link specificity, while the reverse is true for generalist trophic relationships. Link specificity is related to another key concept in the literature on ecological networks, namely, interaction intimacy (i.e., the degree of biological integration between interacting individuals; see Pires & Guimarães, 2013), such that the decision on specificity does not merely involve an analysis of the links in a given network.

pollination and seed-dispersal networks, characterized by lower interaction specificity, and in traditional food webs (Olesen et al., 2007).

Important consequences for conservation may follow from the combined use of an organizational theory of ecological functions and modularity analysis, as it may allow us to ascertain key groups of taxa that need to be conserved for an ecological network to persist, based on the implications of losing them to the network functioning per se, not just on the choices of a scientist in relation to where to focus his or her attention. This is so because such a theoretical-methodological approach can provide us with normativity criteria that can underlie conservation decisions, for instance, about the conservation of biodiversity items that constrain the flow of matter and energy in the ecosystem in such a manner that its resilience and persistence – as aspects of its stability and, accordingly, of its intrinsic teleology – be maintained. These are criteria that depend on the natural normativity of the system and cannot be offered by accounts that fall short from grounding the teleology and normativity of functions. Consider, say, how the network consequences of species extinctions depend, among other factors, on the species role in the topology of the network. For instance, the extinction of a module hub may cause its module to fragment with no or minor cascading impacts on other modules, whereas if connectors are extinct, this may cause the entire network to fragment into isolated modules but with minor impacts on the internal structure of individual modules (Olesen et al., 2007). Accordingly, we can derive criteria, for instance, for choosing conservation priorities from the ascription of functions to different species depending on their topological roles in relation to the modularity of the network, which may be properly captured by interpreting ecological functions in terms of the organizational theory. To briefly mention a central topic discussed by Lean (2021), this will have normative consequences to decisions in invasion biology: for instance, alien invaders of a network may cause, as they are often highly generalist, fusion of modules in an ecological network, with profound, long-term effects on network functioning and selection regime (Olesen et al., 2007). This would be a reason, then, to choose to avoid the establishment of highly generalist invaders in ecological systems.

By considering the modularity of ecological networks, we can conclude, first, that it may provide a first step in the identification of organizationally closed systems individuated as modules in a network (if complemented by approaches to establish that the nodes in a module form a closure of constraints), which do not necessarily correspond to a small set of small networks; second, that if we consider the modularity of entire ecological networks, say, all species interacting through ecological processes such as pollination in a given area, many networks will likely include many modules, and, then, the fact that it may not be possible to describe the whole network as a single organizationally closed system may not hamper functional ascription based on the organizational theory within identified modules; and third, that several different functional roles can be described based on modularity analysis, such as module and network hubs, and that it may be possible to interpret them based on the organizational theory, as related to the constraining actions of biodiversity items on flows of matter and energy through, say, trophic or pollination relationships.[20] Even connectors, which link several modules, can be ascribed functions based on the organizational theory. In particular, this will be so if we consider coupling between modules as a functional role, something that will be possible if these connectors establish a form of mutual dependence between modules, due to their stable connecting interactions, such that one can say the self-maintenance of the modules is related to these interactions. Finally, the roles played by different species in the topology of ecological networks have conservation implications that are not at the discretion of scientists' decisions only but also depend on the nature and structure of the networks, as modularity can even be said to spontaneously evolve in them (Kashtan & Alon, 2005).

13.5 Organizational Functions and Evolution

Dussault and Bouchard (2017) argue that the organizational theory dissociates the concept of ecological function from evolutionary considerations. It is true, on the one hand, that there is much work to be done in order to develop the connection between the causal loop by which functions explain the presence of the function bearer in an organization-based account, which has self-maintenance as its *telos*, and an evolution-based causal loop, which refers to the (past) natural selection of fitness-enhancing traits. But, on the other hand, the organizational theory has the resources to further develop this connection to the etiological dimension, which is part of its elaboration since its inception. The organizational theory proposed by Mossio and colleagues (2009; see also Moreno & Mossio, 2015; Saborido et al., 2011, 2016, among others) aims at accounting for the *explananda* of both etiological and dispositional theories of function. Moreover, in the theory of biological organisms under construction by the ORGANISM-group, which is also an important theoretical framework for our proposal, the concept of function is connected to the principle of organization, and this is in turn integrated with evolutionary thinking through the principle of variation (Montévil et al., 2016; Mossio et al., 2016). This means that in this framework evolutionary changes in organization along time, both qualitative and quantitative, are necessarily integrated into the understanding of biological phenomena, including ecological ones.

Ecological functions have been recently conceived as central for integrating evolutionary and ecological perspectives on ecosystems. As a consequence, even though the concept of function has played an important role in the whole history of ecology, it has become increasingly fundamental to the development of ecological and conservation research in the last three decades (Nunes-Neto et al., 2016a). This happened as a consequence of the biodiversity and ecosystem function (BEF) research program, which aims at establishing a better understanding of the

---

[20] In this sense, our arguments are not affected by the fact that not all flower-visiting animals in the pollination networks are truly pollinators but play different roles, for instance, as nectar-robbers, as observed above.

relationships between biodiversity and the functioning of ecosystems (e.g., Naeem, 2002b; Loreau, 2010a, b). To account for these relationships, this research program attributes to functional diversity the role of a conceptual bridge between community and ecosystem ecology, i.e., between the understanding of biotic communities and, accordingly, of biodiversity, including the interactions among their constituting components and their effects, on the one hand, and their contribution to ecological processes that maintain ecosystems and their properties, on the other. This unification of ecosystem and community ecology is often recognized as an important goal both for the development of ecological knowledge and for meeting the challenges of the current socioenvironmental crisis (e.g., Pickett et al., 2007; Dussault & Bouchard, 2017). They are sought after by BEF researchers through the investigation of how specific traits of organisms and other biological items contribute to the maintenance and functioning of ecosystems. This requires, however, that the understanding of ecological functions be connected with how organismic functions are conceived in evolutionary theory, which is a central component of the theoretical framework in community ecology (Dussault & Bouchard, 2017). Not surprisingly, BEF researchers stressed that a synthesis of community and ecosystem ecology demands that evolutionary considerations be reintroduced into ecosystem studies (e.g., Loreau, 2010a, b).

It is at the purview of the organizational theory of ecological functions to deliver an understanding of functions that is both ahistorical and evolutionarily grounded. To do so, it will be necessary to elaborate more on the relations between the evolution of organisms and the emergence of ecological interactions and functions in organizationally closed ecological systems.[21] A key aspect to bear in mind is that, as ecological systems emerge from interactions (at least part of them functional interactions) between populations that have been selected to a considerable extent, an integration between evolution- and organization-based accounts is a *sine qua non* for understanding ecological functions. But how should this integration take place? A fundamental requisite is to consider how to prioritize functional approaches in organisms or ecological systems. At the level of organisms, an evolution-based functional and teleological understanding should be grounded on an organization-based functional and teleological conception of self-maintaining organisms capable of survival and reproduction (Mossio & Bich, 2017). However, at the ecosystem level, it seems to be the case that organization-based function and teleology should be grounded on the interrelations among organismic functions (and also accidental or fortuitous effects) that emerge in evolution at the population level. That is, at the ecosystem level, organization arises from interactions among populations that have been selected for at the population level, as a kind of by-product of organismic functioning to achieve self-maintenance and increase fitness. After all, other populations are always a relevant part of the environment of any population at stake.

---

[21] A dialogue with the persistence enhancing propensity (PEP) account proposed by Dussault and Bouchard (2017) can be helpful in this effort.

Functions in ecology are relational and contextual, as emphasized by Dussault and Bouchard (2017). They emerge from current interactions between populations that are, at least partly, associated with organismic traits exhibiting functional roles that evolved historically, before a particular ecological system has been formed. This does not mean, however, that such functional traits have evolved for the sake of the ecosystem; rather, they partly evolved due to their fitness-enhancing consequences at the population level, partly due to other evolutionary processes than natural selection, and have been coopted for functional roles in ecological processes within the ecosystem when subject to its closed organization of constraints. For example, in plant-pollinator networks, different populations mutually stabilize each other (allowing for an account in terms of closure of constraints and organizational functions), but it is the evolution at the population level that explains the spread and eventual fixation of functions that are entangled with one another in the ecosystem closure of constraints (e.g., Patiny, 2012). Moreover, the historical constitution of ecosystems involves a "fine-tuning" of functional relations as a result of evolutionary paths, reinforcing the need to integrate evolutionary and organizational perspectives on ecological functions.

13.6   Ascribing Functions to Abiotic Items

For Dussault and Bouchard (2017), the PEP account accommodates the ascription of functions to abiotic components of ecosystems better than the organizational theory, as it allows function ascription to abiotic factors like disturbance regimes and habitat heterogeneity (e.g., Pickett et al., 1999; White et al., 1999). A similar argument is proposed by Odenbaugh (2019) but in a defense of a systemic approach, which is one version of a dispositional theory. But consider how we recently broadened the range of functional items that fall under the umbrella of the organizational theory in order to include abiotic items, provided they are under the control of biodiversity items (El-Hani & Nunes-Neto, 2020). From this perspective, factors like disturbance regimes (say, related to fire) and habitat heterogeneity (for instance, due to the construction of a beaver dam) only play a functional role in an ecological system if they are products of constraints subject to closure in that system and are themselves involved in the production of constraints. If they are not so, then they are not truly functional but just boundary conditions that affect the maintenance of populations and ecological communities in that system (even though they can be functional if we are rather modeling an ecological system at a higher scale). If, say, habitat heterogeneity and disturbance regimes are not under the control of components of an ecological system, they do not have a functional role according to the organizational theory precisely because they are not under the control of the system and do not enable the conditions of existence of other constraints. But this does deny their relevance to the system's dynamics, as external boundary conditions. This relevance seems to be the reason why ecologists ascribe in a number of cases functions to such external entities and processes. A philosophical analysis can offer, then, an important clarification about a conflation between the functional contribution from a system's component and the dynamic relevance of external boundary conditions (at a given scale). To describe the interaction of an external entity or process not controlled by a system but influencing its dynamics in terms of a broadened regime of functional integration is incorrect precisely because in this case the system is not exerting any influence upon the generation of the boundary condition (Bich, 2019). If we consider some ecological systems as showing the same kind of regime of closure as autocatalytic sets (see, e.g., Cazzolla Gatti et al., 2017, 2018), it will be also clear why one should not, from this perspective, ascribe an ecological function to entities or processes external to the system and not under its control. This conflates being a boundary condition to the system's dynamics with playing a function, which is an important drawback, since functions are attributed to components of a given system.

If an external entity or process is under the control of the system, as disturbance regimes or habitat heterogeneity in a number of cases are, then it has a functional role defined in accordance with the closure of constraints defining the system, since by being under the control of the system, it becomes subject to closure, being both a dependent and an enabling constraint. Consider, as a case in point, how fire (as a disturbance regime), when integrated into the dynamics of an ecological system, say, through fire-adapted plant species exhibiting traits that promote flammability and, thus, influence fire frequency (e.g., Mutch, 1970; Schwilk & Ackerly, 2001), is not merely destructive but rather enabling, leading to regrowth processes that are crucial to the system's dynamics. In these cases, vegetation is a driver of fire regimes, and one can even talk about coevolution of fire and biota (McLauchlan et al., 2020).

Dussault and Bouchard (2017) consider, in fact, precisely the argument we are advancing here but refuse it because it would, they argue, run counter to the tendency in contemporary ecosystem ecology to include disturbance regimes into the dynamic of ecosystems irrespective of whether they are under biotic control or not. However, we think there is no real problem in this case, because boundary conditions are part of the dynamics of the system no matter if they are within the closure of constraints or not. This is a clear case in which the concept of closure of constraints is not properly expressed, since it only demands that part of the constraints exerting influence on the system be internal to closure. A boundary condition that affects the system without being within closure is still part of the system's dynamics.

Another argument presented by Dussault and Bouchard concern the difficulty of determining whether a disturbance regime is under the control of the ecological system, since it may lie on a continuum between being biotically controlled and uncontrolled (Pickett & White, 1985, pp. 8–9). First, this is an empirical problem that does not challenge the organizational theory: it is quite common in scientific research that the stipulations of a theory pose empirical challenges for their application to real-world processes. Second, the organizational theory can accommodate through the idea of tendency to closure a situation in which an external process or entity (say, fire) is somewhere on a continuum between being biotically controlled or not. In this case we would in fact ascribe function to fire if it is at least to some extent under biotic control.

Differently from Dussault and Bouchard (2017), who intend to follow ecologists' ascription of ecological function to abiotic entities and processes, no matter if they are under the control of the ecosystem or not, we rather think to be preferable to conceptually clarify the case from a philosophical perspective. It seems to us that – revisionist or not – the philosophical analysis at stake drives home a relevant distinction to ecological research, which we exemplify using fire as an example. If fire is under control of constraints internal to the ecological system, it can be both enabling and dependent, being part of the ecosystem closure of constraints, and, thus, being ascribed function, but if it is not under the control of those constraints, even if fire may be eventually enabling, it will not be dependent on internal constraints, and, thus, it will be just an external boundary condition, which should not be described as functional within that ecosystem, despite their significance to the system's dynamics. We do not see a problem in adopting a "revisionist stance" (as Dussault & Bouchard, 2017, p. 1133, calls it) in relation to some attributions of functions by ecologists. After all, epistemological studies would be quite limited in their utility and contribution to scientific research if we assumed that philosophical analysis could never clarify the uses of concepts by scientists themselves.

It does not matter, then, if some abiotic process has the same effect on an ecosystem as a biotic process to which an ecological function is ascribed, as in the example of nitrogen fixation by lightning or volcanoes. Contra Odenbaugh (2010, p. 251), this does not mean that those abiotic processes should be ascribed a function as well, since what they have in common with the biotic process at stake is just that they are both boundary conditions. Nevertheless, the crucial distinction between being a boundary condition under control or deprived of control by the system still applies and is, in our view, crucial to keep in place the distinction between what is truly functional and what merely affects the system's dynamics. From the perspective of the ecological system, nitrogen fixation by a lightning is merely a boundary condition (which, by accident, can fortuitously affect an ecosystem's dynamics, or eventually become stable enough to affect the dynamics on a steady basis[22]), while the same process carried out by bacteria has an ecological function.

Another case that does not bring as much trouble to the theory as Dussault and Bouchard (2017) think concerns source-sink dynamics (e.g., Pulliam, 1988; Pulliam & Danielson, 1991; Amarasekare & Nisbet, 2001; Loreau et al., 2003), which explains the maintenance of "sink" populations, i.e., populations which would run locally extinct if not maintained by constant immigration from "source" populations, as well as of "source" populations whose abundance would often inflate considerably if there was no emigration to "sink" populations. They correctly argue that the effects of source-sink dynamics are often indistinguishable from those of more conventional density-dependent regulation processes (as discussed by Sterelny, 2006, pp. 219–220) and would thus warrant ecological functional ascriptions just as

---

[22] In this case, it may be that the dynamics of the system eventually turns the boundary condition subject to closure, meaning that the abiotic process become a functional part of that system. This does not affect, however, the core of our argument.

in the latter case. True. But there seems to be no problem, however, in ascribing ecological functions to sink and source populations based on the organizational theory. It is only necessary to describe a higher-level entity of which those populations are part, playing functions within its closure, such as a metacommunity or metapopulation, depending on whether we are dealing with multiple or single species. Therefore, in the case of both density-dependent regulation processes and sink-source dynamics, the closure criterion can be met, and function can be ascribed according to the organizational theory.

13.7   A Word on Pluralism About Ecological Functions

It is worth saying here a few words on pluralism about functions. In the overall literature on biological functions, pluralism has been often regarded as an attractive option. For instance, a number of authors supported a pluralistic solution to the problem of function by advocating that the etiological and dispositional theories offered two complementary concepts (e.g., Millikan, 1989; Amundson & Lauder, 1994; Allen & Bekoff, 1995). Godfrey-Smith (1993) called this solution a "consensus without unity." Currently, one cannot advocate for pluralism about functions without considering also organizational theories, in their several versions, as one of the key players in the debate. Specifically in the ecological domain, Garson (2018) has also defended within-discipline pluralism about functions.

But, when we are dealing with some specific problem, it seems to us that pluralism should be the conclusion we reach once we did our best to find a single, unifying theory. It should be the outcome of an investigation that justifies the principled impossibility of a unified account. If we think that a certain theory about function, ecological or otherwise, cannot be the unique one, a proper justification should be offered. Why is it the case? Moreover, to avoid empty pluralism, we need to identify which kind of phenomena can be accounted for by which models, such that we may in the end reach a theory unified as a family of models, as proposed in a pragmatic view of theories in ecology (Travassos-Britto et al., 2021).

To our understanding, we are not yet at a point in the investigation that allows us to settle the case and conclude for a pluralistic perspective on theories of ecological function. The jury is still out. Thus, rather than assuming pluralism, we will leave for now this possibility open and continue inquiring into the application of the organizational theory to different uses of functional explanations and ascriptions in ecological and conservation research. This does not mean that we are claiming that normative functions will be properly attributed to each and every ecological system. Also, this is not the same as exclusively defending organizational functions as an overarching account for all functional ascriptions in ecological research, as Lean (2021) claims to be our intention. We are simply continuing to pursue our avenue of inquiry, extending the theory as much as we can, but open to the possibility that it may not apply, eventually, to a number of functional ascriptions made by ecologists.

The organizational theory of ecological functions remains expansible to new cases, and we really do not know if the latter may be the case.

As an example of how the domain to which the organizational theory is applied can be extended, we can consider two mechanisms proposed to explain how biodiversity enhances the maintenance and resilience of ecosystems, namely, sampling and compensation effects (e.g., Sterelny, 2005).[23] In the former mechanism, the increased resilience of species-rich ecosystems is attributed to the statistical fact that they have more chances to contain species whose functional performance will not be affected by a range of environmental variations. In this case, the functional contribution to ecosystem maintenance is attributed to items of biodiversity rather than to biodiversity as a whole, and no difficulty is posed for the organizational theory. Compensation effect is, however, a different matter, as the increased resilience of species-rich ecosystems is related in this case to response diversity, i.e., the presence of many species that respond differently to environmental variation but are able to perform similar functional roles in the ecosystem. Response diversity entails, thus, that the species may show compensatory dynamics, i.e., when an ecosystem is subject to variation in its interaction with other systems that leads a formerly dominant species to decrease in abundance, the functional consequence for the ecosystem dynamics can be buffered by the compensation of another species that is functionally equivalent but shows a differential response to the variation at stake. Thus, the likelihood that a variation leads to impacts that may disrupt ecosystem functioning and harm its capacity of maintaining itself is reduced, and, conversely, ecosystem resilience before that variation is maintained or even enhanced.

Compensation effect illustrates a case in which more work is needed to extend the domain of the organizational theory. In the definition of organizational function proposed by Saborido et al. (2011), function is ascribed to a *trait* that exerts a constraint subject to closure in an organization of a given system. Mossio et al. (2013) characterize a constraint as a *configuration* and Moreno and Mossio (2015) as an *entity* that exhibit a symmetry with respect to a process (or set of processes) under its influence. Nunes-Neto et al. (2014) consider *items of biodiversity* as objects of functional ascription in ecology, while El-Hani and Nunes-Neto (2020) recently broadened the set of functional objects in the theory to include *abiotic items*. Biodiversity is a global property or, to put it differently, a distributed feature of an ecological system. The question that arises is as follows: Can a global property be a constraint, such that the organizational theory justifies the ascription of ecological function to biodiversity *per se*? At this point, we do not see any fundamental blockage for formulating the notion of constraint in such a manner that this justification can be done. While it may stretch the concepts too far if we treat biodiversity as an entity in an ecosystem, the use of categories such as trait and configuration in the organizational functional discourse paves the way to encompass biodiversity as an object of functional ascription. This will need some reworking of the notion of

---

[23] The arguments in this paragraph benefited a lot from the discussion made by Dussault and Bouchard (2017, pp. 1133–1134).

constraint in order to include under its extension global properties such as biodiversity, but this will not be some far-fetched conceptual operation. Therefore, while there is still work to do, the organizational theory can be applied to explain both the specific functional contributions of many items of biodiversity to the overall functioning of an ecosystem and the collective stabilizing function of response biodiversity.

13.8 From Organizational Functions to an Integrated Scientific and Ethical Approach to Sustainability

The fortunes of teleological accounts of ecological functions, such as the organizational theory, may have important consequences to conservation ethics. If we show that ecological systems are structured in such a manner that their parts are functional for the whole, we may be able to provide support to the claim that they possess a type of natural value on a naturalistic basis. Such a natural normativity can facilitate objective judgments about the role of populations within ecosystems and about conservation measures, as well as mediate debates in conservation ethics and provide guidance for thorny environmental ethical questions.

What does the idea that ecological (and, possibly, socioecological[24]) systems realize closure of constraints entail, then, for an ethical perspective on such systems? Recognizing that biological systems include constraints that perform functions is to recognize a normative dimension of the very existence of these systems. In this sense we can differentiate between two kinds of systemic state, namely, between organized states, which exist according to the norms of the system's behavior, maintaining conditions of existence that allow its persistence and resilience, and states that work counter to the system's organization, deviating from the norms of its behavior and disrupting its conditions of existence (Moreno & Mossio, 2015; Cooper et al., 2016; Montévil, 2021). That is, the functionality of certain biological features concerns not only a current performance of an ecological system but a performance that the system must do in order to continue to exist. It seems, then, that we may be able to discern in a normative way between *good* and *bad* functioning of ecological systems (Cooper et al., 2016). It is at this point that the descriptive language of biological organization touches on ethical and axiological aspects.

Notice, however, that this is a more demanding normative dimension than that at play when we just speak of functions. If we consider, say, a pumping heart, this organ will be fulfilling the biological norms involved in the performance of its function even when pumping poorly, with consequences to the organism's health. To consider the performance of the heart's function in a healthy condition demands, thus, a second, additional set of norms, establishing that the heart is not only

---

[24] For a brief and initial discussion of the application of the organizational theory of functions to socioecological systems, see Nunes-Neto et al. (2016b).

functioning but also functioning *well*. Accordingly, one thing is ascribing normativity to ecological functions based on the intrinsic purposiveness associated with the realization of closure by an ecosystem, as a causal regime maintaining its own conditions of existence. Another thing is considering whether an ecosystem is functioning poorly or well, as this requires a second source of normativity. What should this source of normativity be is one of the issues to be tackled by an environmental ethics theory.

There are important differences, however, between proposing an organizational view of ecological systems and functions (which is mostly an epistemological stance) and developing an ethical perspective on them (which entails an interpretation based on moral philosophy). Let us begin, thus, by appreciating an important conceptual difference in moral philosophy which is important for our arguments, namely, that between moral agents and moral patients (Warnock, 1971; Goodpaster, 1978; Nunes-Neto & Conrado, 2021). A moral agent is a being capable of emitting moral judgments, which can be – as a consequence – held responsible for its actions. In turn, a moral patient is a being that matters in relation to actions and, accordingly, should be taken into consideration in moral judgments about the latter. When dealing with ecological systems, we are primarily talking about moral patients, rather than moral agents, who could have any kind of moral duty, obligation, or responsibility. When we refer to what an organism or species (say, a bee species) should do in relation to the norms of an ecological system's behavior (say, in a pollination network), we are surely not considering any moral duty, obligation, or responsibility but just manifesting an expectation that a given behavior must happen if those norms are to be observed and conditions needed for the system's resilience and persistence are to be fulfilled. But this expectation may also offer criteria to distinguish between what is good or poor working of the system, providing an ethical perspective on its organization and functioning.

But would we not be committing a fallacy – namely, the naturalistic fallacy – by constructing an ethical perspective on ecological systems from an organizational theory of functions? We cannot simply make inferences from purely factual claims to moral ones, or, to put it differently, *normative claims* about what *ought to be true* can never be validly inferred from *factual claims* about *what is true* (e.g., Kitcher, 1993; Sterelny & Griffiths, 1999). This means that the use of good, bad, well, poorly, or other normative terms in an ethical context does not entail merely an expectation about the natural behavior of systems but also about what we consider that we – human beings as moral agents – *must do* in relation to others (humans or nonhumans), to moral patients, in our everyday life. There is a central difference between developing an organizational theory of ecological functions and an ethical perspective on ecological systems: while the normative language in the organizational theory expresses facts (even if dispositional), the normative language in the ethical field expresses values. The kind of normativity that stems from the organizational theory does not come from the same sources than those at play with ethical and moral human judgments. When we consider an ethical view about ecological systems, we must also recognize, thus, our own (human, thus ethical) perspective, which – at least from our point of view – cannot be reduced to a naturalized outlook about organizational functions only. In this sense the difference between ethics and natural sciences is of central relevance, even if a dichotomous view of facts and values is avoided:[25] ethical issues concern, preponderantly, matters of value, while the natural sciences deal, mainly, with matters of fact, but matters of value and fact, albeit not entailing one another, do interact.

This difference does not mean, however, that it is not possible to build an integrated perspective combining the organizational theory and an ethical theory. It only means that we should not do so by committing a naturalistic fallacy, since one thing is an epistemological (naturalistic) outlook on organisms and ecosystems, and another thing is an ethical standpoint. But these stances are not necessarily in contradiction; they can interact with one another, and, perhaps, in some cases be even conceived in a kind of continuity or complementarity. As Sterelny and Griffiths (1999) argue, even if moral principles cannot be inferred from purely factual biological premises, we can discover morally relevant facts through biological research, which can interact with existing moral principles to produce new practical policies.

In what follows, we are not going to talk about ethics in general, but rather talk about environmental ethics, since ecological systems are our main focus here. As soon as environmental problems gained notoriety (around the 1970s), a new field of ethical reflections was consolidated, environmental ethics, as a way of dealing with a whole range of new issues that could not be well grasped by more traditional ethical frameworks. This was so because those issues concerned a series of beings and processes that had not been commonly considered in previous ethical studies. In its emergence, environmental ethics differed from previous views, which were generally anthropocentric (i.e., focused on human beings). In this sense, environmental ethics broadened the scope of ethical study and reflection to include other natural entities and processes such as animals, plants, rivers, mountains, ecosystems, etc. This was an expansion of the scope of moral considerability (i.e., concerning which beings or entities should be morally considered in our decisions and actions). What was at stake, in short, was which among all the natural beings should we humans (as moral agents) accept as moral patients (Warnock, 1971; Goodpaster, 1978; Vaz & Delfino, 2010).

Kant [1785] 2007) differentiated between direct and indirect moral considerations, depending on the moral status we recognize in other beings. We consider something to be under the purview of indirect moral consideration when its value is not final but rather justified by reference to something else, which is external to it. For example, the value of a hammer comes from the act of hammering, which is external to the hammer itself. In this case, the value of a hammer is merely instrumental. In turn, we generally accept – in accordance with Kant's view – that the importance of a human life is final, in the sense that it has value in itself, without

---

[25] Following Putnam (2002), we do not endorse a dichotomy between facts and values (as assumed, for instance, by logical positivists), but this does not mean that we cannot differentiate between them. Every fact is value-laden, as well as values are connected to facts in the empirical domain. Here, we assume a non-dichotomous difference between facts and values, as well as between science and ethics, recognizing at the same time that there are mutual influences between them.

requiring justification in terms of anything else. Accordingly, a human life cannot be grouped generically together with other entities that might supposedly replace it in fulfilling some external value (as is the case of a hammer, which, when broken, can be replaced by another one, with no harm to the satisfaction of its value). This means a human life is irreplaceable and shows intrinsic value, i.e., a value that is justified in itself. Direct moral consideration results from the recognition of this type of value (Warnock, 1971; Goodpaster, 1978; Vaz & Delfino, 2010).

For Kant ([1785] 2007), only human beings – as rational beings – should have their intrinsic value recognized, being fundamentally different, in ethical terms, from things and other beings. However, the appraisal of this humanist position has changed with the emergence of environmentalism, among other developments. This view came to be regarded as a form of anthropocentrism. Environmental ethics translated moral perceptions that came to the fore with environmentalism into the proposal of expanding moral theories in such a manner that recognition of intrinsic value in other beings could be justified (Warnock, 1971; Goodpaster, 1978).[26]

This expansion of moral theories resulted in a variety of different positions. The sentiocentric current,[27] for instance, attributes intrinsic value to all sentient beings, i.e., to all those that can experience their own life (including humans and a range of nonhuman animals). The biocentric current, in turn, recognizes the intrinsic value of all living beings, whether they are sentient or not (also including, say, bacteria, fungi, etc.). The ecocentric current, finally, ascribes intrinsic value to ecosystems and cannot be regarded as a mere expansion of other moral theories, due to its more holistic character (Vaz & Delfino, 2010; Nunes-Neto & Conrado, 2021).

In the wake of this theoretical expansion, one of the main tasks has been to justify the intrinsic value of nonhuman beings. This means to offer reasons to justify which of these beings (if any) have a purpose of their own. As the organizational theory discussed here naturalizes the concept of function in living systems, it offers a possible contribution to the understanding of this purpose (see, e.g., Holm, 2017; Moosavi, 2019). Biocentrism offers a case in point. In the case of this stance, the justification for ascribing intrinsic value to all living beings stems from the idea that the intrinsic teleology associated with organisms provides a criterion for objective recognition of a good of its own, a good that does not originate from subjective attribution of value (e.g., Taylor, 1986; Varner, 1998).

Holm (2017) investigates whether the biocentric claim can be well justified by the organizational theory in response to what he calls the scope problem. According to this problem, for the biocentric justification to correspond with the moral

---

[26] We chose here, for simplicity, a Kantian way of describing the changes brought about by the emergence of environmental ethics. However, there are other equally important moral theories, such as utilitarian and virtue ethics theories, that would describe the research tradition of environmental ethics differently.

[27] Sentiocentric ethics can be understood not as an environmental ethics *per se* but as an animal ethics, with its own research agenda. However, it is part of the same movement of questioning and overcoming the anthropocentric position and that is why it is described here within the same tradition of environmental ethics.

intuitions of biocentrists, it is necessary that the teleology identified in living beings encompasses all types of possible organisms and be exclusive to them, i.e., not shared with non-organisms (e.g., artifacts and inanimate objects in general). Holm argues that the organizational theory, to a large extent, locates the scope of teleology in the domain of living systems, except for the theoretical possibility that some dissipative systems, such as candle flames and hurricanes, also show a rudimentary sort of constraint closure, resembling the intrinsic teleology described by the organizational theory. He considers, then, that this possibility poses a problem for the defense of a strict view of biocentrism, as it more appropriately points to a defense of a teleocentrism, which acknowledges that beings that are not organisms can also (albeit arguably) be targets of direct moral consideration whenever they show intrinsic teleology.

Moreover, the naturalization of the ascription of functions to biological items, as articulated by the organizational theory, is regarded by Holm as suggesting a potential empirically testable criterion for the biocentric claim. That is, as any system realizing self-determination by means of a closure of constraints will exhibit intrinsic teleology and, hence, a good of its own, the organizational theory enables biocentrists to turn the claim that living systems show such a good into an empirical thesis, without appealing to the contested concept of "life."

Holm's proposal of a teleocentrism points to the possibility that a supraorganismic system be regarded as having a good of its own, provided it shares the same kind of orientation toward the end of self-maintenance exhibited by organisms (which awakens the moral feelings of biocentrists). Once we consider that this is the case of ecosystems, the path is open for an ecocentric argument, such as that developed by Rolston, III (1987), who understands nature as a set of teleologies, ranging from human self-legislation to ecosystem self-maintenance, passing through organic autonomy.

Nunes-Neto et al. (2014) support this understanding by showing how an ecosystem can be treated as an organizationally closed system in which the items of biodiversity (and abiotic items, see El-Hani & Nunes-Neto, 2020), acting as mutually dependent constraints on the flow of matter and energy, give rise to intrinsic teleology, just as we observe in organisms, even though ecosystems typically lack other distinctive features of the latter, such as agency. Once one accepts the organizational theory of ecological functions, it might seem that a strictly biocentric position could not hold, since the same criteria for the good of organisms may be also valid for ecosystems. However, it is not really the case, to our understanding, that the teleological grounding of ecocentrism on the organizational theory of ecological functions denies the epistemological legitimacy of biocentrism. Looking more closely, we must notice that, as Nunes-Neto and Conrado (2021) argue, biocentrism and ecocentrism are not contiguous perspectives on the scope of moral considerability but, instead, are focused on different kinds of entities. While biocentrism lies in the same spectrum as, for instance, sentiocentrism and anthropocentrism, ecocentrism is a response to the lack of moral considerability of nature in general, arguing that holistic entities such as ecological systems should be morally considered as having intrinsic value. This means that biocentric and ecocentric perspectives are not mutually exclusive, even though in some situations there could be tensions between them, such as in the classical example of hunting wild animals for maintaining ecological attributes of ecosystems (see, e.g., the debate between Regan, 2013, and Callicott, 2010) or the example of cutting and removing a tree in order to produce organic matter to maintain an agroforestry system (Miccolis et al., 2019).[28]

Following the argument above, if an ecological (or for that matter, socioecological) system realizes closure of constraints in a similar way to organisms, then it will be also a candidate for the recognition of its own good. What does that mean? Namely, that each and every ecological (and socioecological) system would have its own good, considering only the criteria provided by the realization of closure of constraints and intrinsic teleology. However, this conclusion would lead to serious moral conflicts since, if we dissociate the whole from the parts, it will be possible to conceive the well-being of the whole, even if there is no well-being of one or more parts. For example, it would be possible to think that a socioecological system including slavery might have a good of its own if it showed organizational closure. However, just as ecosystems are formed by items of biodiversity that exhibit their own individual good, so are socioecological systems and, accordingly, the claim that a socioecological system including slavery might have a good of its own would not hold. Rather, we would be facing in this case a conflictive state of affairs. This is analogous, in fact, to a dilemma discussed above: just as there may be conflicts between biocentrism or sentiocentrism, on the one hand, and ecocentrism, on the other, the same is true in the case of socioecological systems. These conflicts will happen whenever there are tensions between the intrinsic goods of individual

---

[28] Another example of tension between biocentric and ecocentric perspectives concerns the implications of redundancy to conservation decisions based on considerations about role functions. For instance, if two species play the same role function in an ecosystem and the extinction of one of them does not impact sustainability (because of redundancy), functional considerations may fail to provide a rationale to preserve it. This is a relevant problem for conservation decisions justified on functional grounds, which does not go away when we propose, from an integrated scientific and ethical point of view, a conception of sustainability that entails our duty as moral agents to support the self-maintenance of ecological systems (see below). This is not the space to engage with this issue in the depth it deserves, but let us just briefly state that, first, pluralism about functions may play an interesting role in this respect, since conservation decisions that seem attractive, but are not justified by some theory of ecological function, may well be justified by another one. Second, that the problems entailed by redundancy for conservation decisions have been recognized and debated in the scientific literature for a while, and one of the outcomes of the discussion has been the requirements of more fine-grained descriptions of ecological role functions, such that what at first may seem to be a redundancy may eventually be shown to be a case of functional complementarity between the roles played by two or more species in relation to ecosystem processes (e.g., Rosenfeld, 2002; Oliver et al., 2015). An ecological community can only be maintained if there is functional complementarity among several anatomic, physiological, behavioral, and other attributes of the populations composing it, as it has been shown, for instance, in several studies on pollination systems (e.g., Brittain et al., 2013; Fründ et al., 2013). Surely, in the case of complementarity, the conservation of all species at stake will be justified.

organisms (humans or not) and the intrinsic goods of whole ecological or socioeco-logical systems. It seems to us that these dilemmas can be avoided by an under-standing of the system's well-being as integrated to the well-being of its parts, which seems reasonable, once the system is composed by the parts and their interac-tions. Cases of conflict between the system itself and its parts may generally involve some kind of malfunctional behavior. However, a more complete evaluation of this problem is out of the scope of this chapter, and we shall leave it for future investigation.

What does the recognition or ascription of intrinsic value to an ecological (or socioecological) system mean? First, that we consider that system as important in itself, that is, as having a purpose of its own, or a value of its own. Second, that we judge we have a duty to the system with regard to its self-maintenance. In short, we must promote the resilience and persistence of the system, and not its destruction. This is equivalent to saying that we must sustain the system, i.e., that we must act toward its being a sustainable system. In short, sustainability, from this perspective, is the realization of the duty to promote the good of an ecological (or socioecologi-cal) system that has its intrinsic value duly recognized by virtuous moral agents integrated into a worldview of respect for nature. This new conception of sustain-ability provides an alternative to the usual anthropocentric and economically based version, associated with the management of natural resources, (social, economic, or ecological) capital, and/or ecosystem services. By combining intrinsic valuation with the self-maintenance of ecological (or socioecological) systems, this new con-ception allows us to use a common "grammar" to refer to respect for nature and responsibility (see Larrère, 2013) in such a manner that the values of technological progress, capital, and the market can be subordinated to what Hugh Lacey (2014, 2016) calls "viable values," associated with the sustainability of socioecological systems, social justice and participation, and universal well-being.

The organizational theory of ecological functions offers a promising way not only to further develop important positions in environmental ethics but also to inte-grate fields of ethical knowledge hitherto pursued in a relatively independent man-ner. This does not mean – it is important to notice – that the organizational theory can ground by itself an ethics. This theory, applied to organisms or ecological sys-tems, offers a naturalized epistemological perspective on their organization and intrinsic teleology, which is not sufficient, in our view, to ground ethical aspects related to the interactions between human beings (as moral agents) and other beings or systems (as moral patients). These ethical aspects demand a consideration, both in theory and practice, of properly ethical and moral perspectives (for instance, theories providing criteria to ground the value of moral actions, or differences in value ascription), which cannot be reduced to a naturalized approach. Nonetheless, there is much to gain from an interaction between morally relevant features of a naturalized approach to the intrinsic teleology of organismic and supraorganismic systems and principles provided by ethical and moral theories.

13.9 Concluding Remarks

We further developed in this chapter the organizational theory of ecological functions by responding to some criticisms that allowed us to sharpen the theory. We argued about the individuation of ecosystems as organizationally closed systems, to which the theory can be applied, provided some comments on how evolutionary considerations may be integrated into an organizational understanding of ecological functions, and took additional steps for elaborating on how functions can be ascribed to abiotic items according to the theory. We expect to have shown how the organizational theory provides a convincing basis for naturalizing the teleological and normative dimensions of ecological functions, as well as for making contributions to the construction of an integrated scientific and ethical approach to sustainability that can avoid an anthropocentric perspective.

Acknowledgments We are indebted to Alan Love, Matteo Mossio, Gaëlle Pontarotti, and an anonymous reviewer for their comments on a previous version of the manuscript, which importantly contributed to its improvement. The work on this chapter has been supported by funding from the John Templeton Foundation (#62220) within the cohort program "Agency, Directionality & Function," coordinated by Alan Love (University of Minnesota, USA). We are also indebted to CNPq (proc. nº 465767/2014-1), CAPES (proc. nº 23038.000776/2017-54) and FAPESB (proc. nº INC0006/2019) for their support to the National Institute of Science and Technology in Interdisciplinary and Transdisciplinary Studies in Ecology and Evolution (INCT IN-TREE), in which the present study has been developed. NNN thanks CNPq for research funding (proc. nº 423767/2018-6). CNEH thanks CNPq for the concession of productivity in research grants (proc. nº 303011/2017-3 and 307223/2021-3) and CAPES/UFBA for senior visiting research grant (proc. nº 88887.465540/2019-00). FRGL is thankful to CAPES for doctoral studies grant (proc. nº 88887633926/2001-00).

References

Adams, F. R. (1979). A goal-state theory of function attributions. *Canadian Journal of Philosophy, 9*, 493–518.

Ahl, V., & Allen, T. F. H. (1996). *Hierarchy theory: A vision, vocabulary, and epistemology.* Columbia University Press.

Allen, C., & Bekoff, M. (1995). Biological function, adaptation and natural design. *Philosophy of Science, 62*, 609–622.

Amarasekare, P., & Nisbet, R. M. (2001). Spatial heterogeneity, source-sink dynamics, and the local coexistence of competing species. *American Naturalist, 158*, 572–584.

Amundson, R., & Lauder, G. V. (1994). Function without purpose: The uses of causal role function in evolutionary biology. *Biology and Philosophy, 9*, 443–469.

Aristotle. (1984). Physics. In J. Barnes (Ed.), *The complete works of Aristotle* (The revised Oxford translation). Princeton University Press.

Babcock, G., & McShea, D. W. (2021). An externalist teleology. *Synthese, 199*, 8755–8780.

Bechtel, W., & Richardson, R. C. (2010). *Discovering complexity: Decomposition and localization as strategies in scientific research.* MIT Press.

Bich, L. (2016). Systems and organizations. Theoretical tools, conceptual distinctions and epistemological implications. In G. Minati, M. R. Abram, & E. Pessa (Eds.), *Towards a post-bertalanffy systemics* (pp. 203–209). Springer.

Bich, L. (2019). The problem of functional boundaries in prebiotic and inter-biological systems. In G. Minati, M. R. Abram, & E. Pessa (Eds.), *Systemics of incompleteness and quasi-systems* (pp. 295–302). Springer.

Bich, L., Mossio, M., Ruiz-Mirazo, K., & Moreno, A. (2016). Biological regulation: Controlling the system from within. *Biology and Philosophy, 31*, 237–265.

Bickhard, M. H. (2000). Autonomy, function, and representation. *Communication and Cognition – Artificial Intelligence, 17*, 111–131.

Bickhard, M. H. (2004). Process and emergence: Normative function and representation. *Axiomathes, 14*, 121–155.

Bigelow, J., & Pargetter, R. (1987). Functions. *Journal of Philosophy, 84*, 181–196.

Brittain, C., Kremen, C., & Klein, A.-M. (2013). Biodiversity buffers pollination from changes in environmental conditions. *Global Change Biology, 19*, 540–547.

Callicott, J. B. (2010). Animal liberation: A triangular affair. *Environmental Ethics, 2*, 311–338.

Christensen, W. D., & Bickhard, M. H. (2002). The process dynamics of normative function. *The Monist, 85*, 3–28.

Clarke, E. (2010). The problem of biological individuality. *Biological Theory, 5*, 312–325.

Clements, F. E. (1916). *Plant succession: An analysis of the development of vegetation.* Carnegie Institution of Washington.

Collier, J. (2000). Autonomy and process closure as the basis for functionality. In J. L. R. Chandler & G. van der Vijver (Eds.), *Closure: Emergent organisations and their dynamics* (Annals of the New York Academy of Sciences) (Vol. 901, pp. 280–290).

Cooper, G. J., El-Hani, C. N., & Nunes-Neto, N. F. (2016). Three approaches to the teleological and normative aspects of ecological functions. In N. Eldredge, T. Pievani, E. Serrelli, & I. Temkin (Eds.), *Evolutionary theory: A hierarchical perspective* (pp. 103–125). University of Chicago Press.

Craver, C. F. (2001). Role functions, mechanisms, and hierarchy. *Philosophy of Science, 68*, 53–74.

Cummins, R. (1975). Functional analysis. *Journal of Philosophy, 72*, 741–765.

Curtis, J. T., & McIntosh, R. P. (1951). An upland forest continuum in the prairie-forest border region of Wisconsin. *Ecology, 32*, 476–496.

Davey, M. E., & O'Toole, G. A. (2000). Microbial biofilms: From ecology to molecular genetics. *Microbiology and Molecular Biology Reviews, 64*, 847–867.

Davies, P. S. (2000). Malfunctions. *Biology and Philosophy, 15*, 19–38.

Delancey, C. S. (2006). Ontology and teleofunctions: A defense and revision of the systematic account of teleological explanation. *Synthese, 150*, 69–98.

Dellinger, A. S. (2020). Pollination syndromes in the 21st century: Where do we stand and where may we go? *New Phytologist, 228*, 1193–1213.

Dussault, A. C. (2019). Functional biodiversity and the concept of ecological function. In E. Casetta, J. M. da Silva, & D. Vecchi (Eds.), *From assessing to conserving biodiversity: Conceptual and practical challenges* (pp. 297–316). Springer.

Dussault, A. C. (2022). Two notions of ecological function. *Philosophy of Science, 89*, 171–179.

Dussault, A. C., & Bouchard, F. (2017). A persistence enhancing propensity account of ecological function to explain ecosystem evolution. *Synthese, 194*, 1115–1145.

Edin, B. B. (2008). Assigning biological functions: Making sense of causal chains. *Synthese, 161*, 203–218.

El-Hani, C. N., & Nunes-Neto, N. F. (2020). Life on Earth is not a passenger, but a driver: Explaining the transition from a physicochemical to a life-constrained world from an organizational perspective. In L. Baravalle & L. Zaterka (Eds.), *Life and evolution – Latin American essays on the history and philosophy of biology* (pp. 69–84). Springer.

Eliot, C. (2007). Method and metaphysics in Clements's and Gleason's ecological explanations. *Studies in History and Philosophy of Biological and Biomedical Sciences, 38*, 85–109.

Eliot, C. (2011). The legend of order and chaos: Communities and early community ecology. In K. deLaplante, B. Brown, & K. A. Peacock (Eds.), *Handbook of the philosophy of science, Vol. 11, Philosophy of ecology* (pp. 49–107). Elsevier.

Elton, C. S. (1927). *Animal ecology*. Macmillan.

Elton, C. S. (1930). *Animal ecology and evolution*. Clarendon Press.

Fenster, C. B., Scott Armbruster, W., Wilson, P., Dudash, M. R., & Thomson, J. D. (2004). Pollination syndromes and floral specialization. *Annual Review of Ecology, Evolution, and Systematics, 35*, 375–403.

Fortuna, M. A., Stouffer, D. B., Olesen, J. M., Jordano, P., Mouillot, D., Krasnov, B. R., Poulin, R., & Bascompte, J. (2010). Nestedness versus modularity in ecological networks: Two sides of the same coin? *Journal of Animal Ecology, 79*, 811–817.

Fründ, J., Dormann, C. F., Holzschuh, A., & Tscharntke, T. (2013). Bee diversity effects on pollination depend on functional complementarity and niche shifts. *Ecology, 94*, 2042–2054.

Garson, J. (2015). *The biological mind: A philosophical introduction*. Routledge.

Garson, J. (2016). *A critical overview of biological functions*. Springer.

Garson, J. (2018). How to be a function pluralist. *British Journal for the Philosophy of Science, 69*, 1101–1122.

Gatti, R. C., Hordijk, W., & Kauffman, S. (2017). Biodiversity is autocatalytic. *Ecological Modelling, 346*, 70–76.

Gatti, R. C., Fath, B., Hordijk, W., Kauffman, S., & Ulanowicz, R. (2018). Niche emergence as an autocatalytic process in the evolution of ecosystems. *Journal of Theoretical Biology, 454*, 110–117.

Gilarranz, L. J., Rayfield, B., Liñán-Cembrano, G., Bascompte, J., & Gonzalez, A. (2017). Effects of network modularity on the spread of perturbation impact in experimental metapopulations. *Science, 357*, 199–201.

Gleason, H. A. (1926). The individualistic concept of the plant association. *Bulletin of the Torrey Botanical Club, 53*, 7–26.

Godfrey-Smith, P. (1993). Functions: Consensus without unity. *Pacific Philosophical Quarterly, 74*, 196–208.

Godfrey-Smith, P. (1994). A modern history theory of functions. *Noûs, 28*, 344–362.

Godfrey-Smith, P. (2013). Darwinian individuals. In F. Bouchard & P. Huneman (Eds.), *From groups to individuals: Evolution and emerging individuality* (Vienna series in theoretical biology) (pp. 17–36). MIT Press.

Goodpaster, K. E. (1978). On being morally considerable. *Journal of Philosophy, 75*, 308–325.

Grilli, J., Rogers, T., & Allesina, S. (2016). Modularity and stability in ecological communities. *Nature Communications, 7*, 12031.

Guimerà, R., Stouffer, D. B., Sales-Pardo, M., Leicht, E. A., Newman, M. E. J., & Amaral, L. A. N. (2010). Origin of compartmentalization in food webs. *Ecology, 91*, 2941–2951.

Hagen, J. B. (1989). Research perspectives and the anomalous status of modern ecology. *Biology and Philosophy, 4*, 433–455.

Hagen, J. B. (1992). *An entangled bank: The origins of ecosystem ecology*. Rutgers University Press.

Holm, S. (2017). Teleology and biocentrism. *Synthese, 194*, 1075–1087.

Hooper, D. U., et al. (2005). Effects of biodiversity on ecosystem functioning: A consensus of current knowledge. *Ecological Monographs, 75*, 3–35.

Huneman, P. (2014a). Individuality as a theoretical scheme. I. Formal and material concepts of individuality. *Biological Theory, 9*, 361–373.

Huneman, P. (2014b). Individuality as a theoretical scheme. II. About the weak individuality of organisms and ecosystems. *Biological Theory, 9*, 374–381.

Jax, K. (2005). Function and "functioning" in ecology: What does it mean? *Oikos, 111*, 641–648.

Jonas, H. (1966). *The phenomenon of life. Towards a philosophical biology*. Harper and Row.

Kant, I. ([1785] 2007). *Fundamentação da Metafísica dos Costumes*. Edições 70.

Kashtan, N., & Alon, U. (2005). Spontaneous evolution of modularity and network motifs. *Proceedings of the National Academy of Sciences of the United States of America, 102*, 13773–13778.

Kauffman, S. A. (2000). *Investigations*. Oxford University Press.

Kitcher, P. (1993). Function and design. *Midwest Studies in Philosophy, 18*, 379–397.

Krause, A. E., Frank, K. A., Mason, D. M., Ulanowicz, R. E., & Taylor, W. W. (2003). Compartments revealed in food-web structure. *Nature, 426*, 282–285.

Krohs, U. (2010). Dys-, mal- et non-: L'autre face de la fonctionnalité. In J. Gayon & A. de Ricqlès (Eds.), *Les Fonctions: Des Organismes aux Artefacts* (pp. 337–351). Presses Universitaires de France.

Lacey, H. (2014). On the co-unfolding of scientific knowledge and viable values. In P. Schroeder-Heister, G. Heinzmann, W. Hodges, & P. E. Bour (Eds.), *Logic, methodology and philosophy of science. Proceedings of the fourteenth international congress (Nancy)* (pp. 269–284). College Publications.

Lacey, H. (2016). Science, respect for nature, and human well-being: Democratic values and the responsibilities of scientists today. *Foundations of Science, 21*, 51–67.

Larrère, C. (2013). Two philosophies of the environmental crisis. In D. Bergandi (Ed.), *The structural links between ecology, evolution and ethics: The virtuous epistemic circle* (pp. 141–149). Springer.

Lean, C. H. (2021). Invasive species and natural function in ecology. *Synthese, 198*, 9315–9333.

Lewinsohn, T. M., Prado, P. I., Jordano, P., Bascompte, J., & Olesen, J. M. (2006). Structure in plant-animal interaction assemblages. *Oikos, 113*, 174–184.

Loreau, M. (2010a). Linking biodiversity and ecosystems: Towards a unifying ecological theory. *Philosophical Transactions of the Royal Society of London. Series B, Biological Sciences, 365*, 49–60.

Loreau, M. (2010b). *From populations to ecosystems: Theoretical foundations for a new ecological synthesis*. Princeton University Press.

Loreau, M., Mouquet, N., & Holt, R. D. (2003). Meta-ecosystems: A theoretical framework for a spatial ecosystem ecology. *Ecology Letters, 6*, 673–679.

McLauchlan, K. K., et al. (2020). Fire as a fundamental ecological process: Research advances and frontiers. *Journal of Ecology, 108*, 2047–2069.

McLaughlin, P. (2001). *What functions explain: Functional explanation and self-reproducing systems*. Cambridge University Press.

McShea, D. W. (2012). Upper-directed systems: A new approach to teleology in biology. *Biology and Philosophy, 27*, 663–684.

McShea, D. W. (2016). Hierarchy: The source of teleology in evolution. In N. Eldredge, T. Pievani, E. Serrelli, & I. Temkin (Eds.), *Evolutionary theory: A hierarchical perspective* (pp. 86–102). University of Chicago Press.

Miccolis, A., Peneireiros, F. M., Vieira, D. L. M., Marques, H. R., & Hoffmann, M. R. M. (2019). Restoration through Agroforestry: options for reconciling Livelihoods with conservation in the cerrado and caatinga biomes in Brazil. *Experimental Agriculture, 55*(S1), 208–225.

Millikan, R. G. (1984). *Language, thought, and other biological categories: New foundations for realism*. MIT Press.

Millikan, R. G. (1989). In defense of proper functions. *Philosophy of Science, 56*, 288–302.

Millstein, R. L. (2020). Functions and functioning in Aldo Leopold's land ethic and in ecology. *Philosophy of Science, 87*, 1107–1118.

Mitchell, S. D. (1993). Dispositions or etiologies? A comment on Bigelow and Pargetter. *Journal of Philosophy, 90*, 249–259.

Montévil, M. (2021). *Disruption of biological processes in the Anthropocene: The case of phenological mismatch*. Available at: https://hal.archives-ouvertes.fr/hal-03574022 Accessed 14 August 2023.

Montévil, M., & Mossio, M. (2015). Biological organisation as closure of constraints. *Journal of Theoretical Biology, 372*, 179–191.

Montévil, M., Mossio, M., Pocheville, A., & Longo, G. (2016). Theoretical principles for biology: Variation. *Progress in Biophysics and Molecular Biology, 122*, 36–50.

Moosavi, P. (2019). From biological functions to natural goodness. *Philosophers' Imprint, 19*, 1–20.

Moreno, A., & Mossio, M. (2015). *Biological autonomy: A philosophical and theoretical enquiry.* Springer.

Mossio, M. (2013). Closure, causal. In W. Dubitzky, O. Wolkenhauer, K.-H. Cho, & H. Yokota (Eds.), *Encyclopedia of systems biology* (pp. 415–418). Springer.

Mossio, M., & Bich, L. (2017). What makes biological organisation teleological? *Synthese, 194,* 1089–1114.

Mossio, M., Saborido, C., & Moreno, A. (2009). An organizational account of biological functions. *British Journal for the Philosophy of Science, 60,* 813–841.

Mossio, M., Bich, L., & Moreno, A. (2013). Emergence, closure and inter-level causation in biological systems. *Erkenntnis, 78,* 153–178.

Mossio, M., Montévil, M., & Longo, G. (2016). Theoretical principles for biology: Organization. *Progress in Biophysics and Molecular Biology, 122,* 24–35.

Mutch, R. W. (1970). Wildland fires and ecosystems: A hypothesis. *Ecology, 51,* 1046–1051.

Naeem, S. (2002a). Functional biodiversity. In H. A. Mooney & J. G. Canadell (Eds.), *Encyclopedia of global environmental change* (The Earth system: Biological and ecological dimensions of global environmental change) (Vol. 2, pp. 20–36). John Wiley & Sons.

Naeem, S. (2002b). Ecosystem consequences of biodiversity loss: The evolution of a paradigm. *Ecology, 83,* 1537–1552.

Neander, K. (1991). Functions as selected effects: The conceptual analyst's defense. *Philosophy of Science, 58,* 168–184.

Nicolson, M., & McIntosh, R. P. (2002). H. A. Gleason and the individualistic hypothesis revisited. *Bulletin of the Ecological Society of America, 83,* 133–142.

Nunes-Neto, N., & Conrado, D. M. (2021). Ensinando ética. *Educação em Revista, 37,* e24578.

Nunes-Neto, N., Moreno, A., & El-Hani, C. N. (2014). Function in ecology: An organizational approach. *Biology and Philosophy, 29,* 123–141.

Nunes-Neto, N. F., do Carmo, R. S., & El-Hani, C. N. (2016a). Biodiversity and ecosystem functioning: An analysis of the functional discourse in contemporary ecology. *Filosofia e História da Biologia, 11,* 289–321.

Nunes-Neto, N. F., Saborido, C., El-Hani, C. N., Viana, B. F., & Moreno, A. (2016b). Function and normativity in social-ecological systems. *Filosofia e História da Biologia, 11,* 259–287.

Odenbaugh, J. (2010). On the very idea of an ecosystem. In A. Hazlett (Ed.), *New waves in metaphysics* (pp. 240–258). Palgrave Macmillan.

Odenbaugh, J. (2019). Functions in ecosystem ecology: A defense of the systemic capacity account. *Philosophical Topics, 47,* 167–180.

Olesen, J. M., Bascompte, J., Dupont, Y. L., & Jordano, P. (2007). The modularity of pollination networks. *Proceedings of the National Academy of Sciences of the United States of America, 104,* 19891–19896.

Oliver, T. H., et al. (2015). Biodiversity and resilience of ecosystem functions. *Trends in Ecology & Evolution, 30,* 673–684.

Patiny, S. (2012). *Evolution of plant-pollinator relationships.* Cambridge University Press.

Pattee, H. H. (1972). Laws and constraints, symbols and languages. In C. H. Waddington (Ed.), *Towards a theoretical biology* (Vol. 4, pp. 248–258). Edinburgh University Press.

Peck, S. L., & Heiss, A. (2021). Can constraint closure provide a generalized understanding of community dynamics in ecosystems? *Oikos, 130,* 1425–1439.

Petchey, O. L., & Gaston, K. J. (2006). Functional diversity: Back to basics and looking forward. *Ecology Letters, 9,* 741–758.

Piaget, J. (1967). *Biologie et Connaissance.* Éditions de la Pléiade.

Pickett, S. T. A., & White, P. S. (Eds.). (1985). *The ecology of natural disturbance and patch dynamics.* Academic Press.

Pickett, S. T. A., Jianguo, W., & Cadenasso, M. L. (1999). Patch dynamics and the ecology of disturbed ground: A framework for synthesis. In L. R. Walker (Ed.), *Ecosystems of disturbed ground* (pp. 707–722). Elsevier.

Pickett, S. T. A., Kolasa, J., & Jones, C. G. (2007). *Ecological understanding: The nature of theory and the theory of nature*. Academic Press.

Pires, M. M., & Guimarães, P. R. (2013). Interaction intimacy organizes networks of antagonistic interactions in different ways. *Journal of the Royal Society Interface, 10*, 20120649.

Polanyi, M. (1968). Life's irreducible structure. *Science, 160*, 1308–1312.

Pulliam, H. R. (1988). Sources, sinks, and population regulation. *American Naturalist, 132*, 652–661.

Pulliam, H. R., & Danielson, B. J. (1991). Sources, sinks, and habitat selection: A landscape perspective on population dynamics. *American Naturalist, 137*, 50–56.

Putnam, H. (2002). *The collapse of the fact/value dichotomy and other essays*. Harvard University Press.

Queller, D. C., & Strassmann, J. E. (2009). Beyond society: The evolution of organismality. *Philosophical Transactions of the Royal Society, B: Biological Sciences, 364*, 3143–3155.

Regan, T. (2013). Animal rights and environmental ethics. In D. Bergandi (Ed.), *The structural links between ecology, evolution and ethics: The virtuous epistemic circle* (pp. 117–126). Springer.

Rolston, I. I. I., & Holmes. (1987). Duties to ecosystems. In J. B. Callicott (Ed.), *A companion to a sand county Almanac: Interpretive and critical essays* (pp. 246–274). University of Wisconsin Press.

Rosenfeld, J. S. (2002). Functional redundancy in ecology and conservation. *Oikos, 98*, 156–162.

Rosindell, J., Hubbell, S. P., He, F., Harmon, L. J., & Etienne, R. S. (2012). The case for ecological neutral theory. *Trends in Ecology & Evolution, 27*, 203–208.

Saborido, C., Mossio, M., & Moreno, A. (2011). Biological organization and cross-generation functions. *British Journal for the Philosophy of Science, 62*, 583–606.

Saborido, C., Moreno, A., González-Moreno, M., & Clemente, J. C. H. (2016). Organizational malfunctions and the notions of health and disease. In É. Giroux (Ed.), *Naturalism in the philosophy of health: Issues and implications*. Springer.

Schleuning, M., et al. (2014). Ecological, historical and evolutionary determinants of modularity in weighted seed-dispersal networks. *Ecology Letters, 17*(454), 463.

Schlosser, G. (1998). Self-re-production and functionality: A systems-theoretical approach to teleological explanation. *Synthese, 116*, 303–354.

Schulze, E.-D., & Mooney, H. A. (1993). Ecosystem function of biodiversity: A summary. In E.-D. Schulze & H. A. Mooney (Eds.), *Biodiversity and ecosystem function* (pp. 497–510). Springer.

Schwilk, D. W., & Ackerly, D. D. (2001). Flammability and serotiny as strategies: Correlated evolution in pines. *Oikos, 94*, 326–336.

Selosse, M.-A., Richard, F., He, X., & Simard, S. W. (2006). Mycorrhizal networks: *Des liaisons dangereuses? Trends in Ecology & Evolution, 21*, 621–628.

Sheykhali, S., Fernández-Gracia, J., Traveset, A., Ziegler, M., Voolstra, C. R., Duarte, C. M., & Eguíluz, V. M. (2020). Robustness to extinction and plasticity derived from mutualistic bipartite ecological networks. *Scientific Reports, 10*, 9783.

Skillings, D. (2016). Holobionts and the ecology of organisms: Multi-species communities or integrated individuals? *Biology and Philosophy, 31*, 875–892.

Sterelny, K. (2005). The elusive synthesis. In K. Cuddington & B. E. Beisner (Eds.), *Ecological paradigms lost: Routes of theory change* (pp. 311–329). Elsevier.

Sterelny, K. (2006). Local ecological communities. *Philosophy of Science, 73*, 215–231.

Sterelny, K., & Griffiths, P. E. (1999). *Sex and death: An introduction to philosophy of biology*. University of Chicago Press.

Sterner, B. (2015). Pathways to pluralism about biological individuality. *Biology and Philosophy, 30*, 609–628.

Stouffer, D. B., & Bascompte, J. (2011). Compartmentalization increases food-web persistence. *Proceedings of the National Academy of Sciences of the United States of America, 108*, 3648–3652.

Strassmann, J. E., & Queller, D. C. (2010). The social organism: Congresses, parties, and committees. *Evolution, 64*, 605–616.

Taylor, C. (1964). *The explanation of behaviour*. Routledge & Kegan Paul.

Taylor, P. W. (1986). *Respect for nature*. Princeton University Press.

Thébault, E. (2013). Identifying compartments in presence-absence matrices and bipartite networks: Insights into modularity measures. *Journal of Biogeography, 40*, 759–768.

Thébault, E., & Fontaine, C. (2010). Stability of ecological communities and the architecture of mutualistic and trophic networks. *Science, 329*, 853–856.

Travassos-Britto, B., Pardini, R., El-Hani, C. N., & Prado, P. I. (2021). Towards a pragmatic view of theories in ecology. *Oikos, 130*, 821–830.

Varela, F. J. (1979). *Principles of biological autonomy*. North Holland.

Varner, G. E. (1998). *In nature's interests? Interests, animal rights, and environmental ethics*. Oxford University Press.

Vaz, S. G., & Delfino, Â. (2010). *Manual de Ética Ambiental*. Universidade Aberta.

Vellend, M. (2010). Conceptual synthesis in community ecology. *The Quarterly Review of Biology, 85*, 183–206.

Vellend, M. (2016). *The theory of ecological communities*. Princeton University Press.

Warnock, G. J. (1971). *The object of morality*. Methuen.

Weber, M. (2005). Holism, coherence and the dispositional concept of functions. *Annals of the History and Philosophy of Biology, 10*, 189–201.

West, S. A., & Toby Kiers, E. (2009). Evolution: What is an organism? *Current Biology, 19*, R1080–R1082.

White, P. S., Harrod, J., Romme, W. H., & Betancourt, J. (1999). Disturbance and temporal dynamics. In N. C. Johnson, A. J. Malk, R. C. Szaro, & W. T. Sexton (Eds.), *Ecological stewardship: A common reference for ecosystem management* (Vol. 2, pp. 281–312). Elsevier Science.

Whittaker, R. H. (1951). A criticism of the plant association and climatic climax concepts. *Northwest Science, 25*, 17–31.

Whittaker, R. H. (1975). *Communities and ecosystems* (2nd ed.). Macmillan, Collier.

Whittaker, R. H., & Woodwell, G. M. (1972). Evolution of natural communities. In J. A. Wiens (Ed.), *Ecosystem structure and function* (pp. 137–159). Oregon State University Press.

Wilson, R. A., & Barker, M. J. (2019). Biological individuals. In E. N. Zalta (Ed.), *The Stanford encyclopedia of philosophy* (Fall 2019 Edition). Available at https://plato.stanford.edu/archives/fall2019/entries/biology-individual/. Accessed 14 August 2023.

Wright, L. (1973). Functions. *Philosophical Review, 82*, 139–168.

Wright, L. (1976). *Teleological explanations: An etiological analysis of goals and functions*. University of California Press.

Index